PARALLEL COMPUTATIONAL FLUID DYNAMICS

PARALLEL COMPUTING AND ITS APPLICATIONS

PARALLEL COMPUTATIONAL FLUID DYNAMICS

PARALLEL COMPUTING AND ITS APPLICATIONS

Proceedings of the Parallel CFD 2006 Conference
Busan city, Korea (May 15-18, 2006)

Edited by

J.H. KWON
Korea Advanced Institute of Science and Technology
Daejeon, Korea

A. ECER
IUPUI, Indianapolis
Indiana, U.S.A.

J. PERIAUX
Dassault-Aviation
Saint-Cloud, France

N. SATOFUKA
University of Shiga Prefecture
Hikone, Japan

P. FOX
IUPUI, Indianapolis
Indiana, U.S.A.

ELSEVIER

Amsterdam – Boston – Heidelberg – London – New York – Oxford – Paris
San Diego – San Francisco – Singapore – Sydney – Tokyo

Elsevier
Radarweg 29, PO Box 211, 1000 AE Amsterdam, The Netherlands
Linacre House, Jordan Hill, Oxford OX2 8DP, UK

First edition 2007

Library of Congress Cataloging-in-Publication Data
A catalog record for this book is available from the Library of Congress

British Library Cataloguing in Publication Data
A catalogue record for this book is available from the British Library

ISBN: 978-0-444-53035-6

For information on all Elsevier publications
visit our website at books.elsevier.com

Transferred to Digital Printing in 2012

PREFACE

Parallel CFD 2006 conference was held in Busan city, Korea, from May 15 to 18, 2006. It was the eighteenth conference in an international series of meetings featuring computational fluid dynamics research on parallel computing. There were over 150 participants from 14 countries.

During the conference, 7 invited papers and 92 contributed papers were presented in two parallel sessions. Nowadays, Korea has strength in IT-based technology and large scale computing became popular in universities, research institutes, and companies. Although CFD is one of basic tools for design procedures in Korea to produce machineries, such as automobiles, ships, aircrafts, etc., large scale parallel computing has been realized very recently, especially for the manufactures. Various applications in many areas could be experienced including acoustics, weather prediction and ocean modeling, flow control, turbine flow, fluid-structure interaction, optimization, heat transfer, hydrodynamics. We believe that this conference gave strong motivation and need for parallel computing to everyone.

These proceedings include about 40 percent of the oral lectures presented at the conference. All published papers have been refereed.

The Editors

ACKNOWLEDGEMENTS

Parallel CFD 2006 was hosted by

Korean Society of Computational Fluids Engineering

Korea Institute of Science and Technology Information

We would like to thank for the generous financial contributions of

Korea Institute of Science and Technology Information

Dynamic Busan City, Busan Convention and Visitors Bureau

Air Force Office of Scientific Research, Asian Office of Aerospace Research and Development

Korea Tourism Organization

Department of Aerospace Engineering, Korea Advanced Institute of Science and Technology

Korea Aerospace Industries, LTD.

KOREAN AIR

Korea Research Foundation

HP & INTEL

CRAY Inc.

IBM

ANST Inc.

FLOW-NOISE LTD.

ATES LTD.

KIMIDATA Corp.

CD-ADAPCO Inc.

KYUNGWON TECH LTD.

The conference could not be successful without the contribution and effort from many people who helped from the beginning to the conference date. We are very grateful for the help and guidance from all the international committee members. Local organizing committee members did great work and we would like to thank for their effort. Special thank should be expressed to Kum Won Cho, the secretary general and Pil Choon Lee of KSCFE for their devotion to the conference.

Jang Hyuk Kwon, Jysoo Lee

Chairmen, Parallel CFD 2006

TABLE OF CONTENTS

3. Parallel Environment

4. Multi-Disciplinary Simulation

Parallel Computational Fluid Dynamics – Parallel Computing and Its Applications
J.H. Kwon, A. Ecer, J. Periaux, N. Satofuka and P. Fox (Editors)

Parallel hybrid particle-continuum (DSMC-NS) flow simulations using 3-D unstructured mesh

J.-S. Wu[a], Y.Y. Lian[a], G. Cheng[b], Y.-S. Chen[c]

[a]Mechanical Engineering Department, National Chio Tung University, Hsinchu 30050, TAIWAN

[b]Mechanical Engineering Department, University of Alabama, Birmingham, AL 35294, USA

[c]National Space Organization, Hsinchu 30048, TAIWAN

1. Introduction

Several important gas flows involve flow fields having continuum and rarefied regions, e.g., hypersonic flows [2], vacuum-pump flows with high compression ratio [3], expanding RCS (reaction control system) nozzle plumes in aerospace and space applications [4], physical vapor deposition processes with heated sources [5], pulsed-pressure chemical vapor deposition processes [6], among other things. Understanding of the underlying physics through simulation and modeling in the above flows are important for the success of these disciplines, in addition to the usual experiemental studies. In general, these flow problems are governed by the Boltzmann equation, which is very difficult to solve numerically or analytically due to the existence of collision integral and high number of phase-space dimensions. It is well known the direct simulation Monte Carlo (DSMC) method [7] can provide more physically accurate results in flows having rarefied and non-equilibrium regions than the continuum flow models such as the NS equations. However, the DSMC method is extremely computational expensive especially in the near-continuum and continuum regions, which prohibits its applications to practical problems with large domains. In contrast, the computational fluid dynamics (CFD) method, employed to solve the Navier-Stokes (NS) or Euler equation for continuum flows, is computationally efficient in simulating a wide variety of flow problems. But the use of continuum theories for the flow problems involving the rarefied gas or very small length scales (equivalently large Knudsen numbers) can produce inaccurate results due to the breakdown of continuum assumption or thermal equilibrium. A practical approach for solving the flow fields having near-continuum to rarefied gas is to develop a numerical model combining the CFD method for the continuum regime with the DSMC method for the rarefied and thermal non-equilibrium regime. A well-designed hybrid scheme is expected to take advantage of both the computational efficiency and accuracy of the NS solver in the continuum regime and the physical accuracy of the DSMC method in the rarefied or thermal non-equilibrium regime.

In the past, there were several efforts in developing the hybrid particle-continuum scheme. Most studies employed structured grid for both the particle and continuum solvers [8–12], in which the location of breakdown interfaces between continuum and rarefied regions was specified in advance [8,9,11,12] or identified after "one-shot" CFD simulation [10]. One immediate disadvantage by employing structured grid is that the pre-specified breakdown interface does not follow faithfully the interface determinde by some breakdown parameters [7,13], which may in turn either increases the runtime or induce inaccuracies of the solution. In addition, some techniques, such as particle cloning [12], overlapped region [8,12] and iteration [8] between particle and continuum regions, are used to reduced the statistical uncertainties in coupling the two solvers. Among these, the hybrid schemes developed by Wang et al. [11] and Roveda et al. [12] are potentially suitable for simulating unsteady flows, while the others were only designed for simulating steady flows. In the above, only one-dimensional and two-dimensional flows were demonstrated and extension to parallel or three-dimensional simulation has not been reported to the best knowledge of the authors. Recently, Wu et al. [1] has developed a parallel hybrid DSMC-NS scheme using 3D unstructured mesh. Parallel implementation is realized on distributed-memory parallel machines, e.g., PC-cluster system. In this method, a domain overlapping strategy, taking advantage of unstructured data format, with Dirichlet-Dirichlet type boundary conditions based on two breakdown parameters, was used iteratively to determine the choice of solvers in the spatial domain. The selected breakdown parameters for this study include: 1) a local maximum Knudsen number, proposed by Wang and Boyd [13], defined as the ratio of the local mean free path and local characteristic length based on property gradient and 2) a thermal non-equilibrium indicator defined as the ratio of the difference between translational and rotational temperatures to the translational temperature. A supersonic nitrogen flow ($M_\infty = 4$) over a quasi-2-D 25° wedge and a nitrogen flow, which two near-continuum parallel orifice jets underexpand into a near-vacuum environment, were simulated to verify its validity in simulating gas flows involving rarefied and continuum regions. In the present paper, this recently developed hybrid DSMC-NS scheme is described briefly for completeness. Improvement of the NS solver by using a pressure-based solver is described next. Then, several practical experiences learned from the development and verifications of the implementation are shown in detail. Finally, simulation results for a RCS (reaction control system) nozzle plume are presented to demonstrate its capability in simulating realistic flow problems.

2. Hybrid DSMC-NS Scheme Using 3D Unstructured Mesh

In the proposed coupled DSMC-NS method [1], steady-state flow calculation is assumed. There were two numerical flow solvers included: one is the 3-D DSMC code for rarefied or continuum breakdown and thermally non-equilibrium regions, named PDSC (Parallel Direct Simulation Monte Carlo Code), developed by Wu's group [14, eg] and the other is HYB3D, a density-based 3-D Euler and Navier-Stokes solver for continuum regions, developed by Koomullil [15, eg]. In the present study for simulating the RCS (reaction control system) nozzle plume issuing from the spacecraft, a pressure-based NS solver, named UNIC-UNS, using 3D unstructured mesh, developed by Chen and his coworkers [16,17, eg] that is applicable at all speeds, is used for continuum regions. It is rather straightforward to exchange the information between the PDSC and the UNIC solver in the proposed DSMC-NS scheme because both methods use the unstructured grid topology with parallel computing. However, proposed coupling procedures between DSMC and NS solvers are not limited to any specific codes, and selection of these two solvers is only for the purpose of demonstration. Both codes are introduced briefly in the following, respectively, for completeness.

2.1. DSMC Solver (PDSC)

Details of the features of the PDSC code can be found in the reference [14] and are only briefly described here for brevity. PDSC features 3-D unstructured grid, parallel computing with dynamic domain decomposition using MPI, variable time-step scheme with adaptive mesh refinement and treatment of high-temperature chemical reacting flows. In addition, iterative pressure boundary treatment is also available for treating internal flows. It can be implemented efficicntly on general distributed-memory parallel machines, such as PC-cluster system.

2.2. Navier-Stokes Solvers

Details of the HYB3D that is an density-based NS code have been described in detail elsewhere [15] and are skipped here for brevity. The present Navier-Stokes solver (UNIC-UNS), developed by Chen and his coworkers [16,17] employs an unstructured-grid topology and has the following important features: 1) Cell-centered finite-volume for the numerical integration of governing equations, 2) An upwind method with linear reconstruction scheme for convective flux evaluation, 3) Modified pressure-velocity-density coupling algorithm of the SIMPLE type with added pressure damping term, 4) Parallel computing based on domain decomposition with message passing interface (MPI), 5) Turbulent flow

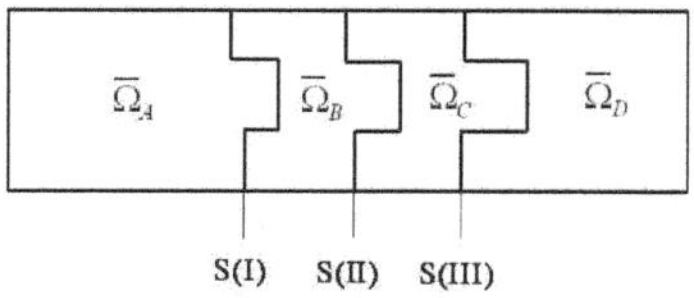

Figure 1. Sketch of numerical domain distribution of the present coulped DSMC-NS methods with overlapping regions and boundaries

simulation capability with the standard and extended k-ϵ turbulence models, and 6) General chemical reacting flow treatment. One of the most important features of this NS solver is the use of pressure-based method which allows accurate simulaiton of the flows at all speeds. Either implicit first-order Euler time-marching or second-order time-centered scheme can be used for time integration. A second-order spatial accuracy is achieved using Taylors series expansion and the gradients of the flow properties are computed using a least-square method. The creation of local extreme during the higher order linear reconstruction is eliminated by the application of limiter proposed by Barth and Jespersen [18]. Parallel computing of the NS solver also incorporates the graph-partition tool, METIS, which is the same as that in the PDSC.

2.3. Breakdown Parameters

A continuum breakdown parameter, proposed by Wang and Boyd [13] for hypersonic flows, is employed in the present hybrid DSMC-NS method as one of the criteria for selecting proper solvers. The continuum breakdown parameter Knmax is defined as

$$Kn_{max} = max\left[Kn_D, Kn_V, Kn_T\right] \tag{1}$$

where Kn_D, Kn_V and Kn_T are the local Knudsen numbers based on density, velocity and temperature, respectively. They can be calculated from the following general formula

$$Kn_Q = \frac{\lambda}{Q}\left|\nabla Q\right| \tag{2}$$

where Q is the specific flow property (density, velocity and temperature) and λ is the local mean free path. If the calculated value of the continuum breakdown parameter in a region is larger than a preset threshold value $Kn_{max}^{Thr.}$, then it cannot be modeled using the NS equation. Instead, a particle solver like DSMC has to be used for that region.

In addition, another breakdown parameter is used to identify regions that exhibit thermal non-equilibrium among various degrees of freedom. In the current study, this thermal non-equilibrium indicator is defined as

$$P_{Tne} = \left|\frac{T_{Tr} - T_R}{T_{Tr}}\right| \tag{3}$$

where T_{Tr} and T_R are translational and rotational temperature, respectively. If the value of the computed thermal non-equilibrium indicator in a region is larger than some preset threshold value P_{Tne}^{Thr} in the current study, then this flow region cannot be modeled correctly by the present thermal-equilibrium NS solver, which assumes thermal equilibrium among various degrees of freedom. Hence, the DSMC method has to be used for that region instead.

2.4. Overlapping regions between DSMC and NS domain

Figure 1 shows the sketch of overlapping regions and boundaries near the interface of the DSMC and NS solvers at an intermediate step (other than the first NS simulation step for the whole domain). The general iterative procedure of the present coupling framework is that running the DSMC solver first after the breakdown regions are identified, and then running the NS solver next with the boundary values calculated from DSMC simulations. Note that all domains mentioned in the following include the boundaries surrounding them. Domain $\Omega_A \cup \Omega_B \cup \Omega_C$ represents the DSMC simulation region, while domain $\Omega_B \cup \Omega_C \cup \Omega_D$ represents the NS simulation region; thus, domain $\Omega_{B-} \cup \Omega_{C-}$ is the designated overlapping region. Boundary conditions (Dirichlet-type) on Boundary-I ($= \Omega_A \cap \Omega_B$) for NS simulation come from part of the previous iterative DSMC simulation, while boundary conditions (Dirichlet-type) on Boundary-III ($= \Omega_C \cap \Omega_D$) for DSMC simulation come from part of the previous iterative NS solution. Location of Boundary-I is determined from strict comparison of breakdown parameters (Kn_{max} and P_{Tne}), computed based on previous iterative solution of domain $\Omega_A \cup \Omega_B \cup \Omega_C \cup \Omega_D$, with the preset criteria. Location of Boundaries-II and -III are then determined by extending from Boundary-I towards the neighboring continuum region. In addition, the thickness (number of cell layers) of domains Ω_B and Ω_C can be adjusted to achieve better convergence for the coupling procedure. Furthermore, in the current coupled DSMC-NS method the choice of solution update for each cell is based on its domain type. Domain $\Omega_A \cap \Omega_B$ is the region where the updated solution comes from the DSMC simulation, while domain $\Omega_C \cap \Omega_D$ is the region where the updated solution comes from the NS simulation.

2.5. Coupling Procedures

In brief summary, major procedures of the present hybrid DSMC-NS method are listed as follows, referring to the overlapped region as shown in Figure 1:

(a) Apply the NS code to simulate the whole flow field as continuum.

(b) Determine the locations of Boundary-I and -III and, thus, the DSMC simulation domain ($\Omega_A \cup \Omega_B \cup \Omega_C$).

(c) Impose Dirichlet-type boundary conditions (velocities, temperature and number density) on Boundary-III, obtained from latest NS simulation, for the next DSMC simulation domain ($\Omega_A \cup \Omega_B \cup \Omega_C$).

(d) Simulate and sample the flow field in the DSMC domain ($\Omega_A \cup \Omega_B \cup \Omega_C$), using the PDSC code, until acceptable statistical uncertainties are reached.

(e) Impose Dirichlet-type boundary conditions (velocities, temperature and density) on Boundary-I, obtained from latest DSMC simulation, for the next NS simulation domain ($\Omega_B \cup \Omega_C \cup \Omega_D$).

(f) Conduct flow simulation in the NS domain ($\Omega_B \cup \Omega_C \cup \Omega_D$), using the NS code, to obtain a converged steady-state solution.

(g) Update solution of the whole computational domain.

(h) Repeat from Steps (b) to (g) until the maximum number of coupling iterations is exceeded or the preset convergence criterion is reached.

2.6. Practical Implementation

Numerical simulations with the hybrid DSMC-NS code are conducted on a distributed-memory PC cluster system (64 Nodes, dual processors, 2GB RAM per node, Gbit switching hub) running under a Linux operating system. 32 processors are used throughout this study, unless otherwise specified. The PDSC and the NS codes are coupled through a simple shell script on Linux system. Thus, the hybrid DSMC-NS code is expected to be highly portable among parallel machines with distributed memory. Most importantly, our experience shows that the I/O time related to the switching of solver and read/write files is negligible as compared to the simulation time used by each solver.

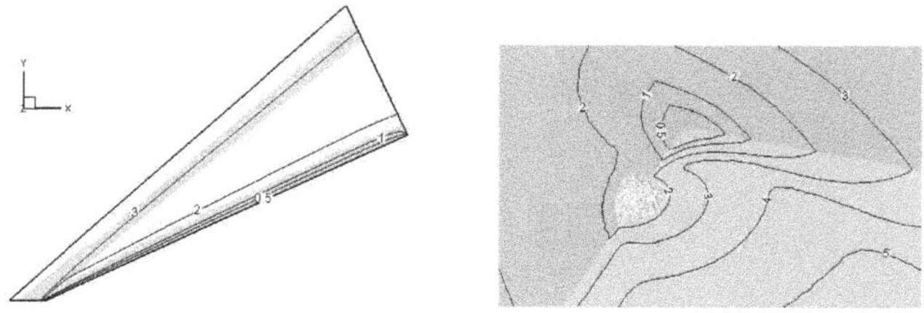

Figure 2. Mach number distribution for the quasi-2D supersonic wedge flow (left) and the two orifice jet (right).(Gray area: DSMC domain; others: NS domain)

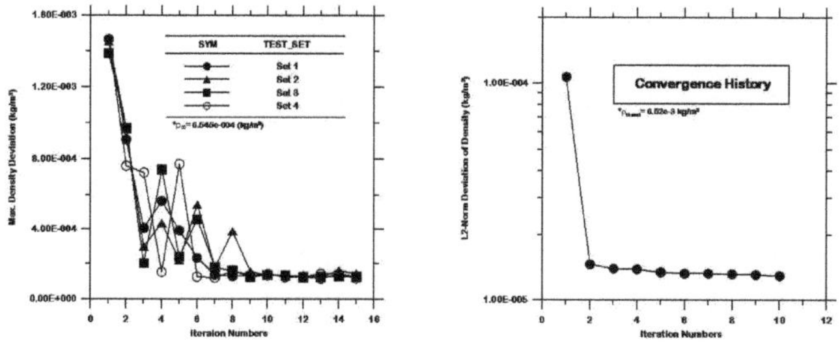

Figure 3. Convergence history of maximum density deviation for:(left) the quasi-2-D 25° wedge flow; (right) two parallel near-continuum orifice free jets flow.

3. Results and Discussions

3.1. Validations

Completed parallel hybrid DSMC-NS code was tested using a supersonic nitrogen flow ($M_\infty = 4$) over a quasi-2-D 25° wedge and a nitrogen flow, in which two near-continuum parallel orifice jets underexpand into a near-vacuum environment, were simulated to verify its validity [1]. Corresponding flow conditions are shown in Table 1 (wedge flow) and Table 2 (orifice jets) for completeness. Simulation data compare excellently with the benchmark simulation results using DSMC for the quasi-2-D wedge flow, and reasonably well with the experimental data for the 3-D orifice jet flow. Details of the description can be found in [1] and are skipped here for brevity.

From these two simulations, we have found that the number of couplings between DSMC and NS solvers strongly depends upon the flow conditions near the breakdown interfaces, which is described in the following.

In the case of supersonic flow past a quasi-2-D wedge (Figure 2a), subsonic flow dominates in the regions normal to the breakdown interface above the boundary layer along the wedge wall that necessitates more number of couplings to exchange the information between two solvers (Figure 3a), although supersonic flow dominates in the regions normal to the breakdown interface around the oblique shock. Note the four sets of data in Figure 3a represent different simulation conditions in the number of cell layers of the overlapped region (1~4 for set 1~4, respectively) and the threshold values

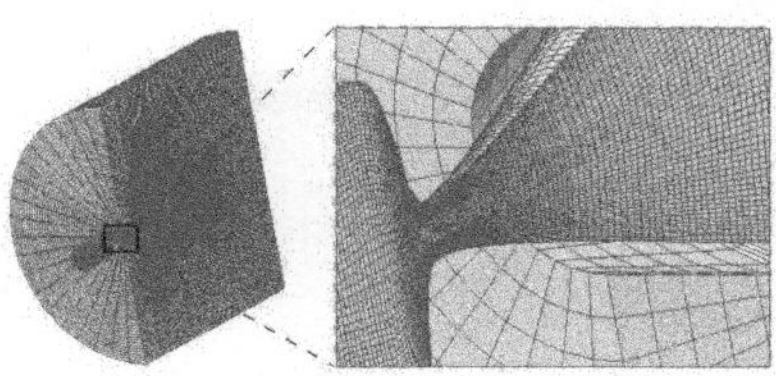

Figure 4. Mesh distribution of the near-field plume simulation.

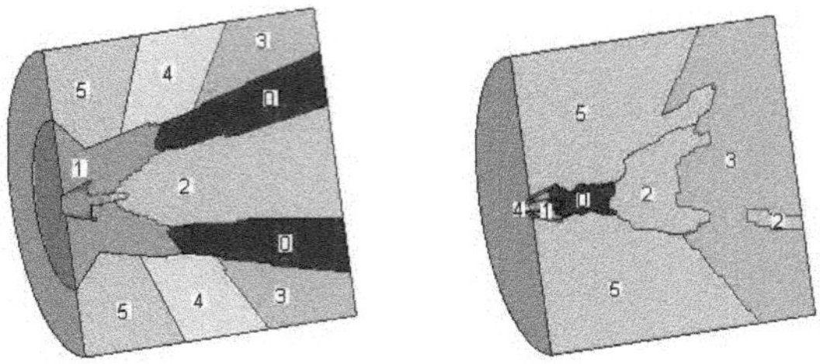

Figure 5. Domain decoposition of DSMC computation. Initial(left) and final (6th) iteration (right).

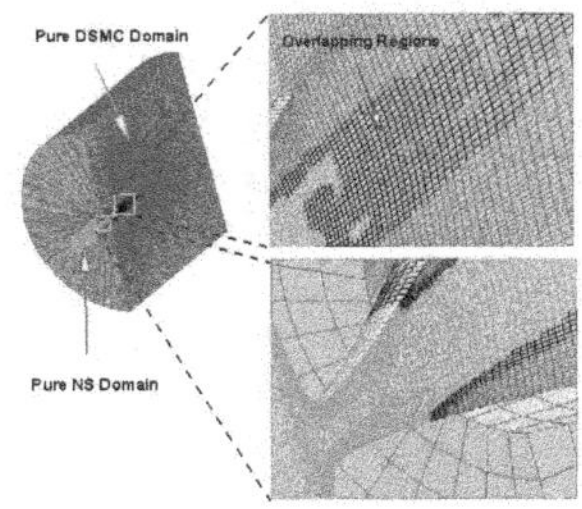

Figure 6. Distribution of the DSMS and NS domains after the 6th coupled iteration.

of breakdown parameters ($Kn_{max}^{Thr.} = 0.02 \sim 0.04$, $P_{Tne}^{Thr.} = 0.03 \sim 0.06$), which all have the similar trend. However, in Figure 2b, supersonic flow dominates normal to the breakdown interface around entrance regime of orifice jets, which greatly reduces the number of couplings (Figure 3b, no. of overlapped cell layer= 2) required for convergence as seen from the simulation. The above observation is very important from the viewpoints of practical implementation. For example, in the early stage of simulation we can determine the number of couplings required for convergence by simply monitoring to which the flow regime (supersonic or subsonic) normal to the breakdown interface belongs. If most flows normal the breakdown interface are supersonic, then two coupling iterations should be enough for a fairly accurate solution. If not, more coupling iterations (e.g. less than ten) are required to have a converged solution. Further investigation in determining the optimum number of coupling iterations is required in practical applications of the current hybrid method.

3.2. RCS Nozzle Plume Simulation

The proposed coupled DSMC-NS method has been verified by the previous two test cases. To demonstrate its applicability, we apply it to simulate a plume issuing from a RCS thruster, which is a very challenging and important problem in designing the ADCS system on a spacecraft. Important flow conditions, which are summarized in Table 3, include: nitrogen gas, stagnation pressure of 0.1 bar, stagnation temperatur of 300K, throat diameter of 4.36 mm, length of the convergent part of 8

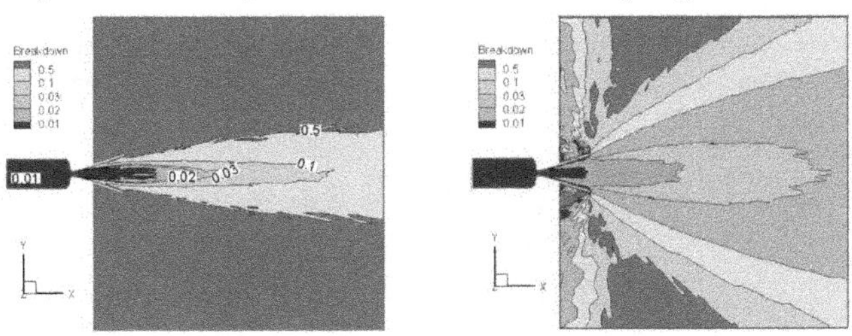

Figure 7. Distribution of continuum breakdown (left) and thermal non-equilibrium ratio (right) at the 6th coupled iteration.

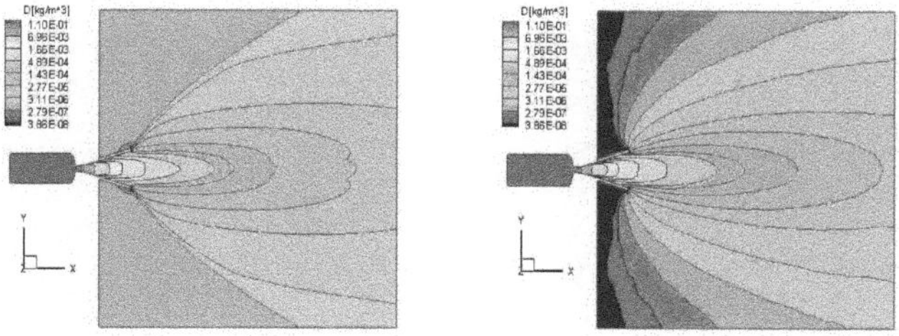

Figure 8. Density distribution of near-field nozzle plume using using pure NS solver (left) and hybrid DSMC-NS solver (the 6th coupled iteration)(right).

Table 1
Free-stream conditions in supersonic flow over quasi-2-D 25° wedge.

Gas	ρ_∞	U_∞	T_∞	M_∞
N_2	$6.545^{-4}\ kg/m^3$	$1111\ m/s$	$185.6\ K$	4

mm, length of the divergent part of 50 mm, and area expansion ratio of 60. Six processors are used in this case. Estimated Reynolds number at the throat using inviscis theory is about 62,000, which necessitates the use of turbulence model in the NS solver. In this study we have used standard k-ϵ model unless otherwise specified. Simulation conditions of the PDSC for coupled method for the RCS nozzle plume are summarized are listed as follows: ~210,000-250,000 cells, 2.2-2.5 million particles, reference timestep size 6.8E-08 seconds and number of sampling timesteps 8000. 3 layers of overlapping region is used. $Kn_{max}^{Thr.}$ and P_{Tne}^* is set as 0.03 and 0.03, respectively. In the current simulation, the downstream boundary condition of the near-field simulation is not predicable in advance of the simulation. Thus, supersonic boundary condition in the NS solver is chosen and applied for those surface boundaries and, furthermore, the NS solver breaks down in the backflow region due to the rarefaction of flow.

Figure 4 illustrates the mesh distribution with exploded view near the C-D nozzle. The three-dimensional (hexahedrons) rather than the axisymmetric mesh is used since we are interested in simulating the far-field plume interaction with the spacecraft in the future. With the present mesh, we can simply add extra grid which includes the spacecraft body to the original 3D mesh. The total number of hexahedral cells is 380,100. Note that length of reservoir in axial direction is about 3.5 times of the inlet diameter of the nozzle, which is necessary to accurately simulate the very low-speed gas flow inside the reservoir.

Figure 5 illustrates the initial and final (6th iteration) distribution of domain decomposition in the DSMC solver by means of dynamic domain decomposition using a multi-level graph-partitioning technique [14]. It is clear that the spatially uniform decomposition at the initial stage has evolved to the spatially non-uniform decomposition, which reflects the correct density distribution at the end. Figure 6 shows the distirbution of DSMC and NS domains after the 6th coupled iteration. Most of the regions inside the nozzle are NS domains, while DSMC domains dominate outside the nozzle exit with intrusion along the nozzle wall from the lip to the mid of divergent part of the nozzle.

Figure 7 shows the distribution of continuum breakdown and thermal non-equilibrium ratio between rotational and translational degrees of freedom at the 6th coupled iteration. The continuum breakdown parameter generally increases along the streamlines from the nozzle and becomes very large due to rapid expansion outside the nozzle. Thermal non-equilibrium ratio shows a similar trend. It is obvious that regions in the reservoir and in the core of the nozzle flow are in continuum and thermal equilibrium

Table 2
Sonic conditions at the orifice exiting plane in two parallel near-continuum orifice free jets flow,

Gas	ρ_{throat}	U_{throat}	T_{throat}	Re_{throat}
N_2	$6.52^{-3}\ kg/m^3$	$314\ m/s$	$237.5\ K$	401

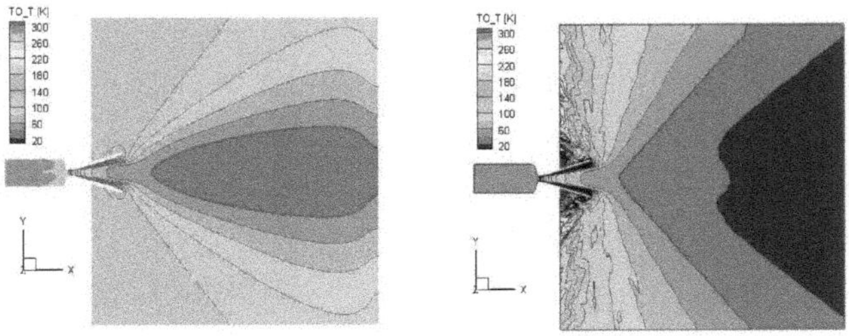

Figure 9. Temperature distribution of near-field nozzle plume using pure NS solver (left) and hybrid DSMC-NS solver (the 6th coupled iteration)(right).

Table 3
Flow conditions of the plume simulation issuing from RCS nozzle.

Gas	P_0	T_0	T_w	D_{throat}	Area Ratio
N_2	0.1 bar	300 m/s	300 K	4.36 mm	60

as expected. In the boundary layer region of the nozzle wall near the exit, value of continuum breakdown parameter is rather large (>0.1) due to the large value of velocity gradient. However, the thermal non-quilibrium ratio is close to unity (between 1 and 1.05), which is suprising. This implies that the in the boundary layer regions may be still in thermal equilibrium although the continuum breakdown parameter is large, which requires further investigation. In addition, the scattered data of the thermal non-equilibrium in the backflow region originates from the statistical uncertainties due to very few simulated particles coming into this area. Based on the data in Figure 7 we can conclude that simulation of the plume flow using either NS or DSMC solver along is practically impossible, which necessitates the use of the hybrid DSMC-NS scheme as developed in the present study.

Figure 8 shows the density contours of the pure NS simulation and the coupled DSMC-NS method after the 6th coupled iteration. It can be found that the density distribution of the coupled method have similar trend to that of the pure NS simulation, although the data differ very much. In addition, only the hybrid DSMC-NS scheme can predict the back flow near the lip of the nozzle. For other flow properties (temperature, streamline and Mach number, as shown in Figures 9, 10 and 11, respectively), the simulation results obtained by the hybrid DSMC-NS scheme are very much distinct from those by the pure NS solver. Similar to density distribution, temperature distribution, streamline distribution, and Mach number dostribution predicted by the coupled DSMC-NS method are more reasonable than those by the the pure NS solver, because the coupled method is capable of handling the continuum breakdown in the rarefied regions and thermal non-equilibrium in the backflow region. Note the temperature scatters in the backflow region can be reduced if more simulated particles in the DSMC solver is used.

Figure 12 shows the L2-norm of number density and temperature as a function of number of it-

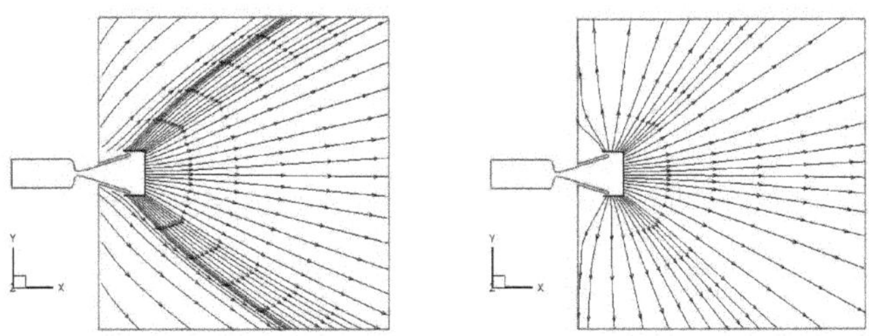

Figure 10. Streamline distribution of near-field nozzle plume using pure NS solver (left) and hybrid DSMC-NS solver (the 6th coupled iteration)(right).

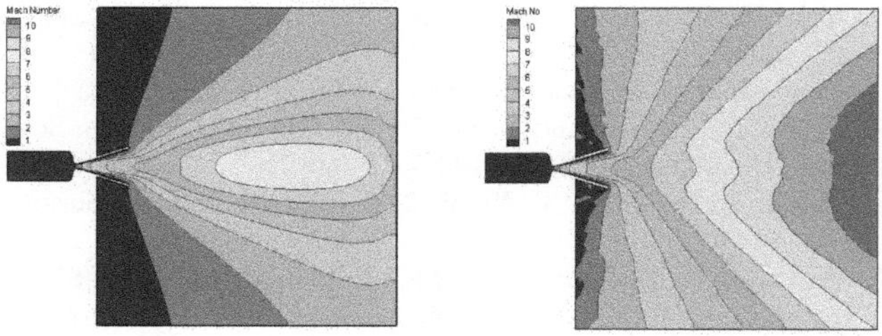

Figure 11. Mach number distribution of near-field nozzle plume using pure NS solver (left) and hybrid DSMC-NS solver (the 6th coupled iteration)(right).

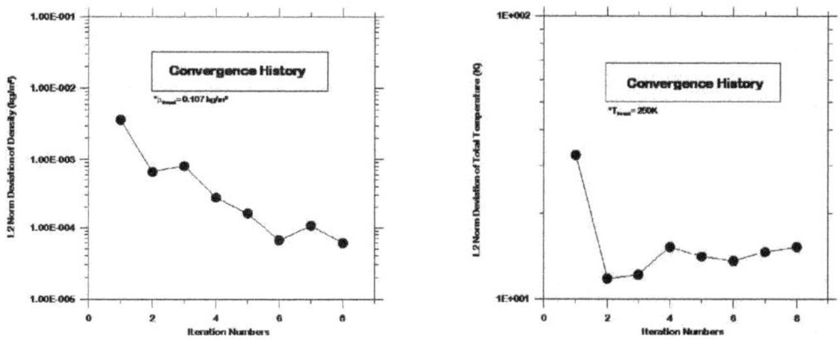

Figure 12. Convergence history of number density(left) and total temperature (right) for the nozzle plume simulation using hybrid DSMC-NS scheme

erations. Results show that the L2-norm of number density decreases 1-2 order of magnitude within 6-8 iterations, while the L2-norm of temperature drops down to a value more than 10K even after 8 iterations, which is too high. The reason of the high L2-norm of temperature originates from the very high statistical uncertainties in the back flow region in the DSMC domain, although the solution in other regions have converged much earlier. Based on this observation and those in the test cases (Figure 3), we propose the the hybrid DSMC-NS simulation converges when the L2-norm of number density decreases down to 0.1 of that resulting from the first iteration, regardless of the magnitude of L2-norm of temperature.

4. Conclusions

A hybrid DSMC-NS approach for steady-state flows using 3-D unstructured mesh is presented to combine the high computational efficiency of the NS solver in continuum and thermal-equilibrium regions with high fidelity of the DSMC method in "breakdown" regions. Flexible overlapping regions between DSMC and NS simulation domains are designed by taking advantage of the unstructured grid topology in both solvers. Two breakdown parameters, including a continuum breakdown parameter proposed by Wang and Boyd [13] and a thermal non-equilibrium indicator, are employed to determine the DSMC simulation and NS simulation domains, in addition to the concept of overlapping regions. Proposed hybrid DSMC-NS scheme was verified using a quasi-2D supersonic wedge flow and a realistic 3D flows with two parallel near-continuum orifice jets expands into a near-vacuum environment. Results show that the number of couplings for convergence between two solvers is approximately 2-3 if supersonic flows dominate near the breakdown interface, while it increases up to 8-10 if subsonic flows dominate near the breakdown interface. At the end, proposed hybrid scheme is employed to simulate a realistic RCS plume to demonstrate it capability in handling realistic challenging problems.

REFERENCES

1. J.-S. Wu, Y.-Y. Lian, G. Cheng, R. P. Koomullil and K.-C. Tseng, Journal of Computational Physics (to appear). (2006)
2. J. N. Moss, J. M. Price,Journal of Thermophysics & Heat Transfer, (1997) 11, 321-329.
3. H.-P. Cheng, R.-Y. Jou, F.-Z. Chen, and Y.-W. Chang, Journal of Vacuum Science & Technology A: Vacuum, Surfaces, and Films,(2000), 18, 543-551.
4. M. Taniguchi, H. Mori, R. Nishihira, T. Niimi, 24th International Symposium on Rarefied Gas Dynamics (2004).
5. S. T. Junker, R. W. Birkmire, F. J. Doyle III, AIChE Journal,(2005), 51, 878-894.
6. V. Versteeg, C. T. Avedisian, R. Raj, U.S Patent Number: 5,451,260 (1994).
7. G. A. Bird,(Oxford Univ. Press, New York, 1994).
8. O. Aktas, N. R. Aluru, J. Comput. Phys.,(2002) 178, 342-372.
9. A. L. Garcia, J. B. Bell , W. Y. Crutchfield , B. J. Alder, J. Comput. Phys.,(1999), 154, 134-155.
10. C. E. Glass, P. A. Gnoffo, NASA report TM-2000-210322 (2000).
11. W.-L. Wang, Q.-H. Sun, I.D. Boyd, 8th AIAA/ASME Joint Thermophysics and Heat Transfer Conference (AIAA Paper 2002-3099, 2002).
12. R. Roveda, D. B. Goldstein, P. L.Varghese, J. Spacecraft & Rockets,(1998), 35, 258-265.
13. W.-L Wang, I. D. Boyd, Physics of Fluids,(2003) 15, 91-100.
14. J.-S. Wu, K.-C. Tseng, U.-M. Lee, Y.-Y. Lian, 24th International Symposium on Rarefied Gas Dynamics (2004).
15. R. P. Koomullil, B. K. Soni, AIAA Journal,(1999), 37, 1551-1557.
16. H. M. Shang, M. H. Shih, Y.-S. Chen, P. Liaw, Proceedings of 6th International Symposium on Computational Fluid Dynamics (1995).
17. H. M. Shang, Y.-S. Chen, AIAA Paper, 97-3183 (1977).
18. T. J. Barth, D. C. Jespersen, AIAA Paper 89-0366, (1989).

Parallel Computational Fluid Dynamics – Parallel Computing and Its Applications
J.H. Kwon, A. Ecer, J. Periaux, N. Satofuka and P. Fox (Editors)

A Parallel CFD-CAA Computation of Aerodynamic Noise for Cylinder Wake-Airfoil Interactions

Sven Peth[a], Jung H. Seo[b], Young J. Moon[b], Marc C. Jacob[c] and Frank Thiele[a]

[a] Technische Universität Berlin, Institut für Strömungsmechanik und Technische Akustik
Müller-Breslau-Straße, 10623 Berlin, Germany

[b] Korea University, Department of Mechanical Engineering
Anam-dong, Sungbuk-ku, Seoul, 136-701, Korea

[c] Ecole Centrale de Lyon Centre Acoustique du LMFA, UMR CNRS 5509
69134 Ecully Cedex, France

Aerodynamic noise from the rod wake-airfoil interactions at M=0.2 and Re_D=46,000 is computed by solving the linearized perturbed compressible equations (LPCE), with the acoustic source and hydrodynamic flow variables computed from the incompressible the incompressible large eddy simulation (LES). The computational results for flow and acoustics are critically validated with the experimental data measured at the Ecole Centrale de Lyon.

1. Introduction

The interaction of vortices with structures is one important factor in industrial design, because it creates noise. And reduction of noise is a sales argument for many products. The way of reducing it is strongly connected with understanding its sources. The rod-airfoil configuration is a simple benchmark setup for the study of vortex interaction noise. The rod acts as vortex generator and creates a Karman vortex street which hits the airfoil. And the properties of the flow past the rod were investigated.

In the present study, a hydrodynamic acoustic splitting method is applied to the rod-airfoil configuration at M=0.2. First, a three-dimensional large eddy simulation (LES) computes the hydrodynamic properties. An incompressible LES is used to obtain the acoustic sources, the total pressure changes in the flow. Second, the two-dimensional acoustic field is calculated by the linearized perturbed compressible equations (LPCE)[1]. Subsequently, there is a 2D Kirchhoff extrapolation to the far-field and a 3D correction after Oberai[2]. An important advantage in the splitting is the computational efficiency. Both solvers can use different grids which are optimized for their particular needs.

The investigation of rod-airfoil configurations has a short history which means there is only little literature, yet, for example by Sorgven et al.[3] and Magagnato et al.[4] or Boudet[5]. This work uses experimental results by Jacob[6] as reference in a similar configuration. The objectives of this study are not only to evaluate critically the computational methodology proposed by Seo and Moon[1] by comparing the hydrodynamic and aeroacoustic results

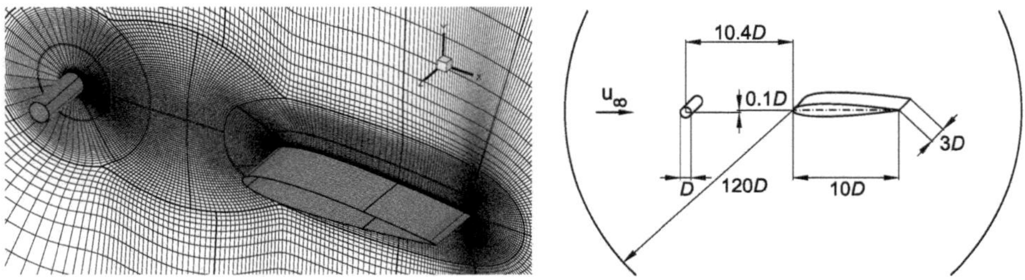

Figure 1. Computational domain, left: 3D mesh of LES with every other grid point plotted; right: sketch of relative dimension.

with the experiment[6] measured at Ecole Centrale de Lyon but also to understand the governing physics to identify the noise sources.

2. Computational Grid

The dimensions of the computational domain aligns to the experimental setup by Jacob[6]. It includes a rod and a NACA0012 airfoil in tandem configuration. The governing flow parameters are Re_D=46,000 and M=0.2. All computations use nondimensional values, but for experimental comparison it is necessary to provide them with units. The rod diameter D is equal to 10 mm, the inflow velocity $u_\infty = 72\,\text{m/s}$, the ambient temperature $T_0 = 293\,\text{K}$ and the ambient pressure $p_0 = 98.9\,\text{kPa}$. There is a small offset between the centers of the rod and airfoil, because in technical applications it is more likely that vortices are not symmetric to the airfoil chord. The trailing edge of the airfoil is rounded with a radius, thus the chord length is 99 mm instead of 100 mm. Different grids were used for hydrodynamic and acoustic calculations. The hydrodynamic computation is performed by an incompressible LES in 3D and employs 3.14 million grid cells with 32 blocks (30 cells used in the spanwise direction). The grid in the wake region is designed to be

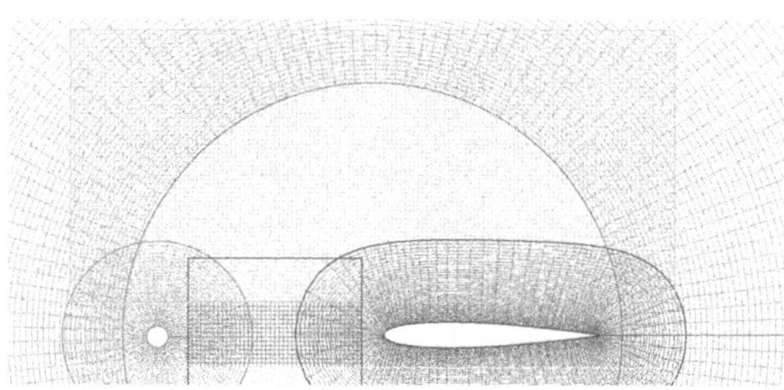

Figure 2. 2D mesh for acoustic calculation, every other grid point depicted.

homogeneous to avoid quick diffusion of the vortices. The grid size is chosen that a vortex is resolved by 20 cells. The nondimensional wall distance y^+ of the airfoil is rather large with a value of 12.5 and some near-wall turbulence physics are not resolved with the present grid. As far as noise prediction is concerned, the present grid resolution may be not too off because main noise comes from the large scale motions of the eddies. Due to the large y^+ the largest ratio of wall cell length to thickness is 47. The hydrodynamic domain has the shape of a cylinder with a diameter of $240D$. All outer boundaries are defined as inflow, the body boundaries are set to no slip, and periodic boundaries are used in spanwise direction.

The acoustic grid consists of 0.16 million cells in 6 blocks. It is an overlaid grid which makes it very easy to fit perfectly to the body contour. The grid in the wake region and around the bodies is coarser than the hydrodynamic grid, because the aeroacoustic and hydrodynamic calculation use the same time step, but the acoustic information propagates with the speed of sound while the flow propagates roughly with u_∞. The domain of the acoustic calculation is a circle of $120D$, and it is relatively fine at the outer boundary compared to the hydrodynamic grid. The outer grid size has been chosen in order to resolve the wavelength of an acoustic wave with 10 kHz with 7 cells.

3. Hydrodynamics

The main idea is to split the whole computation into a hydrodynamic and an acoustic part. The hydrodynamic computation is performed by an incompressible large eddy simulation (iLES) while the acoustic computation uses the linearized perturbed compressible equations[1] (LPCE). Therefore the instantaneous total flow variables are decomposed into the incompressible and perturbed compressible variables.

$$\begin{aligned} \rho(\vec{x},t) &= \rho_0 + \rho'(\vec{x},t) \\ \vec{u}(\vec{x},t) &= \vec{U}(\vec{x},t) + \vec{u}'(\vec{x},t) \\ p(\vec{x},t) &= P(\vec{x},t) + p'(\vec{x},t) \end{aligned} \tag{1}$$

The compressible variables represent the unsteady viscous flow, while the perturbed compressible variables describe the acoustic fluctuations and all other compressible components. The exclusion of compressibility in the Navier-Stokes equations for the iLES leads to the incompressible Navier-Stokes equations. Their filtered form is written as

$$\frac{\partial \tilde{U}_j}{\partial x_j} = 0, \tag{2}$$

$$\rho_0 \frac{\partial \tilde{U}_i}{\partial t} + \rho_0 \frac{\partial}{\partial x_j}\left(\tilde{U}_i \tilde{U}_j\right) = -\frac{\partial \tilde{P}}{\partial x_i} + \mu_0 \frac{\partial}{\partial x_j}\left(\frac{\partial \tilde{U}_i}{\partial x_j} + \frac{\partial \tilde{U}_j}{\partial x_i}\right) - \rho_0 \frac{\partial}{\partial x_j} M_{ij} \tag{3}$$

where the resolved states are marked by ($\tilde{}$). There are many ways to model the sub-grid tensor M_{ij}. One aspect of this work is to reduce the computational effort, consequently the tensor is set to zero, i.e. the sub-scale turbulence is not modeled.

The iLES is solved by an iterative implicit fractional step method (Poisson's equation for pressure). The momentum equations are time-integrated by a four stage Runge-Kutta method and spatially discretized by a sixth-order compact finite difference[7] scheme. Then

Figure 3. Instantaneous vorticity with isosurfaces at $\omega D/u_\infty = -3$ and $\omega D/u_\infty = 3$.

the pressure field is iteratively solved to satisfy continuity, and the velocity is updated by a correction step. The results of the iLES act as source for the LPCE. Consequently, the spanwise average of the results have to be saved for every time step. The computation covers 50 cycles of Karman vortex shedding of the rod to have reasonable spectral-statistics. Two isosurfaces of the instantaneous vorticity are plotted in Figure 3. The three-dimensional character of the flow is clearly visible. There are three sources for vorticity, the rod, the leading edge and the trailing edge of the airfoil. The strongest source is the rod which also creates a Karman vortex street that hits the leading edge of the airfoil. The vortex shedding of the cylinder is very unsteady.

4. Acoustics

The aeroacoustic field at zero spanwise wave number k_z=0 is obtained by solving the linearized perturbed compressible equations (LPCE). The only acoustic source is $\mathrm{D}P/\mathrm{D}t$ from the iLES solution. The interpolation from hydrodynamic to acoustic grid is performed by a Lagrangian interpolation method. The far-field sound pressure is extrapolated by a 2D Kirchhoff method in frequency domain. The acoustic pressure in mid-span plane has to be adapted by Oberai's 3D correction, and a spanwise coherence function of the wall pressure is used to acquire the acoustic pressure for the long span for the experiment.

4.1. Linearized perturbed compressible equations (LPCE)

The linearized perturbed compressible equations can be derived by applying the decomposed state equations (1) to the Navier-Stokes equations, and subtracting the compressible Navier-Stokes equations from them. The LPCE are written in a vector from as,

$$\frac{\partial \rho'}{\partial t} + \left(\vec{U} \cdot \nabla\right) \rho' + \rho_0 \left(\nabla \cdot \vec{u}'\right) = 0 \tag{4}$$

$$\frac{\partial \vec{u}'}{\partial t} + \nabla \left(\vec{u}' \cdot \vec{U}\right) + \frac{1}{\rho_0} \nabla p' = 0 \tag{5}$$

$$\frac{\partial p'}{\partial t} + \left(\vec{U} \cdot \nabla\right) p' + \gamma P \left(\nabla \cdot \vec{u}'\right) + \left(\vec{u}' \cdot \nabla\right) P = -\frac{\mathrm{D}P}{\mathrm{D}t}. \tag{6}$$

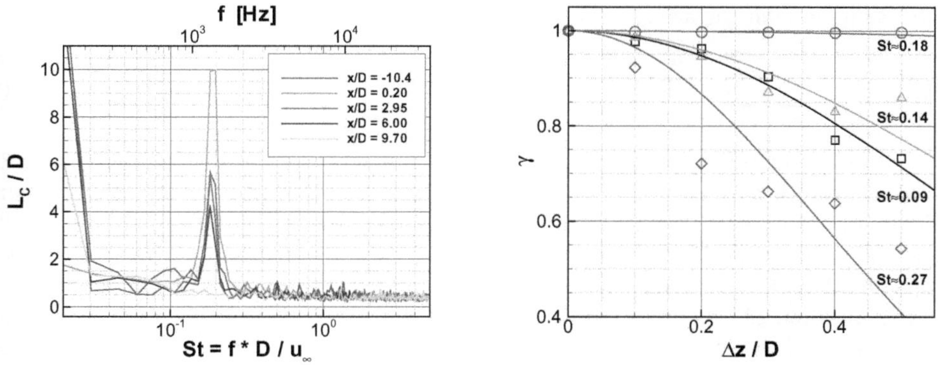

Figure 4. Coherence as function of frequency, left: coherence length at different positions, $x/D = 0$ matches the leading edge; right: regression at $x/D = 2.95$.

The LPCE is solved in 2D only, because a 3D approach needs a large dilatation in spanwise direction for resolving the acoustic waves. A 2D method with subsequent correction is more efficient and produces sufficient results. The LPCE is time integrated by a four-stage Runge-Kutta method and spatially discretized by a sixth-order compact finite difference scheme. For damping numerical instabilities a tenth-order spatial filter by Gaitonde et al.[8] is used.

4.2. Computation of Far-Field Sound

The acoustic grid covers a circular domain with a radius of $60D$. The microphones of the experiment are at a distance of $185D$ from the center of the airfoil. The extrapolation is done by a 2D Kirchhoff method. According to Oberai[2] the 3D radiated far-field acoustic pressure $\hat{p}'_{\text{3D}}$ at the $z = 0$ plane is related to a 2D predicted acoustic pressure $\hat{p}'_{\text{2D}}$ at the zero spanwise wave number $k_z = 0$ by

$$\hat{p}'_{\text{3D}}(r, \theta, 0, \omega) \approx \hat{p}'_{\text{2D}}(r, \theta, \omega) \frac{1+i}{2} \sqrt{\frac{\omega}{c_0 \pi r}} L_{\text{S}}, \tag{7}$$

with L_{S} as the simulated spanwise length of the iLES.

Finally, the acoustic pressure for the simulated span (L_{S}) must be corrected to the actual span L of the experiment. First presume a long span object which is divided into N subsections with each the span of L_s, i.e. $L = N \cdot L_{\text{S}}$. With a statistical homogeneity in spanwise direction and the coherence function which has the properties of a Gaussian distribution[6,9],

$$\gamma_{ij} = \exp\left[-(i-j)^2 \left(\frac{L_{\text{S}}}{L_{\text{C}}} \right)^2 \right], \tag{8}$$

a relation between the acoustic pressure of long span to simulated span is given by

$$|\hat{p}'_{\mathrm{L}}|^2 = |\hat{p}'_{\mathrm{S}}|^2 \sum_{i=1}^{N}\sum_{j=1}^{N} \exp\left[-(i-j)^2\left(\frac{L_{\mathrm{S}}}{L_{\mathrm{C}}}\right)^2\right]. \tag{9}$$

The measuring of γ_{ij} as function of acoustic pressures between subsections is hardly possible. Another available source is a spanwise coherence function of the wall pressure. Figure 4 right shows the coherence at the maximum thickness of the airfoil, $x/D = 2.95$. The coherence length L_{C} as sole parameter in γ can be obtained by a regression. It is apparent that the coherence depends strongly of the frequency/Strouhal number.

The question of where to measure the wall pressure coherence is more interesting. Figure 4 left shows the coherence length distribution over the whole spectra for different locations. All graphs have a peak at the vortex shedding frequency, except for the trailing edge, here is no significant coherence at all. At rod and maximum thickness of airfoil is $L_{\mathrm{C}}/D = 5$, and more downstream the value decreases. The model considers just one distribution for the whole domain. So one has to decide for a distribution that represents the area best. The values of the maximum thickness of the airfoil seem to be a reasonable choice, because the airfoil's laminar part is rather small. The value of 5 fits also well with the data from experiments[6]. The correction has only an impact to a small bandwidth around the shedding frequency. The higher frequencies' coherence length is too short ($L_{\mathrm{C}}/D < 1$ equals to $L_{\mathrm{C}}/L_{\mathrm{S}} < 0.33$) to have any effect.

4.3. Comparison with experiment

The experimental data in this study was obtain by Jacob from Ecole Central de Lyon[6]. Figure 5 depicts a power spectrum density (PSD) of the wall pressure in that region, and it clearly shows peaks for the fundamental and harmonic frequency. The results match well with a LES by Jacob[6].

The computational methodology for obtaining the far-field was described above. The microphones are positioned in a distance of $185D$ from the airfoil center (half chord

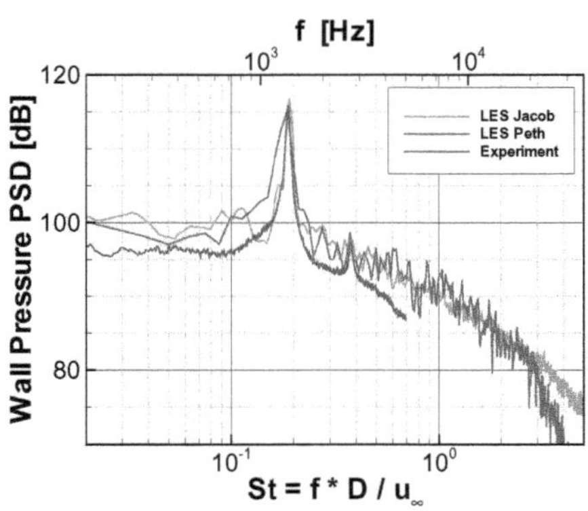

Figure 5. Wall pressure PSD at $x/D = 2$.

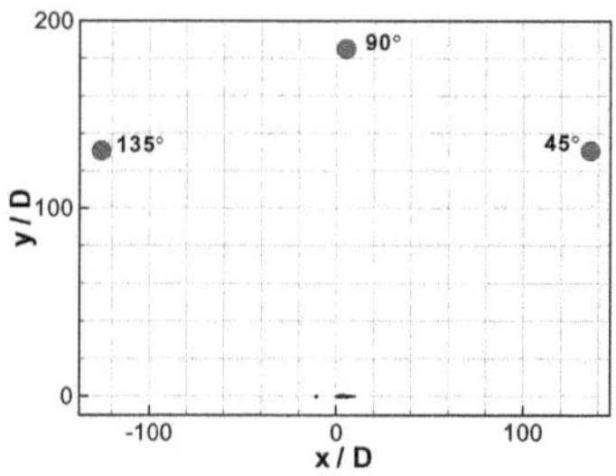

(a) position of microphones

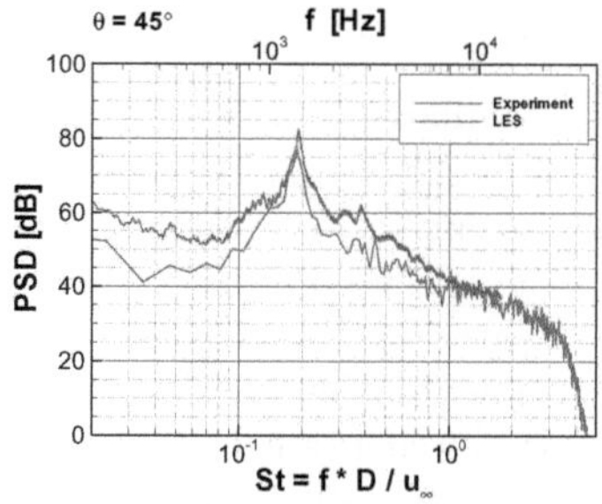

(b) observation at 45°

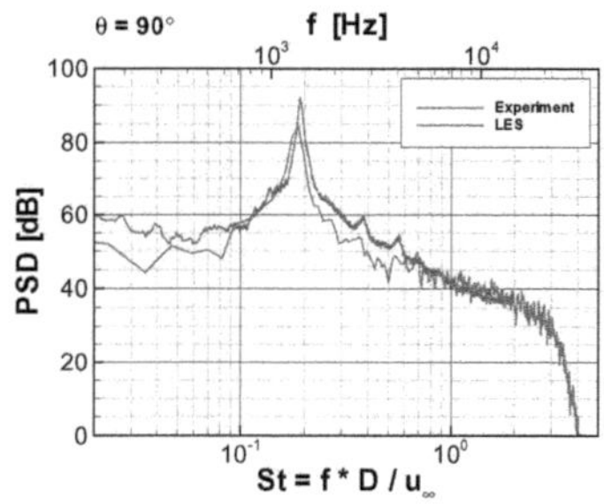

(c) observation at 90°

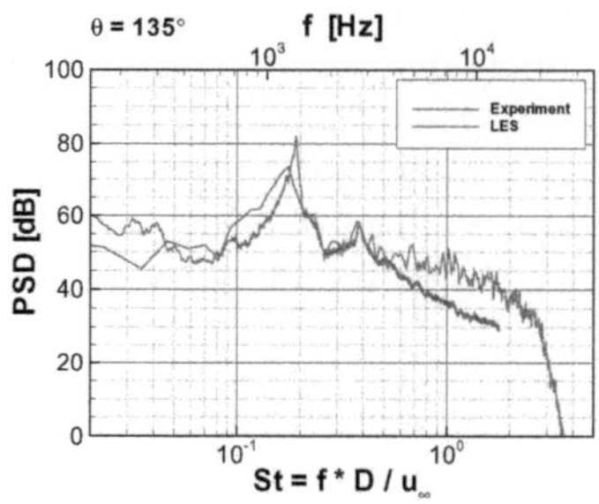

(d) observation at 135°

Figure 6. Far-field acoustic PSD for 3 different observation angles.

downstream of the leading edge) in mid-span plane. The simulation uses just 6 averages and has a spectral resolution of 84 Hz. Figure 6 (a) shows a sketch of the location of the microphones in relation with the rod-airfoil configuration. There are comparisons for three different positions, 45°, 90° and 135°. The LES predicts the peak of the shedding frequency accurately. Its frequency is slightly too small, but its magnitude is always underpredicted by 5 to 7 dB. The overall shape of the distributions agrees well, it just looks like the overall level of the simulation is too low. But this discrepancy is not constant for all angles. At 45° the offset seems to be biggest, while at 90° it is significantly smaller. And at 135° the first harmonic peak is perfectly predicted, but the fundamental peak fits poorly which is maybe caused by the coarse spectral resolution. If all simulated distributions would be shifted to the experimental data then one could see that there is a hump at high frequencies $St > 0.6$. One reason could be that the LES resolves smaller scales than the experiment[6]. Also the signal proceeding is not optimal in terms of spectral resolution and averagings. It is also possible that the quality of the hydrodynamic and acoustic grid cause the deviation. The broadband noise is strongest at 45° and decreases for increasing angles. This is an indication that the dominant source for the broadband noise is at the trailing edge.

5. Conclusions

The various comparisons of hydrodynamic and aeroacoustic results with the experimental data show that the computational methodology used in the present study is reasonably consistent and accurate. Some discrepancy observed in the computational results is primarily due to the grid resolutions for both flow and acoustics. One other possibility is that any form of sub-grid scale model was not used in the present LES. The simulation pointed out that the whole flow field and tonal noise are governed by the vortex shedding of the rod. The generation mechanism of the tonal noise is the swinging of the stagnation point around the leading-edge of the airfoil through periodic interactions of the Karman vortex with the airfoil. The broadband noise is generated by several sources: turbulent wakes between the rod and the airfoil (large amount of volume sources), their interaction with the airfoil leading-edge, and the trailing-edge scattering of the eddies within the boundary layers over the airfoil. The spanwise coherence functions of the wall pressure are rapidly decaying in most frequencies, except at St=0.2. Thereby, the spanwise coherence lengths are smaller than the rod diameter in most cases but at St=0.2, L_c is four or five times the rod diameter widely over the airfoil surface.

REFERENCES

1. J. H. Seo and Y. J. Moon, *Linearized Perturbed Compressible Equations for Low Mach Number Aeroacoustics*, J. Compt. Physics (in review), also 11th AIAA/CEAS Aeroacoustics Conference, AIAA-2005-2927, (2005).
2. A. A. Oberai, F. Roknaldin and T. J. R. Hughes, *Trailing-Edge Noise due to Turbulent Flows*, Technical Report, Boston University, Report No. 02-002, (2002).
3. E. Sorgven, F. Magagnato, M. Gabi, *Acoustic Prediction of a Cylinder and Airfoil Configuration at High Reynolds Numbers with LES and FHW*, ERCOFTAV Bullentin **58**, 47-50, (2003).
4. F. Magagnato, E. Sorgven, M. Gabi, *Far Field Noise Prediction by Large-Eddy Simulation and Ffowcs-Williams and Hawkings analogy*, AIAA paper 2003-3206, (2003).
5. J. Boudet, D. Casalino, M. C. Jacob, P. Ferrand, *Prediction of Broadband Noise: Airfoil in the Wake of Rod*, AIAA paper 2004-0852, (2004).
6. M. C. Jacob, J. Boudet, D. Casalino and M. Michard, *A Rod-Airfoil Experiment as Benchmark for Broadband Noise Modelling*, Theoret. Comput. Fluid Dynamics, Vol. **19**, pp. 171-196, (2005).
7. S. K. Lele, *Compact Finit Difference Schemes with Spectral-like Resolution*, Journal of Computational Physics, Vol. 103, (1992), pp. 16-42.
8. D. Gaitonde, J. S. Shang and J. L. Young, *Practical Aspects of Highorder Accurate Finite-Volume Schemes for Wave Propagation Phenomena*, International Journal for Numerical Methods in Engineering, Vol. 45, No. 12, (1999), pp. 1849-1869.
9. D. Casalino and M. Jacob, *Prediction of Aerodynamic Sound from Circular Rods via Spanwise Statistical Modelling*, Journal of Sound and Vibration, Vol. 262, No. 4, (2003), pp. 815-844.

Parallel Computational Fluid Dynamics – Parallel Computing and Its Applications
J.H. Kwon, A. Ecer, J. Periaux, N. Satofuka and P. Fox (Editors)

CFD Problems Numerical Simulation and Visualization by means of Parallel Computational System

Dmitry A. Orlov, Alexey V. Zibarov, Alexey V. Medvedev, Andrey N. Karpov, Ilya Yu. Komarov, Vladimir V. Elesin, Evgeny A. Rygkov, Andrey A. Parfilov, Anna V. Antonova[a]

[a]GDT Software Group, 27-79, Demonstratsii str., Tula, Russia, 300034

GDT Software Group company provides two software products. One of them – GasDynamicsTool® package (or GDT) – is a numerical simulation package developed to solve various CFD problems [1,2].

Currently, our company mainly focuses on solving various gas dynamics problems. Additionaly, a universal scientific data visualizer – ScientificVR® (or S-VR) is created. Originally, it was capable of working only with GDT simulation results, but we later decided to make it a universal and extensible visualization system [3–5]. This package has a unique feature – it allows the visualization of the simulated data on-the-fly, that is, simultaneously with the computation process. This article will discuss this important feature of the package and the visualization system GDT Software Group offers for CFD simulation data in general.

1. Validation

First of all, it is necessary to mention the long development period of the GDT package, and we think we have a right to be proud of the results we have achieved. All the simulation results have passed numerous precision tests. Figure 1 shows several comparisons with experimental solution of some problems of shock-wave processes simulation. Please, note the high simulation precision in each of the problems.

2. Actual problems

There are some very important problems that all CFD package developers face at this point associated with optimizing for parallel computing systems. We will highlight several very important issues GDT Software Group mainly focus on when developing our software. First of all, this is solving problems with division into a very large number of cells, as well as simulating long-standing processes. Actually, these are the problems that have always been there, at least these are the problems most typical of the simulation of gas dynamics processes. Figure 2 shows only one such problem – this is a simulation of Proton carrier rocket launch, implemented in real geometry of both the carrier rocket itself and the launching pad. Simulation of that kind requires division into tens and hundreds of millions of cells. Besides that, simulation of several seconds of the real process is required,

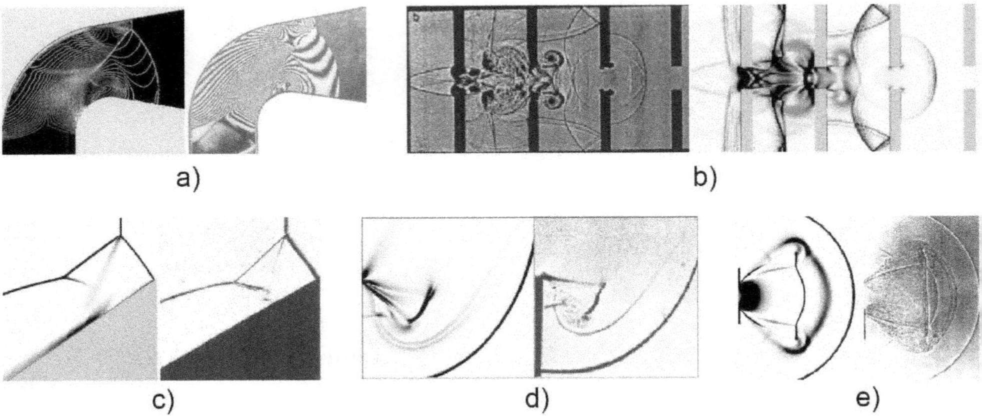

Figure 1. GDT verification examples: a) shock wave diffraction in a duct; b) shockwave propagation through a baffle; c) reflection on a wedge; d) diffraction on a 2D edge; e) Gun Muzzle Blast.

which means you have to calculate tens of thousands of computational scheme steps. It is obvious that you cannot simulate this problem without parallel computing systems with tens or hundreds of processors. As we know, systems with up to 200 processors have become more affordable on the world market of parallel hardware, and we can expect that systems with up to 60 processors will be an ordinary tool for every engineer in the near future. Therefore, GDT Software Group, as simulation software developers will supply the market with the necessary tools that allow efficient simulation of large problems on systems with a huge number of processors. We believe that efficient work in such conditions should be one of our priorities.

Another problem which cannot be underestimated is visualization of computation results. To get those results is just half the work. An engineer has to analyze them in order to make certain conclusions. Semitransparent voxel graphics allows viewing processes from the outside and the inside of a 3D computation domain at the same time. This is especially important when viewing the image stereoscopically. However, this is not the only capability of a modern visualization system. Indeed, solving problems with millions or even billions of cells inevitably results in the need to visualize the obtained computation results, and preferably online.

The last of the most important problems you have to keep in mind when developing a modern parallel simulation package is the problem of the package usability for end users, that is, engineers working on large-scale computation problems on parallel systems with a large number of processors. And here it is all not that easy: the traditional preprocessor-solver-postprocessor scheme does not allow control of the computation process for a large-scale problem because you can get visualization results only after accomplishing all of the three consecutive stages. Further, if it is a really large-scale task, you will have to wait

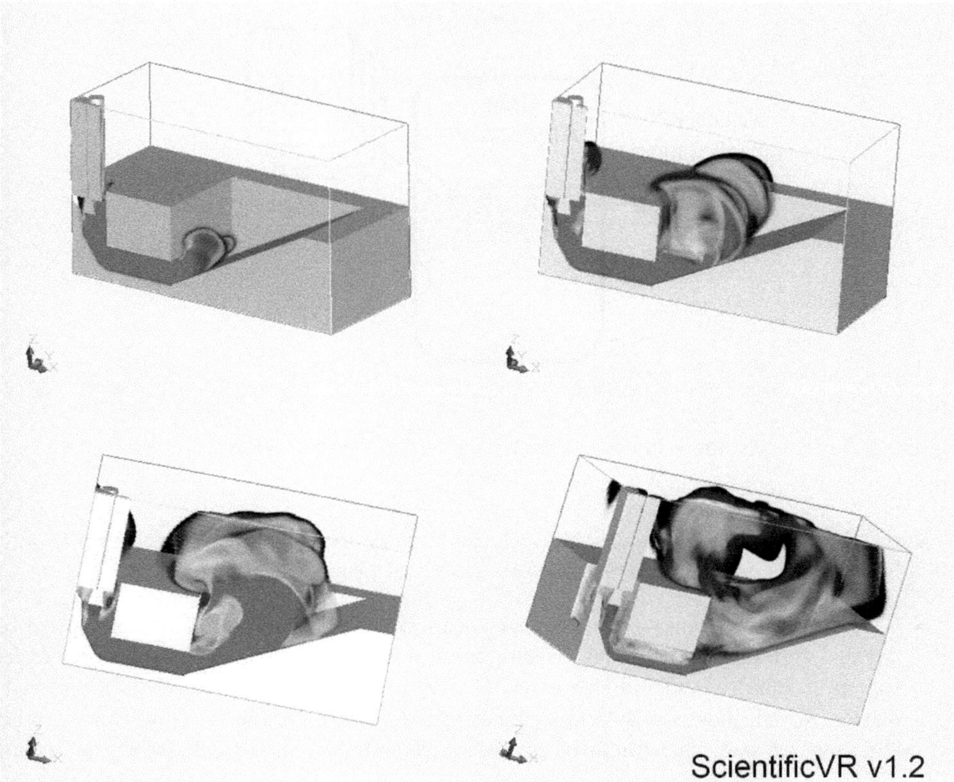

Figure 2. Simulation of the Initial stage of the Proton cargo spaceship start-up on the launch pad in Baikonur. Distribution of products concentration is presented. Simulation was made by GDT package (8-CPU parallel system, 31,6 million cells, Navier-Stokes model, 110000 calculation steps).

for quite a long time before obtaining the final result and you will have to carry out all the three operations from the very beginning in order to correct the problem definition and estimate the new result. Besides that, there are also subjective problems. Many professionals got used to the PC interface and find it very uncomfortable to work with the traditional command-line and batch modes used on large UNIX-computers for parallel applications. GDT Software Group considers package usability to be one of the main factors and believe that a modern simulation package should try to solve these problems and spare the users these inconveniences, as far as possible. Eliminating these problems is also on our priority list.

So these are the three problems GDT Software Group have focused when updating the packages for work on parallel systems:

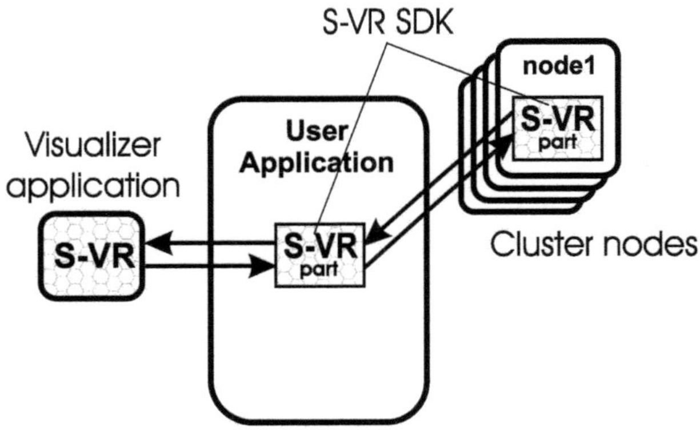

Figure 3. S-VR code embedding in a user parallel CFD application

- GDT Software Group makes its packages capable of solving problems of up to millions of cells on parallel systems with dozens of processors really efficiently;
- GDT Software Group pays a lot of attention to visualization of results as applied to parallel systems and make fast visualization possible even for really large-scale tasks (of up to hundreds of millions of cells). We also provide for the on-the-fly visualization, which means post-processing is accomplished parallel to computation, and the user can view the simulated processes immediately in the course of simulation;
- and finally, GDT Software Group offers the user a high-quality graphic user interface, which provides for all the three simulation processes to be carried out at once – preprocessing, computation (computation control included) and post-processing.

Bringing it all together, we can say that we want to develop a parallel CFD package of a new generation – fast, efficient, user friendly, with unique visualization capabilities, on-the-fly visualization included, and at the same time capable of supporting large parallel systems with various architecture. This is our large-scale goal, and we have already accomplished a lot in this area.

3. On-the-fly visualization on parallel systems

On-the-fly visualization has already been successfully implemented in our package. The process is rather easy: S-VR visualizer is capable of cooperating with an external application within the framework of COM technology. GDT Software Group has implemented the protocol of cooperation between the visualizer and the GDT program for numerical simulation. This gives GDT package users the following opportunities: First, choose the required visualization format and the interval for visualization image refreshing in the course of computation, and the GDT program will automatically transmitt the data to the S-VR visualizer at the preset interval. Thus, the user can follow the simulation process,

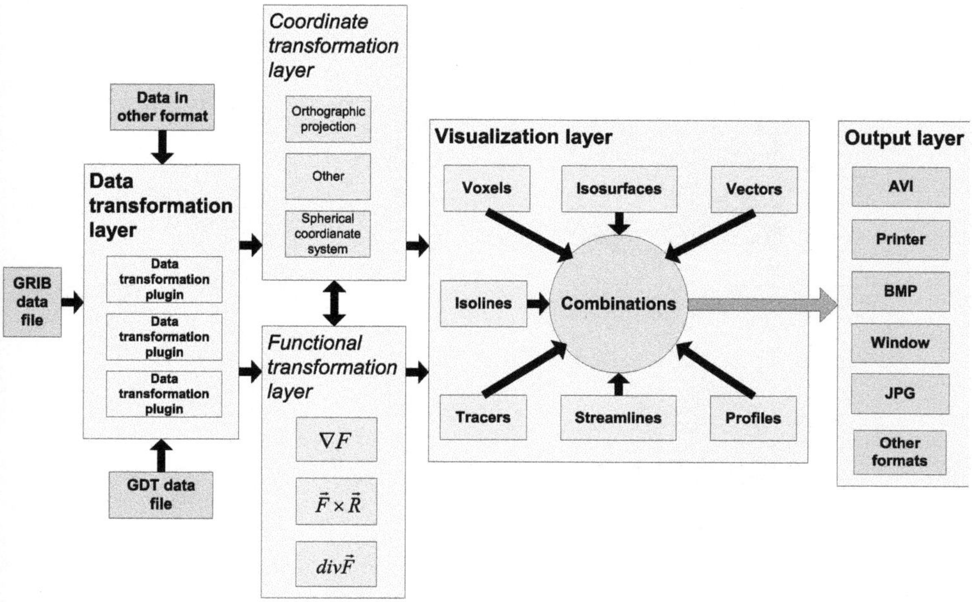

Figure 4. S-VR application structure.

getting a full visual presentation of the chosen parameters. This is how we understand interactive work of the user with a numerical simulation program.

Our process of visualization on parallel systems allows operation with huge data arrays. The point is that part of the visualization process is accomplished on cluster nodes already. Inside the solver part of the GDT application there is a part of the code which reduces data before sending these for visualization. And this data reducing is accomplished according to the algorithm controlled by the S-VR visualizer.

GDT Software Group considers this technology rather advantageous and has decided to develop it so that not only the GDT program, but also any other software with similar data presentation principles, could operate this way. Figure 3 presents our project of a specialized SDK, which will allow developers of parallel simulation applications to use the S-VR visualization pattern for their own needs. Thus, using several plug-ins, developers can get on-the-fly visualization for their parallel applications.

4. S-VR structure

Saying more about the ability of our codes to adapt to user needs, the diagram in Figure 4 shows how flexible the S-VR application is. As you can see, the S-VR application is of modular structure. There are modules responsible for data reading from external sources, modules responsible for data and coordinate systems transformation, modules responsible for graphic presentation of data in this or that form, and modules saving the

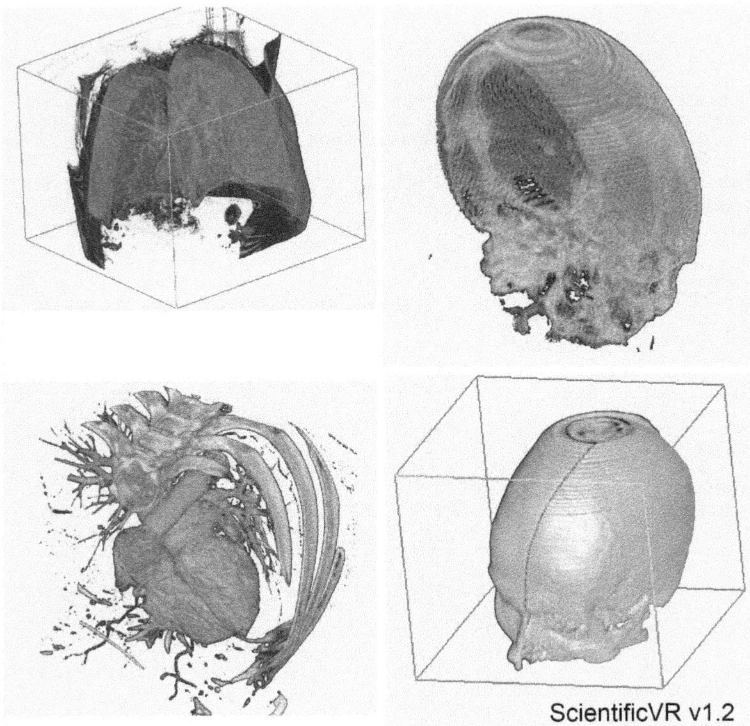

Figure 5. S-VR application for DICOM medical data visualization.

graphic presentation in output formats. What is important, is that any of these modules can be designed as a plug-in. This means that any variety of modules can be written by the user to support his own idea of a visualization method, input or output data format or data transformation. Of course, GDT Software Group supplys a special SDK for this purpose.

Figure 5 shows how efficiently this feature of the S-VR visualizer can be used. Just one example: GDT Software Group has developed a special plug-in which allows opening files in the DICOM format. This is a universal format for medical data. Thanks to this plug-in, the S-VR visualizer can display tomography results, with the data having been previously modified; thus, we can see images of separate organs and systems of human body. This slide presents several examples. So – you write one input format plug-in and S-VR works with quite different data, which has nothing to do with the CFD one.

5. Hybrid UNIX-Windows versions

Returning to the user's comfort again, to simplify the process of working with parallel systems for people used to personal computers, GDT Software Group has developed a

Hybrid Windows - UNIX version of our parallel package. The software package has been divided into two parts. The GUI of the GDT package plus SVR will be started on the user's personal computer under Windows while the calculation part of the package will be automatically started on the UNIX cluster. Saying "automatically" we mean that the user works only with a traditional PC GUI interface, all work with the cluster's command-line, batch/queue will be done by special scripts, absolutely indistinguishable for the user. This working scheme has proved itself very useful and is in great demand, which testifies to the fact that users prefer a simple user interface and do not want to be aware of all the cluster interaction details.

GDT Software Group intends to improve this method of work. Very often the user doesn't need constant control over a long calculation process, because calculation tasks frequently last for several days. In this case the ability to supervise the process of calculation from time to time from different personal computers will be more than useful. We suggest modifying the described scheme so as to allow the user to detach his User Interface part of the application without stopping the calculation process. The calculation will be alternated to the batch mode after that. Then the user will have an opportunity to attach the GUI part back to this calculation process, and have control over the whole application again, as well as implement visualization of the currently available data. What is most important here is that the user can have full control over the process from any PC of the network, from another office, during a business trip, or even from a laptop at home. That is how we see really convenient work with modeling on parallel systems. Your parallel system stays in a special room, and the technical staff looks after it while it solves the problems you have submitted, and meanwhile you can travel all around the world and have full control over the process.

6. GDT package effectiveness

The GDT package is very capable of efficiently solving large-scale gas dynamics problems with a really large number of cells, up to hundreds of millions of cells, as I mentioned before. Here are some figures, which can help you to judge how efficient it is. 1 Gigabyte of parallel system memory can be used to solve 3-dimensional problems of 25 million cells. That means that an ordinary cluster system consisting of only 20 dual-CPU nodes can be used to solve a problem of about 1 billion cells. Calculation speed is outstanding as well. Calculation of 1 million 3D cells problem takes 1 second per step on a Pentium IV 2.8GHz node. Calculation of the above mentioned 1 billion 3D cells problem takes about 30 seconds per step on 40-processor cluster (Xeon 1.4 GHz, 2Gb RAM, Myrinet). This calculation speed is one of highest in parallel CFD industry.

As to scalability indices on parallel systems of various kinds, we can say that they are more than satisfactory. GDT Software Group supports both SMP and NUMA systems and parallel clusters of Beowulf type. There are two implementations of parallelization subsystem in the GDT package: one of them is based on MPI standard, the other on direct memory blocks copies for common or shared memory systems.

It is appropriate to mention the list of supported hardware and software platforms. Our products can be called crossplatform by right, since we support a whole series of modern operating systems (Windows, Linux, Solaris, MacOS X), and we support of microprocessors by the leading vendors – Intel^{tm}, AMD^{tm}, IBM^{tm} – the most significant for the High Performance Computing industry, and we optimize for best performance results on all of them. The GDT package supports various kinds of modern cluster interconnects (Gigabit Ethernet, Myrinet, InfiniBand, SCI) and works well on different SMP and NUMA platforms.

7. Conclusion

In conclusion, we will summarize our company's main priorities in regards to development of parallel modeling and visualization tools.

- First, our package is fully ready to solve large-scale problems on parallel systems with many processors, and there are several examples to prove it.
- Second, GDT Software Group succeeded in developing a tool, or even a system for rapid visualization of large datasets, and moreover the on-the-fly visualization technology has been developed.
- And third, the software products GDT Software Group develop are indeed user friendly. They hide all the difficulties of large cluster system interaction from engineers used to working with personal computers, which saves their time and allows them to focus on the problem itself.

REFERENCES

1. Alexey V. Zibarov, Dmitry B. Babayev, Anton A. Mironov, Ilya Yu. Komarov, Sergei V. Malofeev, Vladimir N. Okhitin Numerical Simulation of 3D Muzzle Brake and Missile Launcher Flowfield in Presence of Movable Objects. // Proceedings of 20th International Ballistics Symposium. September 23-27, 2002, Orlando, USA.
2. A.V. Zibarov, D.B. Babaev, P.V. Kontantinov, A.N. Karpov, I.Ju. Komarov, A.A.Mironov, A.V. Medvedev, Ju.S. Shvykin, Ju. V. Judina, V.M. Kosyakin. Numerical simulation of the finned projectile pass through two chamber muzzle brake. // 21st International Symposium on Ballistics, Adelaide, South Australia, 19-23 April, 2004.
3. Alexey V. Zibarov, Dmitry B. Babayev, Anton A. Mironov, Ilya Yu. Komarov, Pavel V. Konstantinov Modern Visualization Techniques in ScientificVR(R) Package. // Proceedings of 10th International Symposium on Flow Visualization. August 26-29, 2002, Kyoto, Japan.
4. D.B. Babayev, A.A. Mironov, A.N. Karpov, I.Yu.Komarov, P.V. Konstantinov, A.V. Zibarov. Semitransparent voxel graphics realization in the ScientificVR visualisation package // 4 Pacific Symposium on Flow Visualization and Image Processing, 3-5 June 2003, Chamonix, France.
5. Zibarov, A.V., Karpov, A.N., Medvedev, A.V., Elesin, V.V., Orlov, D.A., Antonova, A.V. Visualization of Stress Distribution in a Solid Part // Journal of Visualization, Vol. 9, No. 2 (2006) 134.

Parallel Computational Fluid Dynamics – Parallel Computing and Its Applications
J.H. Kwon, A. Ecer, J. Periaux, N. Satofuka and P. Fox (Editors)

Markov Prefetching in Multi-Block Particle Tracing

Marc Wolter[a], Andreas Gerndt[a], Torsten Kuhlen[a] and Christian Bischof[b]

[a]Virtual Reality Group, RWTH Aachen University

[b]Center for Computing and Communication, RWTH Aachen University

Visualization of flow phenomena in large, time varying data sets demands extensive graphical, computational and memory resources. Recent research in the field of parallel post-processing algorithms and out-of-core data management shows that a good utilization of given resources together with a considerable speed-up may be achieved. One of the main problems is the I/O bottleneck commonly alleviated by data caching and overlapping I/O with computation. Standard prefetching techniques suffice for predictable data demands but fail for more unpredictable requests, e.g. in particle tracing. Therefore, we introduce a Markov prefetcher on application level for predicting data requests in multi-block data sets. Our prefetcher uses former request behavior as well as various information about the flow field and the data set's topology graph to provide required data almost without delay. We achieve a reduction of data access and time lags when applying Markov prefetching with tracing particles in different data sets.

1. Introduction

In general, post-processing of large-scale data sets is a time-consuming task. In order to enable explorative analysis of Computational Fluid Dynamics (CFD) data in interactive environments, a common approach for a considerable speed-up is parallelization. However, quite often the size of the main memory is also a strong constraint, which led to the development of out-of-core strategies. The general idea behind this approach is that only that data portion is loaded into main memory which is really needed for current computation tasks. A rather straightforward out-of-core algorithm is based on multi-block (MB) data sets. In such data structures, the flow field is already spatially clustered into disjunctive regions. Therefore, only the block that contains the currently needed data cells must be loaded from the file system. But loading data on demand can be one of the most fundamental bottlenecks in a post-processing system as the algorithm must stop the computation and wait for new data. In a parallel environment, these interruptions typically result in scaling and balancing problems.

In order to alleviate this problem for complex parallel stream- or pathline computations, this paper presents a new prefetching approach for particle tracing that uses Markov processes to predict the data needed next. This prediction is initiated by probabilities determined either by random seed samples applied offline or by topological heuristics. During the online integration, the actually computed trajectory of the particle is continuously evaluated in order to modify the probability graph. Using this kind of meta-information, in an optimum overlapping case, particle integration can now work without delay, as selected data blocks are prefetched in time.

The following section presents previous work in the area of particle tracing and prefetching strategies. Thereafter, in section 3, multi-block data sets and the use of their topology for parallel particle integration is briefly explained. Section 4 introduces the Markov prefetcher for multi-block CFD data sets, which has been implemented in our framework to overlap particle tracing computation with loading of predicted blocks. Finally, the achieved results are presented and evaluated in section 5.

2. Previous Work

Visualization algorithms for large disk-resident data sets have been studied by many research groups. In particular, out-of-core solutions for various feature extraction methods like isosurfaces or particle tracing have been developed.

Prefetching techniques are primarily designed and analyzed for computer or file systems but may also be adapted to most application level caching systems. A Markov prefetching for computer systems that can be easily integrated in existing systems and prefetches multiple predictions is described in [6]. Doshi et al. present adaptive prefetching techniques for visual data exploration [2]. Different prefetching schemes are implemented and the appropriate combination of strategies is dynamically selected according to the user's exploration pattern.

Visualization of time-varying fluid flows using large amounts of particles is facilitated by the UFAT (Unsteady Flow Analysis Toolkit) software [7]. UFAT provides instantaneous stream-, path- and streaklines for unsteady single and multi-block data sets. Depending on the underlying disk system, the I/O time accounts up to 25% of the execution time. Application controlled demand paging was investigated with UFAT by Cox and Ellsworth [1]. They point out trashing problems when relying on virtual memory and describe a paged segment out-of-core technique to take advantage of existing main memory.

Interactive out-of-core computation of streamlines is done by Ueng et. al. [9]. Unstructured grids are restructured by an octree, and required data in the main memory is extremely reduced by on-demand loading of new octant blocks. Block extents are bound by a fixed set of different sizes, so-called block size levels.

3. Multi-Block Topology

The type of grids generated for CFD is not only important for the quality of the simulation but also for the efficient design of post-processing algorithms. In this paper, we focus on multi-block data sets consisting of several structured grids, each stored separately in a file. For inter-block connectivity, we furthermore apply topology information in order to speed-up the determination of neighbors. This topology is also stored as meta-data in a separate file [4].

Additionally to simple block-to-block connectivity, the knowledge which cells are actually adjacent is considered. In general, a group of cell sides of one block borders to another group of cell sides of a second block. Therefore, not each cell link but a so-called *connection window* is stored. It is defined by means of i-j-k-indices of two grid nodes lying diagonally at corners of that window, which must be restricted to one side of a hexahedral block (see figure 1).

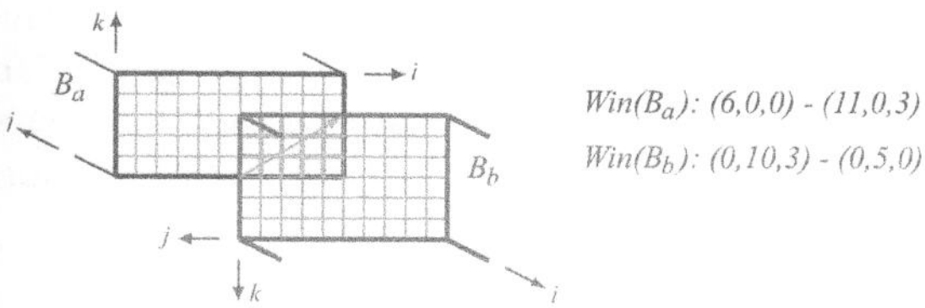

Figure 1. Connection window between block B_a and B_b.

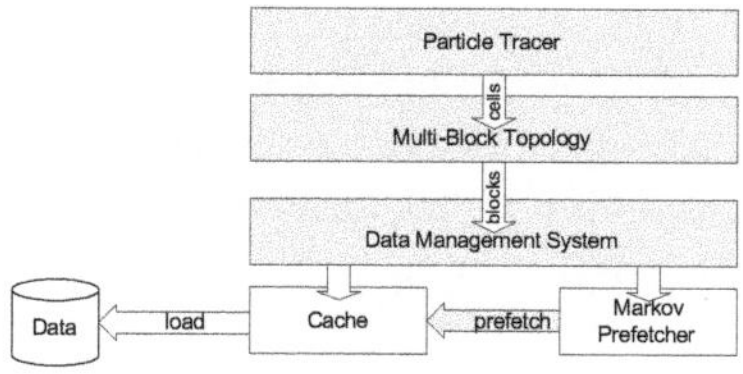

Figure 2. Software scheme of the postprocessing toolkit.

As the Markov prefetcher is a general approach, it is embedded into an existing data management system. For interactive exploration of unsteady CFD data sets in virtual environments, the software framework ViSTA FlowLib [8] is applied. The extraction of flow features from large CFD data sets is done on a remote parallel post-processor called Viracocha [3], connected with ViSTA FlowLib via a standard TCP/IP network. This decoupling allows the use of specialized hardware for different tasks: high-end graphics cards on the visualization workstation, and high-performance, large memory computer systems for feature extraction. The software scheme of the particle tracer on each parallel node is depicted in figure 2. All block requests from the multi-block topology are answered by Viracocha's data management system, which prefetches files into the cache according to the markov prefetcher's predictions.

4. Markov Prefetching

With predictable access patterns, Viracocha achieves quite good results by the use of simple OBL prefetching [3]. But when algorithms with unpredictable access patterns are applied, the sequential prefetcher has almost no effect on the reduction of cache misses. To provide useful prefetching for unpredictable patterns, we now introduce the Markov prefetcher for multi-block data sets.

A Markov predictor of ith order uses the i last queried blocks to predict the next request, i.e. it utilizes the probability $P(X_t = j|X_{t-i} = i_{t-i}, \ldots, X_{t-1} = i_{t-1})$. The simplest variant of first order chooses the next prefetch only on basis of the last block requested.

The prefetcher may be an independent unit of the software framework. As input, it works on the stream of block requests from an algorithm. Using this stream, the prefetcher builds a probability graph for the succession of blocks. The output of the prefetching unit is at every time a set of predicted blocks which are possible successors for the current block. Since we are dealing with multi-block data sets, in the current implementation only the most likely successor is used as prefetch prediction. Other Markov prefetchers, e.g. in [6], prefetch several proposed predictions, but they work on small, equal sized blocks. Prefetching a set of blocks increases the I/O load in our system considerable due to the size and imbalance of our blocks.

While the Markov probability graph provides good results after a certain runtime, which is needed to adopt to the overall request behavior, the probability graph is empty after the system starts. In this case, the implemented Markov prefetcher makes simply no predictions. This results in non-overlapping of computation and I/O. Even sequential prefetching works better in the starting phase, since it makes predictions and some are actually hits. This drawback leads to the introduction of the external initializing for Markov prefetchers.

4.1. External Initializing

Initializing the Markov prefetcher is either done by simulating a request stream with an external resource or by loading a Markov graph filled with topology information. This section deals with the second possibility how to use existing or precomputed topology information to obtain a "good" Markov graph, i.e. a graph which results in many successful and accurate prefetches.

We introduce two approaches for multi-block CFD data: a connection window based heuristic and precomputed flow statistics. Both methods' goal is to generate a probability distribution that describes the flow field in an appropriate way.

The **connection window heuristic** uses the already existent topological information. The multi-block structure described in section 3 contains connection windows for every neighboring block. A particle movement between blocks has to be performed via a connection window, which may be accessed using the topology meta-data. Therefore, we use the block topology graph as Markov graph, where each topology edge is divided into two opposed directed edges, as connection windows may be passed from both sides. The graph's edges are labeled with the transition probability proportional to the size of the connecting windows. We define the size of a connection window with window coordinates $i_{min}, j_{min}, k_{min}$ and $i_{max}, j_{max}, k_{max}$, respectively, as product of the two non-zero terms $(i_{max} - i_{min})$, $(j_{max} - j_{min})$, $(k_{max} - k_{min})$. This Markov graph is easy to compute but involves only the topological information of the grid. The behavior of the flow is completely disregarded, which leads to the second approach.

Precomputed flow statistics are generated using an iterative algorithm. In an initialization phase, the probabilities are set according to a uniform distribution, i.e. every neighboring block has the same probability to be chosen. For each block, a number of seed points is generated and traced until they leave the current block. Thereafter the algorithm determines whether the particle left the whole data set or entered a new block. In the latter case, the probability distribution will be updated. The number of seed points depends on two termination conditions: first, a total maximal number of seed points may not be exceeded in order to have some control over the total runtime of the preprocessing algorithm. Second, the algorithm terminates if the probability distribution stays unchanged for several seeding steps. This criterion stops once the distribution has reached some kind of balanced state, being the favored result.

Both offline computed probability distributions describe the flow field of one timelevel in a heuristical way. For initializing streamline computations, these probabilities are inserted directly in the Markov prefetcher's graph. When computing pathlines, we consider the time-dependency by inserting weighted edges to the next timestep using fixed weights for each data set.

5. Results

In this section, we present and discuss the results of the methods introduced before. For the evaluation of the presented approaches, we currently use two different multi-block data sets. The nose data set is used to compute 50 particle traces in 30 timelevels (see figure 3). 8 trajectories in 9 timelevels are integrated in the engine data set (see figure 4). All results are computed on a SunFire v880z with 4 GB main memory and four Ultra

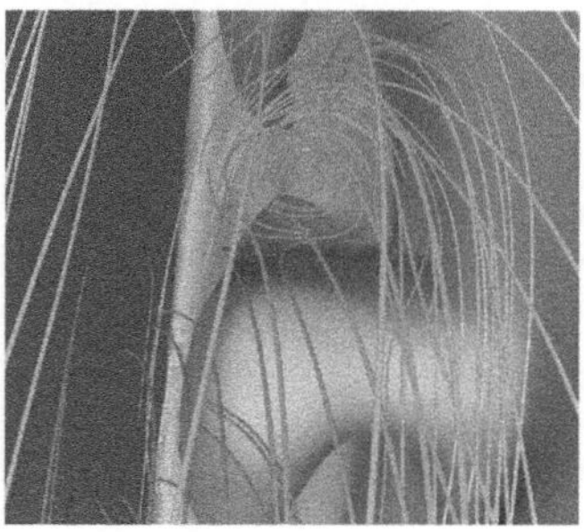

Figure 3. Inside view of streamlines through the nose data set.

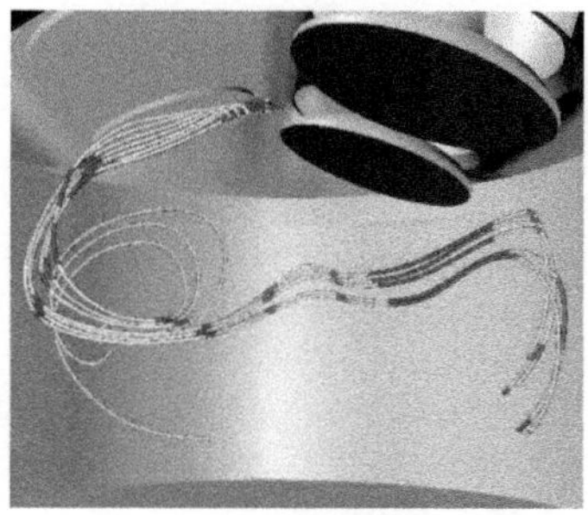

Figure 4. Pathlines in the engine data set, coloring specifies block changes.

Sparc III Cu processors. While both data sets are small enough to fit into main memory completely, their different flow fields are suitable to show the behavior of the markov prefetcher.

The nose data set represents the flow field inside an artificial human nasal cavity [5]. The domain decomposition is time-invariant, i.e. the geometrical definition of used grids does not change over time. Furthermore, during the inspiration and expiration phase, the air flow within a nose shows a dominant direction. Each of the 165 timelevels is divided into 34 blocks containing 443,000 points. The engine data set depicts the inflow and compression phase of a four-valve combustion engine. In contrast to the multi-block geometry of the nose, this data set is a moving grid, i.e. the geometry and dimension of defining blocks change now with each successive timelevel. Besides, an obvious flow direction is hardly determinable. An average of 200,000 points is contained in 23 blocks over 62 timelevels.

5.1. Multi-Block

The on-demand multi-block method's main goal is to reduce the number of loaded blocks. The values in figure 5 indicate the percentage of blocks loaded on demand compared to the number of total blocks.

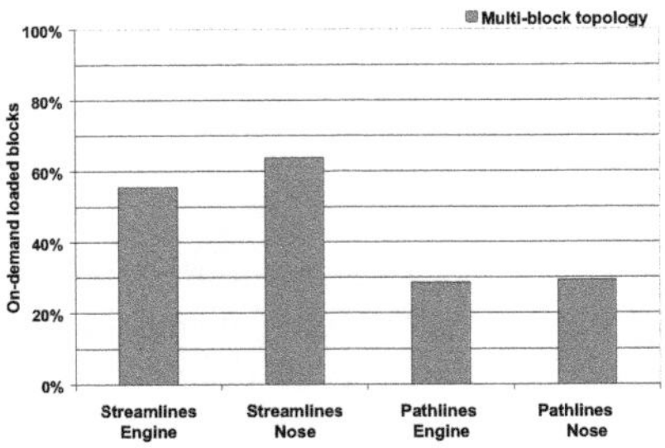

Figure 5. Percentage of on-demand blocks using multi-block topology.

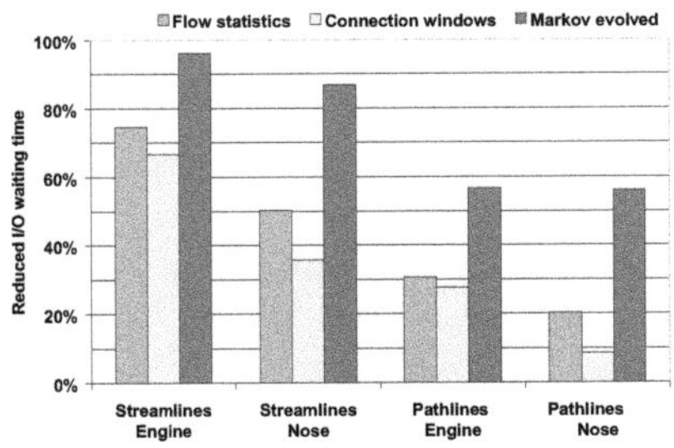

Figure 6. Saved loadtime (LT) for evolved and externally initialized Markov graphs.

When computing streamlines, we saved nearly half of the loading operations. With pathlines, the achieved results are even better as the absolute amount of saved blocks is higher in pathline than in streamline extraction. This is because the total amount of possible blocks is $block \cdot timesteps$ and therefore grows with used timelevels instead of the fix number of blocks in one timelevel.

5.2. Prefetching

To judge the efficiency of our implemented prefetching methods, we use several measures:

1. Prefetch coverage $= \frac{\text{useful prefetches}}{\text{useful prefetches} + \text{cache misses}}$ indicates the fraction of removed cache misses due to prefetching.

2. Prefetch efficiency $= \frac{\text{useful prefetches}}{\text{total prefetches}}$ describes the prefetching accuracy, i.e. how many prefetches could effectively be used.

3. Loadtime (LT) saving refers to the saved time the computation thread has to wait for I/O. This is a criterion for the quality of prefetch overlapping.

While the first two values are more structural, the latter value measures the real gain using the prefetching method. With reduced loadtime (i.e. computation waiting for I/O), the parallel algorithms are better balanceable and the user's waiting time decreases.

Three test series were performed. The first one named *Markov runtime evolved* starts with an empty Markov graph. Then it evolves from consecutive particle computations with similar seed points. This behavior resembles an exploring user during runtime examining a special part of the flow field. The measurement itself results from a single particle trace at the end of the exploring phase. The prefetching measures for the runtime-evolved Markov prefetcher are shown in table 1. Efficient prefetches are made, which reduces most cache misses. Efficiency is highly fluctuating with used seed points.

The *Flow Statistic* and *Connection Windows* series present values from a single particle trace with a respectively initialized Markov graph. The initialization has been executed

Table 1
Prefetching measures for all three series.

	Engine Streamline	Engine Pathline	Nose Streamline	Nose Pathline
Markov runtime evolved				
Coverage	92 %	57 %	79 %	72 %
Efficiency	100 %	74 %	90 %	87 %
Flow statistic series				
Coverage	46 %	23 %	41 %	26 %
Efficiency	60 %	47 %	53 %	24 %
Connection windows series				
Coverage	17%	18 %	27 %	29 %
Efficiency	40 %	38 %	55 %	35 %

on an empty graph. Side-effects from previous particle traces are excluded, so that the series depict the pure benefit from an external initialization. As shown in table 1, the efficiencies for both external initialized particle traces are at most 60 %. However, up to 46 % of all cache misses are reduced by the flow statistics method. This is a quite good result regarding the simple heuristic we used. Tracing in the more turbulent engine data set, the flow statistics approach beats the heuristical method. This may be due to the larger connection windows (more than 72 %) in the engine's topology which do not correspond to the main flow direction.

The savings in loading time for particle tracing using just initialized graphs are depicted in figure 6. Regarding waiting time for I/O, both methods reduce loading time, in particular when applied to streamline calculation. The inferior I/O time reduction with pathlines in contrast to streamlines may be ascribed to the time-dependent nature of pathlines, while the used topology characterizes a single timelevel only.

6. Conclusion and Future Work

We presented a Markov prefetcher as an optional component of the parallel post-processor Viracocha. The prefetcher is used for predicting data requests in multi-block particle tracing. By the exploitation of auxiliary multi-block topology meta-data, the particle integration algorithm already diminishes the amount of blocks to be loaded. The prefetcher takes care of an optimum overlapping of I/O and computation.

The remaining I/O waiting time is reduced considerably when using the runtime evolved Markov prefetcher. To be efficient even when the post-processing framework is started or new data sets are selected, different approaches of external Markov graph initializations are integrated. This yields a substantial improvement in comparison to uninitialized Markov prefetching or sequential prefetching strategies.

One shortcoming of the applied MB meta-data is that it merely considers topology information on timelevels separately. This can occasionally result in inadequate Markov initializations for time-variant algorithms. Therefore, in the future we will also regard MB topologies between timelevels, which could result in improved predictions especially for moving grids. Currently we are working on more efficient storage of and access to simulation data. Tree based indexing structures for fast loading of cells or meta-cells instead of blocks allow for a more granular access. We will evaluate our prefetcher for these structures, too.

Acknowledgements

The authors are grateful to the Institute of Aerodynamics, Aachen University, for the combustion engine and nose data set kindly made available. We would also like to thank the German Research Foundation (DFG), who kindly funded parts of the methodical work under grand WE 2186/5.

REFERENCES

1. M. Cox and D. Ellsworth. Application-controlled demand paging for out-of-core visualization. In *Proceedings IEEE Visualization '97*, pages 235–244, 1997.
2. P. R. Doshi, G. E. Rosario, E. A. Rundensteiner, and M. O. Ward. A strategy selection framework for adaptive prefetching in data visualization. In *Proceedings of the 15th International Conference on Scientific and Statistical Database Management*, pages 107–116, Cambridge, Massachusetts, 2003.
3. A. Gerndt, B. Hentschel, M. Wolter, T. Kuhlen, and C. Bischof. VIRACOCHA: An Efficient Parallelization Framework for Large-Scale CFD Post-Processing in Virtual Environments. In *Proceedings of the IEEE SuperComputing (SC2004)*, November 2004.
4. A. Gerndt, M. Schirski, T. Kuhlen, and C. Bischof. Parallel calculation of accurate path lines in virtual environments through exploitation of multi-block CFD data set topology. *Journal of Mathematical Modelling and Algorithms (JMMA 2005)*, 1(4):35–52, 2005.
5. B. Hentschel, T. Kuhlen, and C. Bischof. VRhino II: Flow field visualization inside the human nasal cavity. In *Proceedings of the IEEE VR 2005*, pages 233–236, March 2005.
6. D. Joseph and D. Grunwald. Prefetching using markov predictors. *IEEE Transactions on Computers*, 48(2):121–133, 1999.
7. D. A. Lane. UFAT: a particle tracer for time-dependent flow fields. In *Proceedings IEEE Visualization '94*, pages 257–264, Washington, D.C., 1994.
8. M. Schirski, A. Gerndt, T. van Reimersdahl, T. Kuhlen, P. Adomeit, O. Lang, S. Pischinger, and C. Bischof. ViSTA FlowLib - a framework for interactive visualization and exploration of unsteady flows in virtual environments. In *Proceedings of the 7th Internationial Immersive Projection Technology Workshop and 9th Eurographics Workshop on Virtual Environment*, pages 241–246. ACM SIGGRAPH, May 2003.
9. S.-K. Ueng, C. Sikorski, and K.-L. Ma. Out-of-core streamline visualization on large unstructured meshes. *IEEE Transactions on Visualization and Computer Graphics*, 3(4):370–380, 1997.

Parallel Computational Fluid Dynamics – Parallel Computing and Its Applications
J.H. Kwon, A. Ecer, J. Periaux, N. Satofuka and P. Fox (Editors)

A Parallel 2-D Explicit-Implicit Compressible Navier–Stokes Solver

C. B. Velkur[a]*, J. M. McDonough[a]

[a]Department of Mechanical Engineering, University of Kentucky.,
P.O. Box 40506-0503, Lexington, KY, USA

In this paper we present a 2-D compressible Navier-Stokes solver which is based on a predictor-corrector methodology. The predictor part of the scheme consists of a half time step explicit forward Euler time integration of the Euler equations followed by a full time step implicit backward Euler time integration of the complete Navier–Stokes equations. Fairly standard numerical techniques have been used to develop an algorithm that is easily parallelizable. We validate the solver for a supersonic flow scenario; in particular we use a shock tube problem with a pressure ratio of 100. We further explicitly show that the solver is capable of capturing boundary layer effects. Finally we present speedups obtained by parallelization using the OpenMP programming paradigm.

Key Words: Hybrid methods, Shock-tube problem, OpenMP

1. INTRODUCTION

It is well known that solutions to the compressible Navier–Stokes (N.–S.) equations admit discontinuities (shock waves). To compute flows with shocks one can proceed along two different lines of thought resulting in two different methods/approaches. The first, which are known as "shock capturing" methods relies on the proven mathematical legitimacy of weak solutions. These methods are designed to have shock waves appear naturally within the computational domain as a direct result of the overall flow field solution, *i.e.*, as a direct result of the general algorithm, without any special treatment to take care of the position of the shocks themselves. This is in contrast to the alternate "shock fitting" approach, where shock waves are explicitly introduced into the flow field solution; the exact Rankine–Hugoniot relations for changes across a shock are used to relate the flow immediately ahead of and behind the shock, and the governing equations are used to calculate the remainder of the flow field. Though both these methods have inherent advantages and disadvantages, shock capturing methods are best suited for complex flow problems involving shock waves for which we do not know either the location or number of shocks—a typical scenario for engineering applications. The interested reader is referred to the review by Moretti [1] for further details regarding these two classes of methods.

Use of shock capturing methods allows us to develop algorithms to solve only the governing system of PDEs over the entire domain. Generally time dependent PDEs are

*The authors would like to thank the University of Kentucky's Center for Computational Sciences for their extended support during the course of the present work.

solved using time-marching procedures. Such procedures fall into one of two different approaches, explicit or implicit. The time step in an explicit scheme is governed by the Courant number for high Reynolds number problems, which must not be greater than unity for stable calculations. The stability limit for explicit schemes is set by regions in the domain where wave speeds are high. These regions drastically reduce the time step possible for explicit schemes. Implicit schemes, on the other hand, can maintain stability with much larger time steps when compared to their explicit counterparts. For coupled nonlinear PDEs such as the compressible N.–S. equations, the use of implicit schemes results in having to iteratively solve a coupled system of linearized equations at each time step. Hence, a reduction in the number of time steps may be outweighed by increase in the number of arithmetic operations at each time step. Hybrid schemes containing both implicit and explicit approaches have also been developed to abate the disadvantages of the above mentioned approaches.

In 1981 MacCormack [2] presented an explicit-implicit predictor-corrector method that involved the inversion of bi-diagonal matrices in an effort to reduce computer time. The algorithm was second-order accurate in space and time. While marching in time, the order of finite differencing in the explicit predictor and corrector steps was cycled from one step to the next, whereas that in the implicit steps was taken as forward differencing in the predictor and backward differencing in the corrector. Fourth-order artificial dissipation, expressed in terms of pressure, is used in the explicit part. Implicit damping was also added to prevent oscillations around sharp discontinuities such as shocks (see [2] for details). An implicit-explicit hybrid scheme was developed by Fryxell *et al.* [3] which extended Godonov schemes to the implicit regime. Dai *et al.* implemented an iterative version of the scheme developed in [3] to solve the Euler equations. The regimes are determined based on the Courant number. If the Courant number is less than unity the regime is explicit, and the equations are solved explicitly; whereas if the Courant number is greater than unity the equations are solved implicitly (see [4] for further details).

Our solver is based on a predictor-corrector methodology. The predictor part of the scheme consists of a half time step explicit forward Euler time integration of the Euler equations followed by a full time step implicit backward Euler time integration of the complete N.–S. equations. Spatial discretization is second-order centered for both predictor and corrector parts of the scheme, with dependent variables being evaluated at cell centers and fluxes evaluated at cell walls. In the corrector stage of the algorithm we implicitly solve the complete N.–S. equations; hence there is a need to linearize the nonlinear terms in the governing equations. The nonlinearities in the N.–S. equations are handled iteratively by δ-form quasilinearization; and δ-form Douglas–Gunn time splitting [5] is used to solve the linearized equations leading to an easily parallelizable algorithm.

All second-order schemes generate oscillations around sharp discontinuities. In order to remove the unavoidable high frequency oscillations, we need to use the concept of artificial viscosity. This is nothing but additional terms that simulate the effects of physical viscosity on the scale of the mesh locally around discontinuities, and are negligible (that is of an order equal to or higher than the dominant truncation error) in smooth regions of the flow. We apply a third-order artificial viscosity developed by MacCormack and Baldwin. We will follow [6] for the formulation of this dissipation term D. This term is made proportional to a second derivative of the pressure field in order to enhance the effects of dissipation in the presence of strong pressure gradients (shocks, expansion waves) and to reduce it in smooth flow regions.

Our ultimate goal is to have 3-D capability including turbulence effects in the context of a large-eddy simulation methodology, but the present study is in two space dimensions. Moreover, in calculations of high-*Re* viscous flows where changes in the flow field occur close to a surface, finer gridding is required to capture these effects, especially the boundary layer. Though LES significantly reduces the amount of arithmetic when compared with DNS, the required arithmetic still can scale as badly as Re^2. This in turn results in very long run times, and hence the need for using parallelizable algorithms for such simulations.

We introduce the governing equations and the mathematical formulation of the problem in the next section of the paper. This is followed by a brief description of the standard test problem, the shock tube employed to validate the solver. Finally we present numerical solutions and the speedups obtained by parallelization via OpenMP.

2. MATHEMATICAL FORMULATION

Since our solver is of a hybrid type having a predictor step solving the Euler equations and a corrector step solving the full compressible N.–S. equations, we will write both the systems of equations in their generic form

$$\frac{\partial \mathbf{U}}{\partial t} + \frac{\partial \mathbf{F}}{\partial x} + \frac{\partial \mathbf{G}}{\partial y} = 0\,. \tag{1}$$

Equation (1) represents the entire system of governing equations in conservation form if **U**, **F**, and **G** are interpreted as column vectors given by

$$\mathbf{U} = \begin{bmatrix} \rho \\ \rho u \\ \rho v \\ \rho\left(e + \frac{\mathbf{V}^2}{2}\right) \end{bmatrix}, \tag{2}$$

$$\mathbf{F} = \begin{bmatrix} \rho u \\ \rho u^2 + p \quad \tau_{xx} \\ \rho v u \quad \tau_{xy} \\ \rho\left(e + \frac{\mathbf{V}^2}{2}\right) u + p u \quad k\frac{\partial T}{\partial x} \quad u\tau_{xx} \quad v\tau_{xy} \end{bmatrix}, \tag{3}$$

$$\mathbf{G} = \begin{bmatrix} \rho v \\ \rho u v \quad \tau_{yx} \\ \rho v^2 + p \quad \tau_{yy} \\ \rho\left(e + \frac{\mathbf{V}^2}{2}\right) v + p v \quad k\frac{\partial T}{\partial y} \quad u\tau_{yx} \quad v\tau_{yy} \end{bmatrix}. \tag{4}$$

where ρ, p and T are the usual density, pressure and temperature which are related via the equation of state for an ideal gas; $\mathbf{V} \equiv (u, v)^T$ is the velocity vector; the quantity $\left(e + \frac{\mathbf{V}^2}{2}\right)$, which will be represented as E in the remainder of the paper, corresponds to total energy (per unit mass). Elements of the stress tensor are given by:

$$\tau_{ij} = \mu\left(\frac{\partial u_i}{\partial x_j} + \frac{\partial u_j}{\partial x_i}\right) + \delta_{ij}\lambda\frac{\partial u_k}{\partial x_k}\,. \tag{5}$$

Finally, k is thermal conductivity; μ is dynamic viscosity; λ is second viscosity, and δ_{ij} is the Kronecker delta. The corresponding Euler equations are obtained by setting $\mu, \lambda, k \equiv 0$.

As alluded to in Sec. 1 either an inherent transient solution or a time-dependent solution leading to steady-state require a time-marching solution method. For such time-marching methods, we isolate $\partial \mathbf{U}/\partial t$ by rearranging Eq. (1) as

$$\frac{\partial \mathbf{U}}{\partial t} = \quad \frac{\partial \mathbf{F}}{\partial x} \quad \frac{\partial \mathbf{G}}{\partial y}. \tag{6}$$

2.1. Predictor Stage

The predictor stage consists of a half time step forward time integration of the Euler equations with a centered discretization in space. On discretizing Eq. (6) we obtain a discrete formula for the unknown $\mathbf{U}_{i,j}$ in the form

$$\mathbf{U}_{i,j}^{n+\frac{1}{2}} = \mathbf{U}_{i,j}^{n} \quad \Delta t d_{0,x}\left(\mathbf{F}_{i,j}\right) \quad \Delta t d_{0,y}\left(\mathbf{G}_{i,j}\right). \tag{7}$$

The difference operator $d_{o,x}$, is defined as $d_{o,x}\left(u_{i,j}\right) = \left(u_{i+1,j} \quad u_{i,j}\right)/\Delta x$, with a corresponding definition for $d_{o,y}$. The value of $\mathbf{U}_{i,j}^{n+\frac{1}{2}}$ now serves as an initial guess for the corrector step.

2.2. Corrector Stage

In the corrector stage of the algorithm we implicitly solve the complete N.–S. equations; hence there is a need to linearize the nonlinear terms in the governing equations. This is achieved by employing the Newton–Kantorovitch procedure. The form that we shall employ here is often termed quasilinearization (even though it is actually linearization and not quasilinearization). The simplest form of quasilinearization, from a conceptual standpoint, is that in which all dependent variables, as well as their derivatives are viewed as independent variables in a Fréchet–Taylor expansion of any nonlinear terms present in the problem. We comment that though the preceding formulation is a very general and effective way to treat nonlinear PDEs, there is considerable arithmetic involved in evaluating the resulting equations. We therefore consider an alternative formulation, often called δ-form quasilinearization. Notice that the fluxes are nonlinear functionals of the solution vector $\mathbf{U}$. We first linearize the fluxes by constructing Fréchet–Taylor expansions in δ-form. For the general m^{th} iteration we have

$$\begin{aligned}
\mathbf{F}(\mathbf{U})^{(m+1)} &= \mathbf{F}(\mathbf{U})^{(m)} + \left(\frac{\partial \mathbf{F}}{\partial \mathbf{U}}\right)^{(m)} \delta \mathbf{U}, \\
\mathbf{G}(\mathbf{U})^{(m+1)} &= \mathbf{G}(\mathbf{U})^{(m)} + \left(\frac{\partial \mathbf{G}}{\partial \mathbf{U}}\right)^{(m)} \delta \mathbf{U},
\end{aligned} \tag{8}$$

where $\delta \mathbf{U} = \mathbf{U}^{(m+1)} \quad \mathbf{U}^{(m)}$ is the difference between two successive iterates. In general for a time dependent calculation $\delta \mathbf{U} = \mathbf{U}^{n+1(m+1)} \quad \mathbf{U}^{n+1(m)}$. Substituting Eq. (8) into Eq. (6) and discretizing the linearized system leads to

$$\begin{aligned}
\mathbf{U}^{n+1(m+1)} &= \mathbf{U}^{n} \quad \Delta t d_{o,x}(\mathbf{F}^{(m)}) \quad \Delta t d_{o,y}(\mathbf{G}^{(m)}) \quad \Delta t D_{o,x}\left(\frac{\partial \mathbf{F}^{(m)}}{\partial \mathbf{U}} \delta \mathbf{U}\right) \\
&\quad \Delta t D_{o,y}\left(\frac{\partial \mathbf{G}^{(m)}}{\partial \mathbf{U}} \delta \mathbf{U}\right).
\end{aligned} \tag{9}$$

The centered-difference operator $D_{o,x}$ is defined as $D_{o,x}(u_{i,j}) = (u_{i+1,j} \quad u_{i\ 1,j})/2\Delta x$, with a corresponding definition for $D_{o,y}$. We now observe that the unknown time level $n+1$ solution occurs in two different ways in the above equation, namely both as $\mathbf{U}^{n+1}$ and as $\delta\mathbf{U}$. We need the expression to be entirely in terms of $\delta\mathbf{U}$, so we write $\mathbf{U}^{n+1}$ as $\mathbf{U}^{n+1} = \delta\mathbf{U} + \mathbf{U}^{(m)}$ and substitute into Eq. (9). The solution vector now becomes $\delta\mathbf{U}$ instead of $\mathbf{U}$. $\partial\mathbf{F}/\partial\mathbf{U}$ and $\partial\mathbf{G}/\partial\mathbf{U}$ result in two 4×4 matrices which are the Jacobian matrices for the 2-D compressible N.–S. equations. The corresponding equation can be rearranged as follows:

$$
\begin{aligned}
&\left\{I + \Delta t D_{o,x}\left(\frac{\partial F_i}{\partial U_i}\,\cdot\right)^{(m)} + \Delta t D_{o,y}\left(\frac{\partial G_i}{\partial U_i}\,\cdot\right)^{(m)}\right\}\delta U_i \\
&\quad = U_i^n \quad U_i^{n+1(m)} \quad \Delta t d_{o,x}(F_i^{(m)}) \quad \Delta t d_{o,y}(G_i^{(m)}) \\
&\qquad \Delta t D_{o,x}\left(\frac{\partial F_k}{\partial U_k}\delta U_k\right)^{(m)} \quad \Delta t D_{o,y}\left(\frac{\partial G_k}{\partial U_k}\delta U_k\right)^{(m)}, \qquad \forall\, i,k \in \{1,2,3,4\}, k \neq i\,.
\end{aligned}
\tag{10}
$$

There are several remarks to be made regarding Eq. (10). The obvious one is that it contains four unknowns. We should observe that as $U_i^{n+1(m)} \to U_i^{n+1}$, $\delta U_i \to 0$. Thus, if the Newton-Kantorovitch iterations converge, they converge to a solution of the original nonlinear problem. Moreover in the context of using time-splitting methods within this solution process, the splitting errors will also be iterated out of the solution. Using compact notation Eq. (10) can be rewritten as

$$
[I + Ax_i + Ay_i]\,\delta U_i = S^{(m)}\,, \tag{11}
$$

with obvious definitions for Ax_i, Ay_i, and $S^{(m)}$.

An important consequence of the above rearrangement is that we obtain a form to which time-splitting procedures can be directly applied. We comment at the outset that the reason for choosing a time-splitting method such as Douglas–Gunn, is its efficiency in solving multi-dimensional problems. The efficiency is achieved by decomposing multi-dimensional problems into sequences of 1-D problems. For the first split step we obtain

$$
[I + Ax_i]\,\delta U_i^{(1)} = S^{(m)}\,, \tag{12}
$$

and the second split step is

$$
[I + Ay_i]\,\delta U_i^{(2)} = \delta U_i^{(1)}\,. \tag{13}
$$

The update formula,

$$
U_i^{(m+1)} = \delta U_i^{(2)} + U_i^{(m)} \tag{14}
$$

provides the new approximation to the solution. We further notice that the time-splitting procedure, along with the second-ordered centered discretizations employed herein, leads to a tridiagonal system at each split step which can be solved efficiently using LU decomposition. Further, the dimensional splitting of the equations leads to an algorithm that is easily parallelized.

3. TEST CASE

We validate our scheme using the standard shock-tube problem described by Sod in [7]. The shock-tube problem is an interesting test case because the exact time-dependent solution is known for the Euler equations in 1-D, and hence we can compare our computed viscous solution at least qualitatively to the exact inviscid solution. The initial data for the shock-tube problem are composed of two uniform states (generally known as left and right states) separated by a discontinuity, physically a diaphragm. When the diaphragm is broken, two pressure waves appear (a shock wave and an expansion fan) and propagate into the initial fluid states, resulting in two uniform states 2 and 3 as shown in Fig. 1. The final states 2 and 3 are separated by a contact surface (discontinuity in first derivatives) implying that pressure and velocity of these states are equal, but density jumps across the discontinuity. The governing equations Eq. (1) are solved on a domain $\Omega = [0, L] \times (0, W) \equiv [0, 1.0m] \times (0, 0.2m)$ with boundary conditions consisting of the no-slip condition imposed at $y = 0$ and $y = W$, an inflow condition $\partial \mathbf{U}/\partial n = 0$ and outflow condition $\partial \mathbf{U}/\partial n = 0$ applied at $x = 0$ and $x = L$, respectively. The initial conditions in the left and right sections of the shock tube are given in Table 1 below (all entries are in SI units).

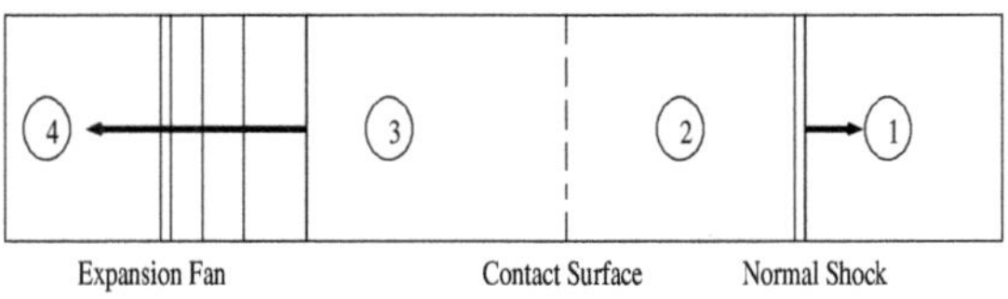

Figure 1. Shock Tube

Parameters	Left	Right
Velocity	0	0
Density	1	0.125
Pressure	10000	1000
Total Energy	2.5e5	2.5e4

Table 1: Initial conditions

4. RESULTS

To compare our 2-D results with the 1-D case the density, u-velocity, pressure and Mach number profiles at the horizontal centerline of the domain were used. Calculations reported here were performed on a 401 × 401 grid with a time step $\Delta t = 1 \times 10^{-6}s$. The following figures suggest that this is adequate. In Fig. 2 we present comparisons between computed and exact solutions. Part (a) of the figure shows the density profile; part (b) displays the u-velocity profile; (c) displays the pressure profile, and in (d) we present the Mach number profiles. It should be noted that the inviscid discontinuities are now transformed to sharp but continuous variations due to physical viscosity and heat conduction effects, but these cannot be resolved using typical gridding; hence, artificial dissipation still must be used to capture these discontinuities. Resolution of the shock front and the expansion fan are satisfactory; in particular, position of the shock front and of the head and tail of the expansion fan are predicted with fairly good accuracy. However, in part (a) of the figure we notice that there is a visible smearing at the contact discontinuity. It is evident from part (b) of the figure that the Gibbs oscillations are still

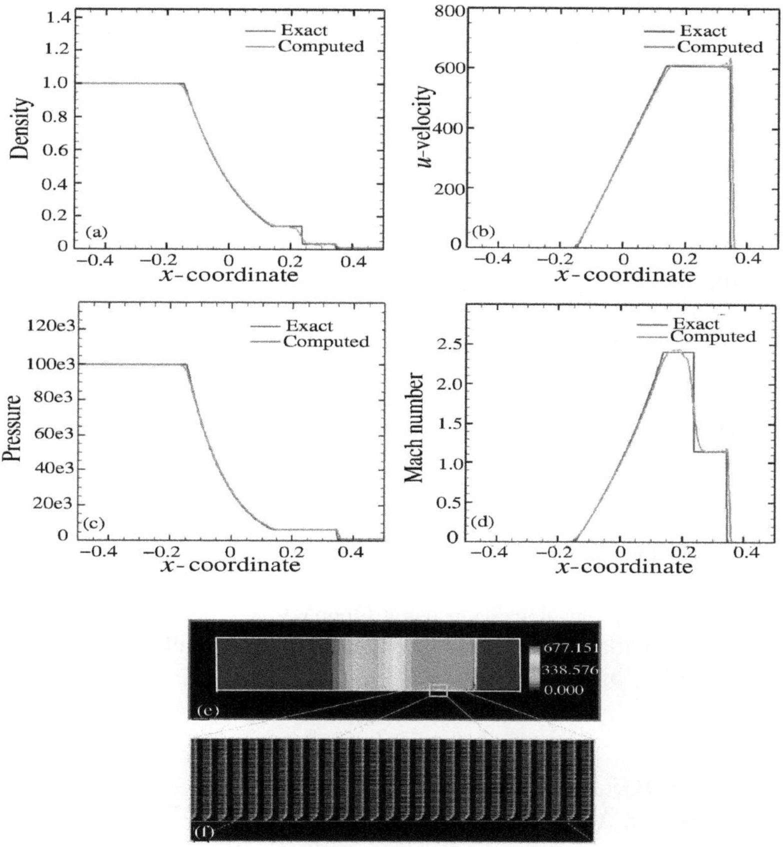

Figure 2. (a)-(d) Exact vs Computed results, (e) u-velocity contour (f) Boundary layer

present, but their magnitudes are greatly reduced. The Mach number profile is also in good agreement with the exact solution, but there is some smearing at the discontinuities.

Figure 2e presents a contour plot of u-velocity magnitude variation in the x-direction showing the final states at $t = 0.39$ms. In Fig. 2f we display a zoomed vector plot of the computed boundary-layer velocity profile. It is clear that the vertical grid spacing is sufficient to provide reasonably good resolution of the boundary-layer.

The code was parallelized using OpenMP on the HP Superdome SMP at the University of Kentucky. The DO loops of the Douglas–Gunn time splitting were parallelized by issuing compiler directives such that the line solves were distributed among the processors. To study the speedup obtained, we used a range of number of processors from 2 to 32 to execute the parallel algorithm. Results of the speedup are presented in Fig. 3. It should be noted that the speedups are sub-linear and not especially good; moreover, no improvement was seen in going to 32 processors, so these results are not shown. These results are fairly consistent with those obtained from several other pieces of software parallelized with OpenMP on Hewlett Packard hardware and suggest that MPI should be used.

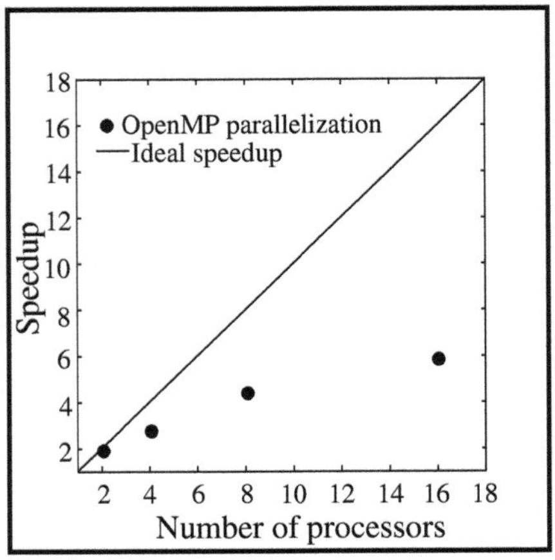

Figure 3: Speedup

5. SUMMARY

A hybrid N.–S. solver with an explicit Euler equations predictor and an implicit corrector solving the full viscous N.–S. equations was introduced. We demonstrate relatively good accuracy between computed and exact inviscid solutions for Sod's shock tube problem. We are able to resolve the boundary-layer profile upon using sufficiently fine grids. We remark that employing time-splitting methods such as Douglas–Gunn to solve multi-dimensional problems results in solution algorithms that are easily parallelizable. Finally, we presented the speedups obtained by parallelization using OpenMP. We note that the speedups are sublinear, but as noted above the current speedups are not very different from earlier ones obtained with OpenMP by other investigators.

REFERENCES

1. G. Moretti, Ann. Rev. Fluid Mech. 19 (1987) 317.
2. R. W. MacCormack, AIAA Paper. 81-0110 (1981).
3. B. A. Fryxell, P. R. Woodward, P. Colella, and K. H. Winkler, J. Comput. Phys. 63 (1986) 318.
4. Wenglong Dai and Paul Woodward, J. Comput. Phys. 124 (1996) 229.
5. J. Douglas, Jr. and J. E. Gunn, Numer. Math. 6 (1964) 453.
6. Charles Hirsch, Numerical Computation of Internal and External Flows, Vol. 2, John Wiley & Sons (1988).
7. G. A. Sod, J. Comput. Phys. 27 (1978) 31.

Parallel Computational Fluid Dynamics – Parallel Computing and Its Applications
J.H. Kwon, A. Ecer, J. Periaux, N. Satofuka and P. Fox (Editors)

A numerical analysis on the collision behavior of water droplets.

H.W. Nam[a], E.J. Kim[a], J.H. Baek[a]

[a]Department of Mechanical Engineering
Pohang University of Science and Technology(POSTECH)
San 31, Hyoja-Dong, Nam-Gu, Pohang, KyungSangBukDo, South Korea
E-mail:jhbaek@postech.ac.kr

Keywords: Droplet collision; Stretching separation; Reflexive separation; Satellite droplet

1. Introduction

The droplet collision is an important phenomenon that can be seen in nature, such as atmospheric raindrop formation, spray characteristic, etc. In the process of spray formation, distributions of droplet size can be affected by the collision angle and the relative velocity of the droplet collision. Recently, the process of spray formation is used frequently in a lot of industrial field applications such as an internal-combustion engine, surface handling, etc. Therfore it is very important to fully understand the droplet collision phenomenon.

Very complex flow phenomena appear after two droplets collide each other. Droplet collision behavior is normally divided into four categories; bouncing, coalescence, stretching separation and reflexive separation. For example, the behavior of droplet collision is affected by a species of working fluid; usually bouncing occurs at collision of hydrocarbon droplets, barely at collision of water droplets. And the droplet collision behavior is normally affected by Weber number, droplet size-ratio and impact parameter. Ashgriz and Poo[1] and Brazier-smith at al[2] conducted experiments on the collision of water droplets and found some critical conditions where the separation occurs. Jiang et al[3] and Qian and Law[4] provided various experimental data on the collision of hydrocarbon droplets. Schelkle and Frohn[5] performed numerical simulations using Lattice Boltzmann method. And Rieber and Frohn[6] did also on the central and non-central collision for droplet size ratio of 1 using 3D VOF method.

In this paper, we performed a series of numerical simulations on the droplet collision for various conditions of Weber number, impact parameter and droplet size-ratio, and compared the results with some experimental results and other theoretical predictions.

2. Theories of droplet collision

2.1. Droplet collision phenomena and dimensionless number

One of the important parameters associated with the collision outcome is the relative velocity of colliding droplets. The relative velocity is defined as follows,

$$U=(U_L^2+U_S^2-2U_LU_S\cos\alpha)^{1/2} \qquad (1)$$

where α is the collision angle, U_L the velocity of larger droplet and U_S the velocity of smaller droplet. Another important parameter that governs collision outcomes is the impact parameter. Impact parameter, X, is defined as the distance from the center of one droplet to relative velocity vector located on the center of the other droplet. Using these two parameters, the following dimensionless numbers can be defined,

$$Re=\rho d_L U/\mu \qquad (2)$$

$$We=\rho d_s U^2/\sigma \qquad (3)$$

$$\Delta=d_s/d_L \qquad (4)$$

$$x=2X/(d_L+d_s) \qquad (5)$$

where Re is the Reynolds number, We the Weber number, Δ droplet size-ratio, and x the dimensionless impact parameter. ρ and μ are the density and viscosity of the droplets respectively and σ is the surface tension coefficient. d_L is the diameter of larger droplet and d_S is the diameter of the smaller droplet.

The range of Reynolds number in this paper is between 500 and 4000. According to previous investigations, Reynolds number in the range of 500 to 4000 does not play a significant role on the outcome of the collision. Therefore main parameters that can influence the collision outcome are Weber number, droplet size-ratio and impact parameter. Weber number is controlled by changing the relative velocity and the impact parameter by changing the collision angle

2.2. Reflexive separation

When two droplets are merging, the overall surface area is decreased, resulting in the reduced surface energy. Reflexive separation occurs when the surface energy of new droplet is insufficient to contain the internal flow of droplet. Usually, reflexive separation occurs at a large Weber number and a small impact parameter. Small impact parameter means that the collision is close to the head-on collision.

Ashgriz and Poo[1] assumed nominal spherical droplet which has same volume with two droplets. Using the relation between the kinetic energy and the surface energy of the nominal spherical droplet, they found theoretical prediction separating the reflexive separation region from the coalescence region.

2.3. Stretching separation

Stretching separation occurs when the region of overlap between two colliding droplets is small. While the interface of two colliding droplets is merging, the remaining volumes, except the interaction region, continue to move in the opposite directions, stretching the interface and ultimately severs the connection.

Ashgriz and Poo[1] found a criterion for the stretching separation. Earlier Park[7] considered an equilibrium between the surface energy and momentum on the colliding region, and derived a criterion for the stretching separation. Brazier-Smith et al[2] assumed that the rotational energy excess the surface energy for stretching separation and derived a relation. Arkhipov et al[8] assumed the rotating system which has a constant angular velocity and derived a criterion of stretching separation using the variation quantity of the minimal potential energy.

3. Numerical method

VOF method is used for free surface tracking, in which, each phase is identified by using volume fraction. The volume fraction and the volume evolution equation are defined as following equations.

$$f = V_{fluid1}/V_{cell} \quad (6)$$

$$\frac{\partial f}{\partial t} + \nabla \cdot (u_c f) = 0 \quad (7)$$

In this paper, two working fluids are specified as air and water, assumed as incompressible and insoluble. The Navier-stokes equations are

$$\nabla \cdot u_c = 0 \quad (8)$$

$$\rho[\frac{\partial \mathbf{u}_c}{\partial t} + \nabla \cdot (\mathbf{u}_c \mathbf{u}_c)] = -\nabla p + \nabla \cdot \mu \nabla \mathbf{u}_c + \mathbf{s} \quad (9)$$

where $\mathbf{u}_c$ is the fluid velocity, p is the pressure. Governing equations are solved by using fractional step method on the cell centered grid. And time marching scheme for convection term is Adams-Bashforth method and, for viscous term, Crank-Nicolson method is applied. CSF(Continuum Surface Force) model[9] is used for the application of surface tension, where the surface tension is represented as a volumetric force in the source term of the Navier-Stokes equation.

4. Numerical results

In the previous investigations, a few representative experimental results were numerically analyzed in case of the drop-size ratio, 1, for two droplets. In this paper, performed are numerical simulations on the droplet collision for different size ratio with the various boundary conditions. The results are compared with previous experimental results as well as theoretical predictions.

4.1. Collision at the size ratio of Δ =1

Results of numerical simulation for droplet size ratio of 1 will be discussed in this section with Weber number of 5 to 100 and dimensionless impact parameter of 0 to 1.

Figures 1 to 3 show results of reflexive separation. Usually reflexive separation occurs at large Weber number and small impact parameter. While Fig.1(a) and Fig.2(a) are the experimental results of Ashgriz and Poo[1], Fig.1(b) and Fig.2(b) is numerical ones. There could be some time lag between the experiments and the numerical results respectively, but in generally, numerical results agree well with experiments. Results at the condition of We=40 and We=83 are shown in Fig.2 and Fig.3 respectively. In Fig.2, one satellite droplet is formed but in Fig.3, two satellite droplets are formed. Numerical results show that the number of satellite droplet increases with Weber number which coincide with previous experiments. Figure 4(a) shows various theoretical predictions dividing the regions of coalescence, reflexive separation and stretching separation. While a solid line is the theoretical prediction of Ashgriz and Poo[1] differencing the region of reflexive separation from stretching separation region, the inverted triangles indicate reflexive separation region of numerical results. Our numerical results shows good agreement with theoretical prediction of Ashgriz and Poo[1].

Figure 5 shows results of stretching separation. Usually stretching separation occurs at the conditions of large Weber number and large impact parameter. Since stretching separation is affected by the interaction area of collision, the interaction region reduces if impact parameter increases, resulting in a higher stretching energy. In Fig.5, the drop size after the collision is found to be different from each other and this reveals the mass transfer during the collision. Results on the number of satellite droplets after stretching separation is shown in Fig.6(a) and the experimental results of Ashgriz and Poo[1] and the numerical result of Ko[10] are shown in Fig.6(b). Both our numerical results and previous investigaton result show the same tendency that the number of satellite droplets is increased up to critical impact parameter which has the maximum number of satellite droplets and then decreases. The maximum number of satellite droplets appears at the range of impact parameter 0.4 to 0.6. In Fig.4(a), various theoretical predictions dividing the region of coalescence from stretching separation region are shown by several lines. While Ashgriz and Poo[1] and Brazier-Smith et al[2] derive theoretical prediction using the assumption that the combined mass at any phase of the collision process can be transformed into a nominal spherical droplet, Park[3] and Arkhipov et al[4] derived theoretical prediction using the energy equations of the volume in the interaction region. In Fig.4(a), symbols of diamond are numerical results indicating stretching separation region and those of square indicating coalescence region. Our numerical results show a large discrepancy with the theoretical prediction by Park[3] and Arkhipov et al[4]. Numerical results show good agreement with the theoretical prediction of Ashgriz and Poo[1] and Brazier-Smith et al[2], as well as the experimental results of Ashgriz and Poo[1]. From the comparison, the theoretical predictions of Ashgriz and Poo[1] and Brazier-Smith et al[2] are found to be more reasonable.

4.2. Collision at the size ratio of Δ =0.75

The results of numerical simulation for droplet size-ratio of 0.75 will be discussed in this section.

In Fig.4(b), the reflexive separation region of numerical results agree well with the theoretical prediction of Ashgriz and Poo[1] as in the results of droplet size ratio of 1. However, the reflexive separation region for droplet size ratio of 0.75 become smaller and the coalescence region is extended compared with the regions for droplet size ratio of 1. This is due to the decrease of the repulsion force. Decreasing of the repulsion force makes coalescence occur more often and therfore under the same Weber nuber condition, maximum impact parameter x for reflexive separation for droplet size ratio of 0.75 is lower than that for droplet size ratio of 1. When two droplets with different size collide as shown in Fig.7, the larger droplet becomes smaller after the collision, but smaller drop becomes larger because it takes some momentum from the larger drop. It seems that the mass transfer affect to decrease of reflexive separation region, resulting in the decrease of repulsion force.

The stretching separation region for droplet size ratio of 0.75 as shown in Fig.4(b) becomes smaller as compared with the regions for drop-size ratio Δ=1. It is found that under the same Weber number, stretching separation for droplet size ratio of 0.75 occur at larger impact parameter than that for drop-size ratio of 1. In case of the collision of two droplets with different size, pressure of the smaller droplet is higher than that of larger droplet, and therefore mass transfer occurs from the smaller droplet into the larger droplet. This flow disturbs stretching separation, resulting in smaller stretching separation region. And stretching separation region of the numerical results agrees well with those of the theoretical prediction of Ashgriz and Poo[1] and Brazier-Smith et al[2] as in the result of drop-size ratio Δ=1. Also, the numerical results show a more considerable difference compared with the prediction of Park[7] and Arkhipov et al[8] than those for drop-size ratio of 1 . From the results of the collision for droplet size ratio of 0.75, the theoretical predictions of Ashgriz and Poo[1] and Brazier-Smith et al[2] are found to be more reasonable.

5. Conclusions

A series of numerical simulations on the droplet collision for various conditions of Weber number, impact parameter and droplet size-ratio are performed and compared with some experimental results and other theoretical predictions. From the results, three categories of collision (coalescence, reflexive separation, stretching separation) show a good agreement with the previous theoretical predictions of Ashgriz and Poo[1] and

Brazier-Smith et al[2] and with the experimental results by Ashgriz and Poo[1], But not with the theoretical predictions of Park[7] and Arkhipov et al[8] . Results of stretching separation show the same tendency that the number of satellite droplets is increased up to critical impact parameter which has the maximum number of satellite droplets and then decreases to zero. Compared with the regions for droplet size ratio of 1, regions of reflexive and stretching separation for droplet size ratio of 0.75 become smaller and the region of coalescence is extended in result.

REFERENCES

1. N.Ashgriz and J.Y.Poo, 1990, "Coalescence and separation in binary collisions of liquid drops."*Journal of Fluid Mechanics*,Vol.221, pp.183-204.
2. P.R.Brazier-Smith,S.G.Jennings & J.Latham, 1972, "The interaction of falling water drops:coalescence,"*Proc.Roy.Soc.Lond.(A)*,Vol.326, pp.393-408.
3. Y.J.Jiang, A.Umemura and C.K.Law "An experimental investigation on the collision behavior of Hydrocarbon droplets,"*Journal of Fluid Mechanics*, Vol. 234,(1992), p.171-190
4. J.Qian and C.K.Law "Regimes of coalescence and separation in droplet collision," *Journal of Fluid Mechnics*, Vol. 331, (1997), p.59-80
5. M. Schelkle, A. Frohn "Three-dimensional lattice Boltzmann simulations of binary collisions between equal droplets,"*Journal of Aerosol Science*, Vol. 26, (1995) p.145-146
6. M. Rieber, A. Frohn "Three-dimensional Navier-Stokes simulations of binary collisions between equal droplets of equal size,"*Journal of Aerosol Science*, Vol. 26, (1995) p.929-930
7. R.W.Park, 1970, "Behavior of water drops colliding in humid nitrogen,"Ph.D.thesis, Department of Chemical Engineering, The University of Wisconsin.
8. V.A.Arkhipov, I.M.Vasenin & V.F.Trofimov , 1983, "Stabillity of colliding drops of ideal liquid,"*Zh.Prikl.Mekh.Tekh.Fiz*,Vol 3, pp.95-98.
9. J. U. Brackbill, D. B. Kothe and C. Zemach, 1992, "A continuum method for modeling surface tension," *Journal of Computational Physics*, Vol.100, pp.335-354.
10. G.H.Ko, S.H.Lee, H.S.Ryou, and Y.K.Choi "Development and assessment of a Hybrid droplet collision model for two impinging sprays,"*Atomization and Sprays*, Vol. 13, (2003), p.251-272

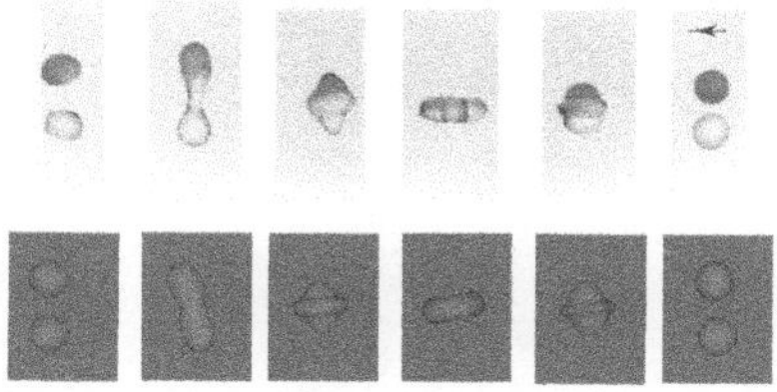

(a)

(b)

Fig.1. Reflexive separation with no satellite for $\Delta = 1$, $We = 23$ and $x = 0.05$ (a : experimental results[1], b : numerical results)

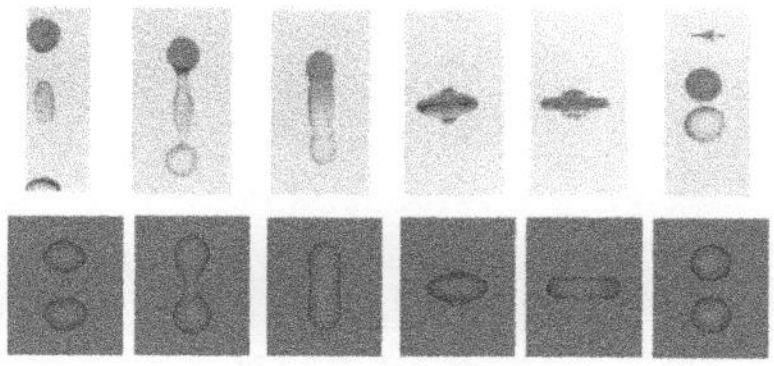

(a)

(b)

Fig.2. Reflexive separation with three satellites at $\Delta = 1$, $We = 40$, $x = 0$ (a : experimental results[1], b : numerical results),

Fig.3 Reflexive separation at $\Delta = 1$, $We = 83$, $x = 0$

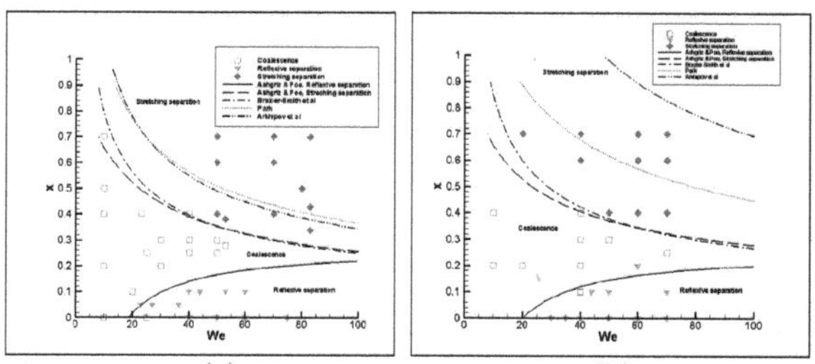

(a) (b)

Fig.4. Comparisons of the numerical results with the analytic ones(a; droplet size ratio of 1, b; droplet size ratio of 0.75)

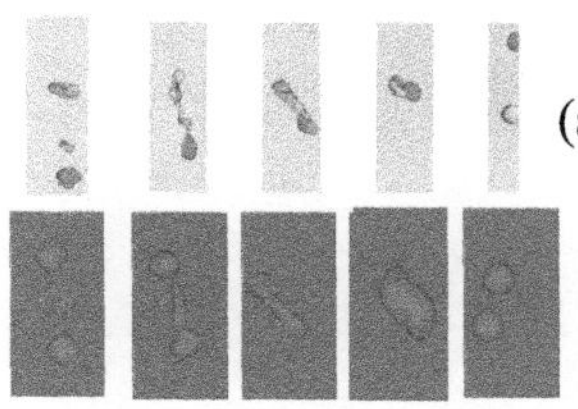

(a)

(b)

Fig.5. Stretching separation at $\Delta = 1$, $We = 83$, $x = 0.34$ (a : experimental results[1], b : numerical results)

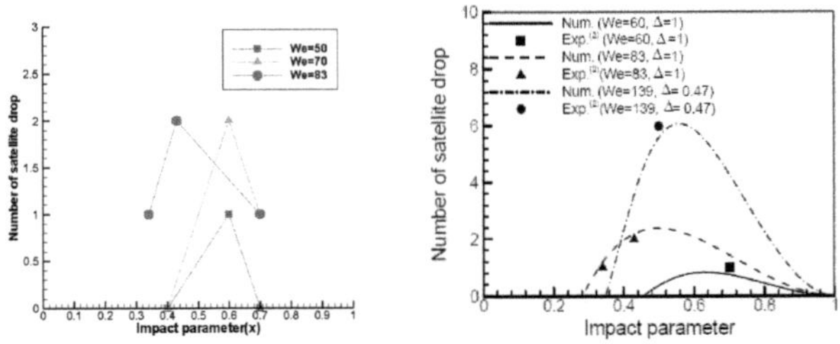

(a) (b)

Fig.6. Number of the satellite droplets after stretching separation of water droplets for Δ=1

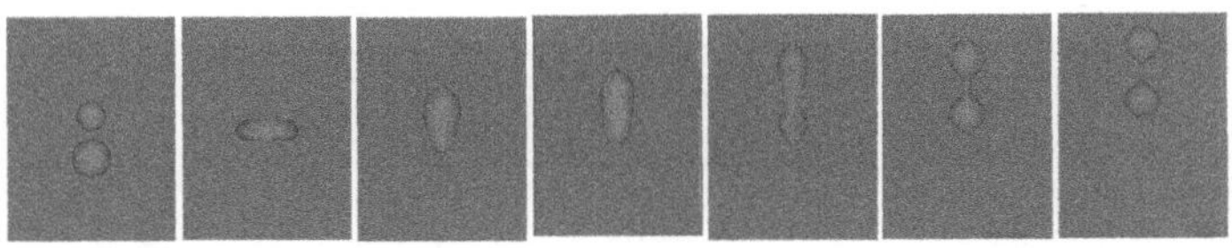

Fig.7. Reflexive separation at $\Delta = 0.75$, $We = 70$,and $x = 0$

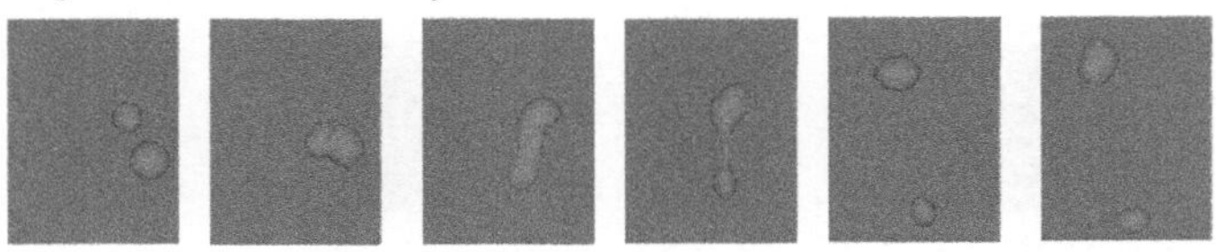

Fig.8. Stretching separation at $\Delta = 0.75$, $We = 60$,and x=0.6

Parallel URANS Simulations of an Axisymmetric Jet in Cross-Flow

R. K. Agarwal* and J. Cui
Department of Mechanical and Aerospace Engineering, Washington University, One Brookings Drive, St. Louis, MO 63130-4899

e-mail: rka@me.wustl.edu

An Unsteady Reynolds-Averaged Navier-Stokes (URANS) solver - WIND is employed to compute the flow field of an axisymmetric synthetic jet in cross-flow. This flow field corresponds to case 2 of NASA Langley Research Center Workshop "CFD validation of synthetic jets and turbulent separation control" held at Williamsburg, VA in March 2004. Parallel Three-dimensional simulations have been performed by employing the appropriate boundary conditions at the diaphragm of the actuator using Shear-Stress Transport (SST), Spalart-Allamaras (SA) and Detached Eddy Simulation (DES) turbulence models. The numerical simulations are compared with the experimental data; fair agreement is obtained.

Key Words: Flow Control, Computational Fluid Dynamics, Synthetic Jets

Nomenclature

ρ	=	density of air
μ	=	dynamic viscosity of air
p	=	pressure of air
T	=	temperature of air
B	=	diaphragm width
L	=	diaphragm length
D	=	jet orifice diameter
f	=	frequency of the synthetic jet
ω	=	$2\pi f$
U_J	=	maximum jet velocity
U_∞	=	freestream velocity
Re	=	jet Reynolds number, $U_J D\rho/(\pi\mu)$
Sr	=	jet Strouhal umber, $\pi fD/U_J$
S	=	jet Stokes number, $(\omega D\rho/\mu)^{1/2}$
(x, y, z)	=	Cartesian coordinates (x is freestream direction; y is spanwise; z is vertical)
(u, v, w)	=	velocity components in (x, y, z) directions

INTRODUCTION

Recently, a promising approach to the control of wall bounded as well as free shear flows, using synthetic jet (the oscillating jet) actuators, has received a great deal of attention. A synthetic jet (SJ) actuator is driven by a piezoelectric diaphragm or a piston in a periodic manner with zero net mass flux. A variety of impressive flow control results have been achieved experimentally including the vectoring of conventional propulsive jets, modification of aerodynamic characteristics of bluff bodies, control of lift and drag of airfoils, reduction of skin-friction of a flat plate boundary layer, enhanced mixing in circular jets etc. Kral et al.[1, 2] peformed the two-dimensional simulation of a synthetic jet by RANS with boundary conditions applied at the orifice of the synthetic jet (not including the effect of the cavity). For two adjacent synthetic jets, Guo et al.[3, 4] have performed detailed RANS computations and parametric studies; the flow inside the actuator cavity was also included. These results agree with the experiments of Smith and Glezer [5, 6] near the exits of the jets, however the difference between the simulations and the experiments becomes larger in the far flow field away from the exits. Lee and Goldstein[7] performed the 2-D Direct Numerical Simulation (DNS) simulation of the synthetic jet. For studying the behavior of synthetic jets in cross-flow, Mittal et al.[8] and Cui et al.[9] have performed the numerical simulations of the interaction of one and two synthetic jets respectively, with a flat plate boundary layer. All these simulations were performed for 2-D synthetic jets emanating in quiescent air or in cross-flow. Rizzetta et al.[10] investigated the flow field of a slot synthetic jet by DNS. The dimension of the synthetic jet employed in their simulation was the same as in the experiment of Smith and Glezer[11]. Their 2-D solutions produced fluctuations that were clearly larger than the experimental values. However, the 3-D computations captured the spanwise instabilities that led to a breakup of the coherent vortex structures, comparing more favorably with the measured data; nevertheless the 3-D computations agreed only qualitatively with the data. Lee et al.[12] have also studied the slot jet MEMS (Micro-Electronic Mechanical Systems), to alter the fine scale flow structures within a boundary layer by 3-D simulation and the feedback control. Their results show that although the jets eliminated the targeted structures, however new structures developed by blowing pulses of fluid leading to a 1-2% increase in overall drag for a single pulse.

Several experimental studies have been conducted for both 2-D, axi-symmetric and 3-D synthetic jets with and without cross-flow. Experiments for an isolated 2-D synthetic jet[5] and two adjacent synthetic jets[5, 6] have been performed by Smith et al. For an isolated axi-symmetric synthetic jet, experiments and computations have been performed by Ahmed and Bangash[13]. Recently experiments have also been performed for circular and elliptic synthetic jets in cross-flow[14, 15]. Honohan[16] studied the interaction of slot synthetic jets with cross flow. NASA LaRC[17] recently held a workshop on CFD validation of synthetic jets in March 2004. In the workshop, numerical results using a variety of computational methods were compared against the experimental data for three cases: (1) an isolated high aspect ratio synthetic slot jet emanating in quiescent air, (2) an axisyemetric synthetic jet in crossflow and (3) the separated flow behind a 3-D hump and its control with a steady and synthetic jet. This paper presents the results of CFD validation for case 2 using the URANS CFD solver WIND 5.[19]

SOLUTION METHODOLOGY

CFD code WIND is a product of the NPARC Alliance[19, 20], a partnership between the NASA Glenn Research Center (GRC) and the Arnold Engineering Development Center (AEDC). WIND computes the solution of the Euler and Navier-Stokes equations, along with supporting equation sets for turbulent and chemically reacting flows by employing a variety of turbulence models and chemistry models respectively. WIND is coded in Fortran 77, Fortran 90, and C programming languages. The governing equations are solved in conservation form. Explicit viscous terms are computed employing the second-order central differencing and the convection terms are discretized using the third-order accurate Roe upwind algorithm. In all the simulation results presented here, the overall spatial accuracy of the algorithm is second-order. The time-implicit convection terms are computed using an approximate factorization scheme with four-stage Runge-Kutta time-stepping. WIND uses externally generated computational grids. The solution is executed iteratively on this grid.

WIND has various turbulence models: Spalart-Allmaras (SA) one-equation model[21] and Menter's Shear Stress Transport (SST) two-equation model[22, 23] etc. For three-dimensional unsteady flows, the Spalart Detached Eddy Simulation (DES) model[24] is also available. This reduces to the standard Spalart-Allmaras model near viscous wall, where the grid is fine and has a large aspect ratio; it acts like a Large Eddy Simulation (LES) model away from the boundary, where the grid is coarser and has an aspect ratio of order one. It is intended to improve the results for unsteady and massively separated flows. An input parameter C_{des} (the default value 0.65 is used here) in the SA-DES model specifies the size of the RANS and LES zones. Increasing C_{des} increases the size of the region in which the combined model reduces to the standard SA model.[19] SST, SA and DES turbulence models have been tested in the simulations presented in this paper.

IMPLEMENTATION AND CASE SPECIFIC DETAILS

The details of flow configuration and actuator specifications for case 2 can be found in Ref. 17. In case 2, a synthetic jet issues from a circular orifice and interacts with the turbulent boundary layer that has developed on a splitter plate mounted in an atmospheric wind tunnel; the boundary layer thickness 50.8 mm upstream of the center of the jet is approximately 20 mm. The throat of the orifice is smoothly tapered from a diameter of 15.2 mm on the inside cavity wall to $D = 6.35$ mm diameter at the exit, as shown schematically in Fig. 1. The volume changes in the internal cavity are accomplished by moving the bottom wall with a piston by a sinusoidal voltage at a forcing frequency of 150Hz. The tunnel medium is air at sea level with tunnel operating at Mach number M=0.1. The tunnel dimensions at the test section are 381 mm (width) * 249 mm (height). The detailed flow conditions for this case are given in Table 1.

Figure 1: Schematic of the synthetic jet actuator with a circular orifice

Table 1: Flow Parameters

Param-eters	P (pa)	T (K)	ρ (kg/m^3)	μ kg/(m-s)	U_∞ (m/s)	U_J (m/s)	B (mm)	L (mm)	f (Hz)	D (mm)	Re	Sr	S
case 2	1.0e5	297	1.185	1.84e-5	34.6	50	51	50.4	150	6.35	6510	0.060	621

Based on the STRUCTURED 3D GRID #1 provided by CFDVAL2004 website[17], 3-D grid for this simulation is generated, as shown in Fig. 2. It comprises of 7 zones, with a total number of about half million points. The grid extends 0.508 m upstream of the center of the orifice, and 0.1016 m downstream of the center of the orifice. The height is 0.076m above the floor, and extends from $y = -0.38$ m to 0.38 m in the span-wise direction, same as in the grid provided by CFDVAL2004 website. Zone 1 has 49∗129∗29 mesh points, with an O-topology, as shown in Fig. 2(b). The surface mesh is generated by ZONI3G, and the internal volume grid is then obtained by Grid MANipulation (GMAN)[25] code. Zonal connectivity information is also computed by GMAN, and is stored in the grid file used by WIND. During the course of a solution, WIND maintains continuity in flow properties across zone boundaries through a process known as zone coupling.[26]

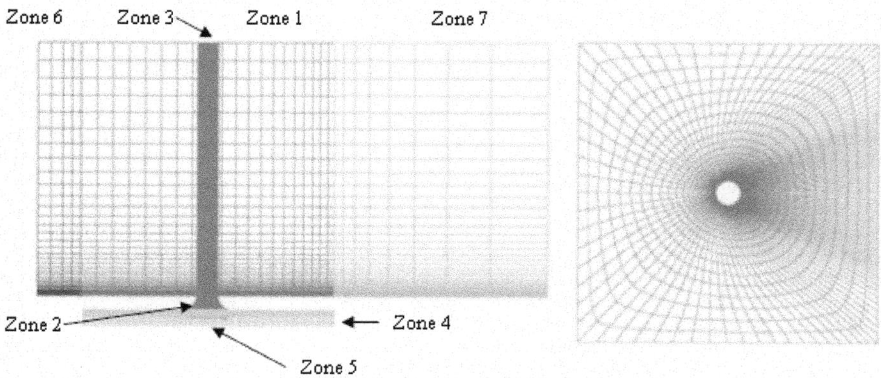

(a) The overview of the grid (b) Top view of zone 1 (I=IMAX)

Figure 2: The grid employed in the simulations

All the boundary conditions are also specified in GMAN. The floor (comprising of zones 1, 6 and 7, I = IMAX = 49) is specified as viscous wall. The steady turbulent boundary layer velocity profile (interpolation of the experiment data) is specified at the left boundary of zone 6 (x = 0, J = 1) as inflow. All other sides of the external flow region are subsonic outflow (top, side and right). For the actuator, the side walls are modeled as slip walls (k=1, k=KMAX of zone 4), and top walls of the cavity and the jet orifice (zone 2, 4 and 5) are viscous. The boundary condition used at the bottom of the cavity (including both the elastic diaphragm and the solid plate, i.e. bottom of zone 3,4 and 5) is as follows: The pressure is taken as a constant, and the w-velocity is specified as in Eq. (1). The same approach as in Ref. 18 is employed here: the pressure is tuned by satisfying the zero-net-mass-flux condition during one synthetic jet period. And then the amplitude of the velocity is tuned to obtain the desired maximum jet velocity of $1.45U_\infty$ (about 50m/s) at the exit. Thus the appropriate pressure at the bottom wall is 100867 Pa; and the velocity amplitude W_{amp} is 5.5 m/s. There are 10,000 time steps for one synthetic jet period, i.e. the time step is 0.6667 μs.

$$w(x, y, z = const, t) = W_{amp} \sin(\omega t) \tag{1}$$

SIMULATION RESULTS AND DISCUSSION

Two simulations are performed on the same grid, shown in Fig. 2. First, SST turbulence model is employed for the whole flow field. Another simulation is then conducted by employing the DES model in zones 1, 2 and 3, where synthetic jet interacts with the cross-flow, and the SA model is employed in all other zones. Both results are discussed in this section. Washington University Center for Scientific Parallel Computing (CSPC)[27] provides the host "chico", which has 64 x R12000 MIPS processors with 400MHz clock, which is employed in the computations. The 3-D simulation runs parallel on 4 CPUs. It takes about 45 hours to calculate one synthetic jet cycle. There are 10,000 time steps per jet cycle. At least 10 cycles are needed to get a periodic solution. After that, one or more cycles are calculated to obtain output data for post-processing. 100 files are saved for one jet cycle to get the phase-locked and long-time averaged flow field information.

A. Phase-locked flow field

Figures 3, 4 and 5 show the time history of the velocity components at three locations: the center of the jet exit (50.63mm, 0, 0.4mm), $1D$ downstream of the jet at the center plane (57.15mm, 0, 10mm), and $2D$ downstream of the jet at the center plane (63.5mm, 0, 10mm). The criterion to align the simulation results with the experiment in phase is provided on the workshop website[17], where 50° phase location is defined such that the w-velocity has reached the average of its maximum and minimum during one synthetic jet cycle and is increasing. The velocity was measured by Laser Doppler Velocimetry (LDV). Comparing with the experiment, both simulations agree well with the LDV u-velocity data during [0, 100°] as seen in Fig. 3(a). The simulation with SST model agrees with the LDV measurement very well during [100°, 200°], while DES model predicts lower u-velocity. Both simulations predict an earlier increase in u-velocity during the suction stroke. For the spanwise v-velocity shown in Fig. 3(b), LDV measurement has a peak nearly equal to U_∞. Neither simulation is able to capture it. As reported in the workshop, no CFD simulation has been able to capture this feature[17]. In Fig. 3(c), SST and DES models predict the same w-velocity. The amplitude of w-velocity is well captured. However, both models fail to predict the sharp peak and the additional hump in the suction stoke. Same was the case for all other results presented in the workshop.[17] Figure 4 gives the plots of velocity history $1D$ downstream of the jet center. For u-velocity, the LDV data shows an increase during [100°, 200°], while the simulations predict a decrease instead. The LDV data again shows a large spanwise velocity close to 0.25 U_∞. The DES model is able to simulate this asymmetry somehow, with the peak $v = 0.15\ U_\infty$. However, SST model still gives zero v-velocity at all time. All the simulations reported in the workshop also did not obtain this experimentally observed large variation in v-velocity.[17] In Fig. 4(c), both simulations predict a weaker w-velocity and a zigzag shape during [100°, 200°]. The SST simulation performed by Rumsey[17] has a similar behavior. $2D$ downstream of the center of the jet, the simulations agree with the LDV data for u-velocity fairly well in Fig. 5(a). DES model again predicts much better v-velocity compared to the experiment than SST model. The w-velocity obtained by the simulations qualitatively agrees with the experiment except during [100°, 200°], where LDV data is noticeably smaller than the computed values. In general, SST and DES models obtain the same w-velocity history, however DES model is superior in capturing the spanwise asymmetry. Both simulations only qualitatively agree with the experimental data.

B. Long-time averaged flow field

The long-time averaged flow field is obtained by averaging 100 data files during one jet cycle, after the solution becomes periodic. Figure 6 shows the velocity profiles $2D$ downstream of the center of the synthetic jet in the center plane, i.e. in the plane $y = 0$. At $2D$ downstream, both the SST and the DES simulations give identical u-velocity, as seen in Fig. 6(a). The LDV data has a slightly fuller inner layer than the simulations. In Fig. 6(b), the SST and the DES models predict significant different v-velocity profiles, where the DES model predicts the same trend as the LDV measurement. Two simulations have almost same w-velocity profile, with similar distortion as the LDV measurement, but slightly larger values. The velocity profiles $8D$ downstream of the jet are given in Fig. 7. Both simulations agree with the LDV data for u-velocity profile very well, while the LDV data once again has a fuller inner layer. As seen in Fig. 7(b), the DES shows similar amplitude of v-velocity as in the experiment. SST model still predicts almost zero v-velocity from the floor all the way up in the far field. In Fig. 7(c), the DES model predicts larger w-velocity, with a similar twist as the SST simulation. w-velocity in both the simulations is larger than the LDV measurement. Generally speaking, the computed velocity profiles only agree qualitatively with the experiment. The DES model gives a better prediction for the spanwise velocity.

PARALLELIZATION

The computations reported in this paper were performed on four processors of a SGI Origin 2000, 64 R12000 MIPS processor supercomputer (known as Chico) at Washington University. Each processor has 400MHz clock, 8MB Cache and a theoretical peak performance of 800MFLOPS. The grid employed in the calculations has approximately half million points. The CPU time required is 2.481e-08seconds/time step/grid point. The time step for time accurate calculations is 6.667e-04 seconds. There are 10000 time steps per jet cycle. At least ten jet cycles are required to get a periodic solution. It required 45 hours of CPU time on 4-processors to calculate one jet cycle. All the cases were computed using 4-processors. However, one calculation with SST model was performed on 16-processors. Almost linear speed up with parallel efficiency of 93% was obtained.

CONCLUSIONS

3-D numerical simulations for case 2 of NASA Langley CFD validations workshop[17] have been presented by employing SST, SA and DES turbulence models, and compared with the experimental data. The computed phase-locked velocity history qualitatively agrees with the experimental data. The long-time averaged flow field results are able to capture the global nature of the velocity field. DES model is better in predicting the spanwise *v*-velocity changes both in time and space. The computations reported were performed on four and sixteen processors of a SGI Origin 2000, 64 R12000 MIPS processor supercomputer (known as Chico) at Washington University. Almost linear speed up with parallel efficiency of 93% was obtained.

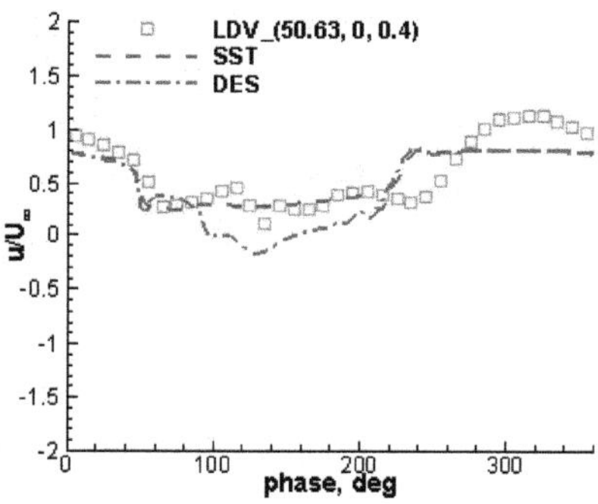

(a) *u*-velocity

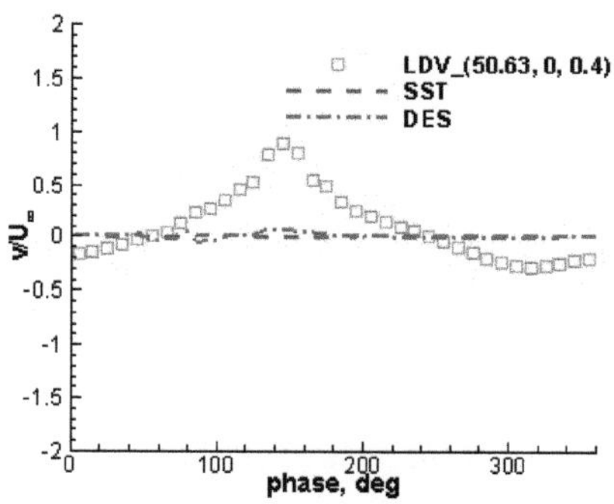

(b) *v*-velocity

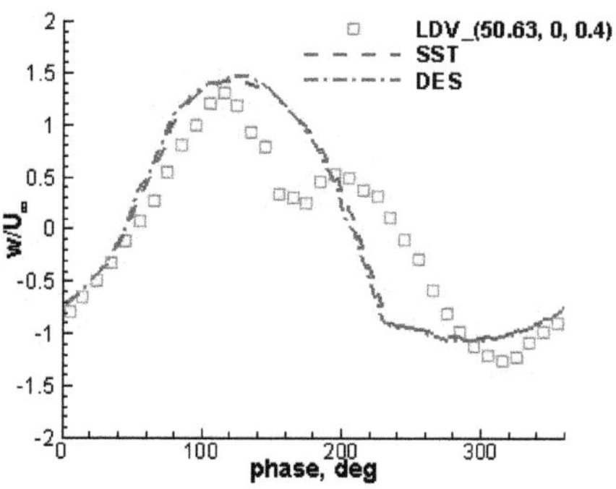

(c) *w*-velocity

Figure 3: Time history of velocity components at (50.63mm, 0, 0.4mm)

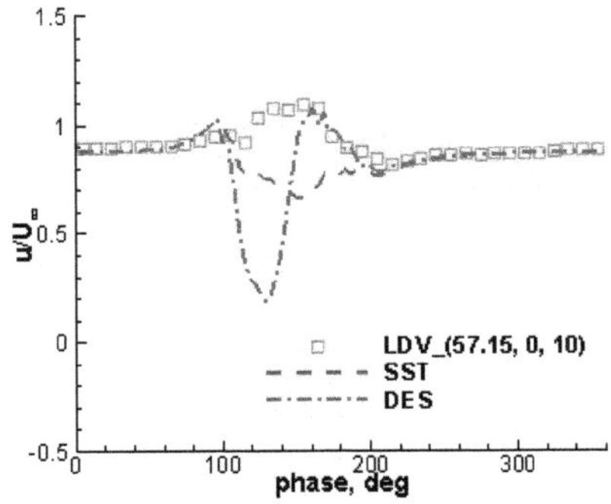

(a) *u*-velocity

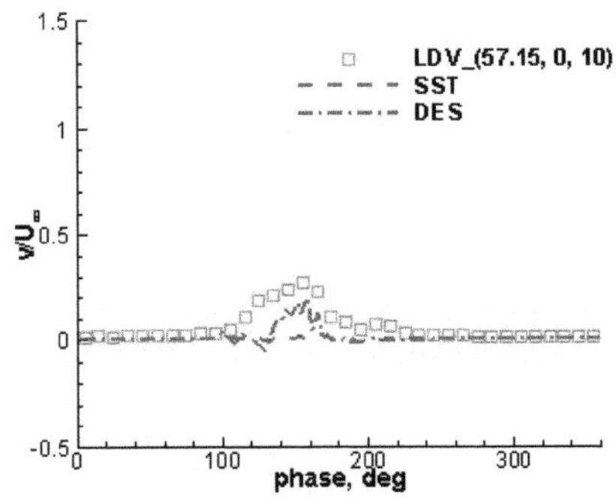

(b) *v*-velocity

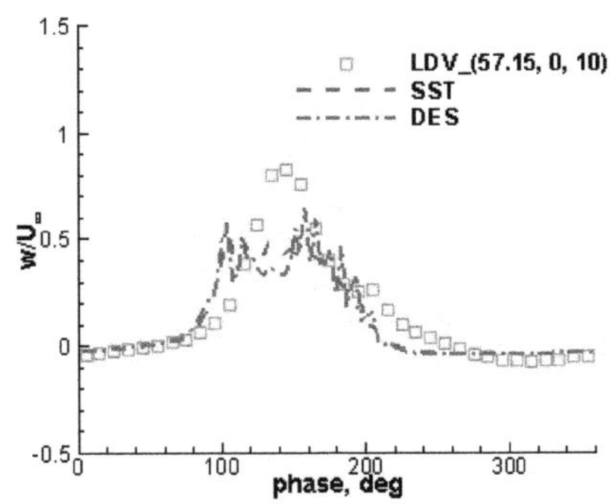

(c) *w*-velocity

Figure 4: Time history of velocity Components at (57.15mm, 0, 10mm)

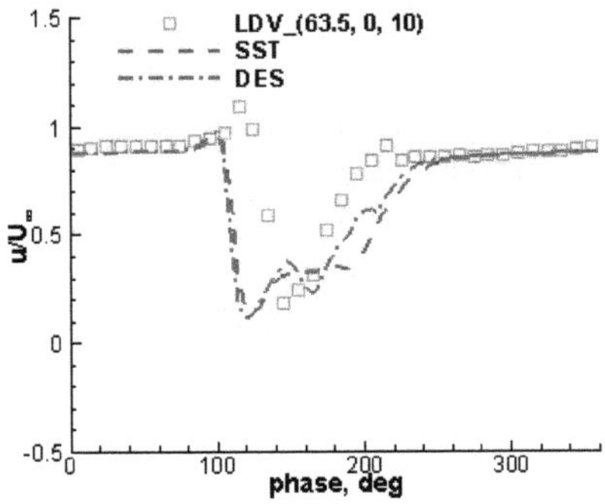

(a) *u*-velocity

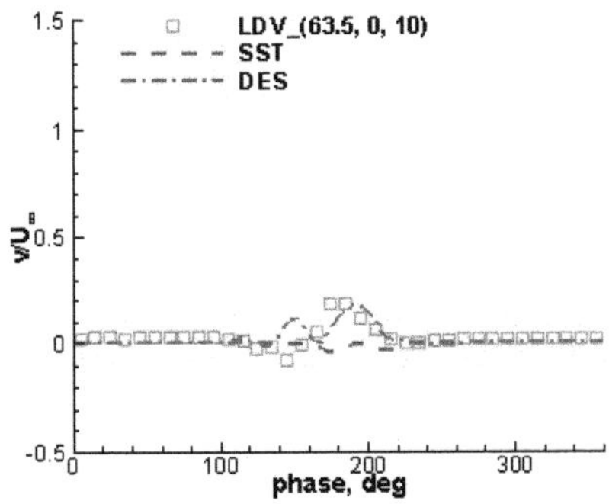

(b) *v*-velocity

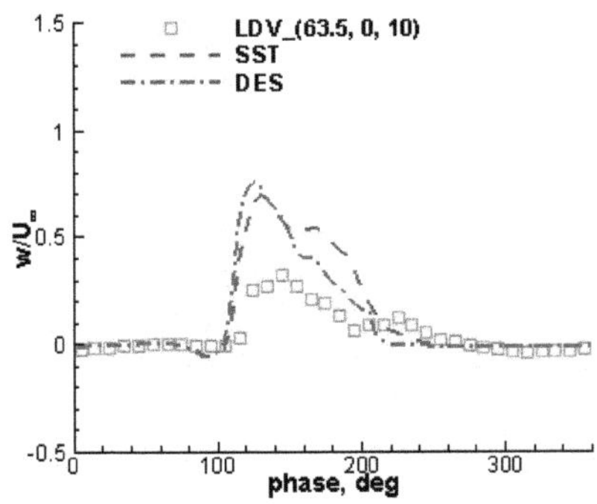

(c) *w*-velocity

Figure 5: Time history of velocity components at (63.5mm, 0, 10mm)

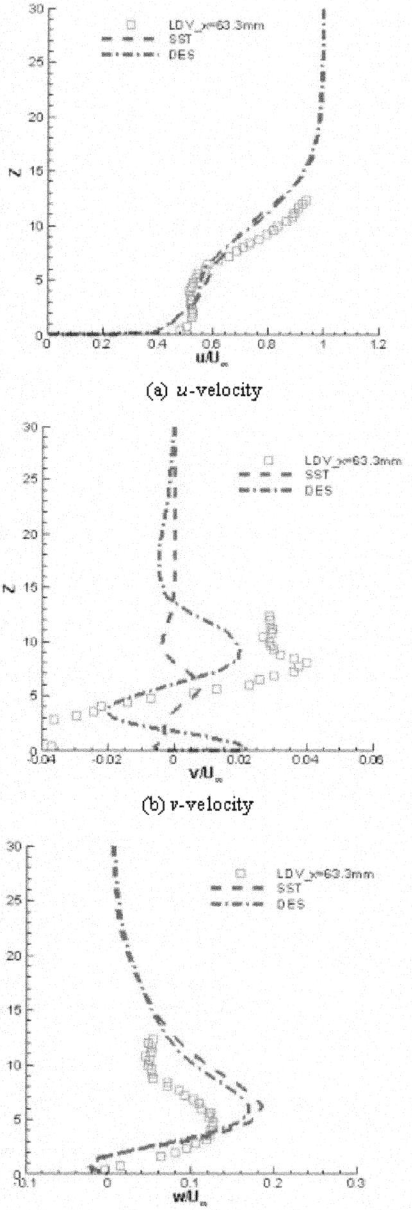

Figure 6: The averaged Velocity at the center plane: $v = 0$, x = 63.3mm, 2D downstream of the center of SJ

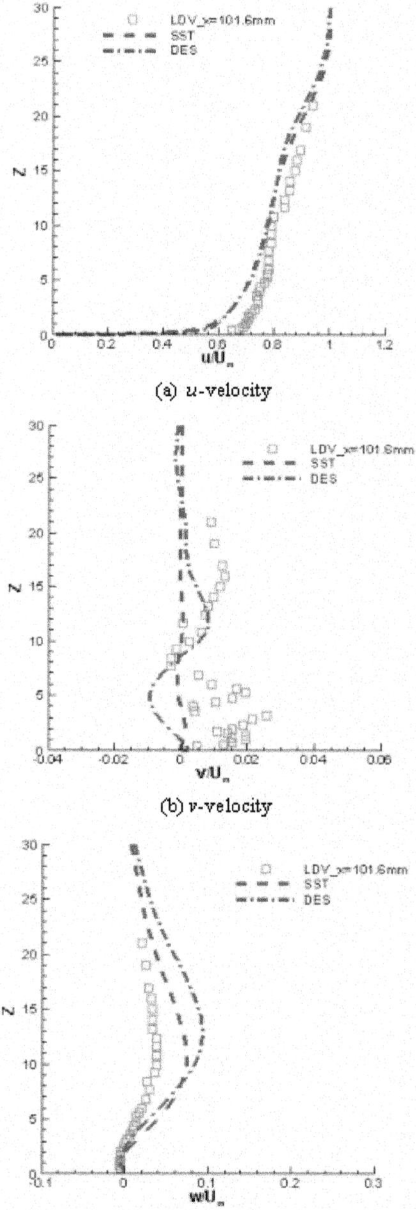

Figure 7: The averaged velocity at the center plane $v = 0$, x = 101.6mm, 8D downstream of the center of SJ

REFERENCES

[1]Kral, L. D., Donovan, J. F., Cain, A. B., and Cary, A. W., "Numerical Simulations of Synthetic Jet Actuators," AIAA Paper 97-1284, 28th AIAA Fluid Dynamics Conference, 1997.

[2]Kral, L. D. and Guo, D., "Characterization of Jet Actuators for Active Flow Control," AIAA Paper 99-3573, AIAA 30th Fluid Dynamics Conference, Norfolk, VA, June 28-July 1, 1999.

[3]Guo, D., Cary, A. W., and Agarwal, R. K., "Numerical Simulation of the Interaction of Adjacent Synthetic Jet Actuators," Proc. of the Int. Conf. on Computational Fluid Dynamics (ICCFD2), Sydney, Australia, 15-19 July 2002.

[4]Guo, D., Cary, A. W., and Agarwal, R. K., "Numerical Simulation of Vectoring Control of a Primary Jet with a Synthetic Jet," AIAA Paper 2002-3284, 1st AIAA Flow Control Conference, St. Louis, MO, 24-27 June 2002.

[5]Smith, B. L. and Glezer, A., "The Formation and Evolution of Synthetic Jets," *Physics of Fluids*, Vol. 10, No. 9, 1998, pp. 2281-2297.

[6]Smith, B. L., Trautman, M. A., and Glezer, A., "Controlled Interactions of Adjacent Synthetic Jets," AIAA Paper 99-0669, 1999.

[7]Lee, C. Y. and Goldstein, D. B., "Two-Dimensional Synthetic Jet Simulation," *AIAA Journal*, Vol. 40, No. 3, 2002, pp. 510-516.

[8]Mittal, R., Rampunggoon, P., and Udaykumar, H. S., "Interaction of a Synthetic Jet with a Flat Plate Boundary Layer," AIAA Paper 2001-2773, 2001.

[9]Cui, J., Agarwal, R., Guo, D., and Cary, A. W., "Numerical Simulations of Behavior of Synthetic Jets in Cross-Flow," AIAA Paper 2003-1264, 2003.

[10]Rizzetta, D. P., Visbal, M. R., and Stanek, M. J., "Numerical Investigation of Synthetic Jet Flowfields," *AIAA Journal*, Vol. 37, 1999, pp.919-927.

[11]Smith, B. L., and Glezer, A., "Vectoring and Small-Scale Motions Effected in Free Shear Flows Using Synthetic Jet Actuators," AIAA Paper 97-0213, 1997.

[12]Lee, C. Y. and Goldstein, D. B., "Simulation of MEMS Suction and Blowing for Turbulent Boundary Layer Control," AIAA 2002-2831, 1st Flow Control Conference, 24-26 June 2002, St. Louis, MO.

[13]Ahmed, A. and Bangash, Z., "Axi-symmetric Coaxial Synthetic Jets," AIAA Paper 2002-0269, 40th AIAA Aerospace Sciences Meeting and Exhibit, Reno, NV, 14-17 January 2002.

[14]Zaman, K. and Milanovic, I., "Synthetic Jets in Cross-Flow, Part I: Round Jet," AIAA Paper 2002-3714, 33rd AIAA Fluid Dynamics Conference, Orlando, FL, 23-26 June 2002.

[15]Milanovic, I. and Zaman, K., "Synthetic Jets in Cross-Flow, Part II: Jets from Orifices of Different Geometry," AIAA Paper 2002-3715, 33rd AIAA Fluid Dynamics Conference, Orlando, FL, 23-26 June 2002.

[16] Honohan, A.M., "The Interaction of Synthetic Jets with Cross Flow and the Modification of Aerodynamic Surfaces," Ph.D. Dissertation, School of Mechanical Engineering, Georgia Institute of Technology, Atlanta, Georgia, May 2003.

[17]Gatski, T. and Rumsey, C., "CFD Validation of Synthetic Jets and Turbulent Separation Control," NASA Langley Research Center Workshop, 29-31 March, 2004. (*http://cfdval2004.larc.nasa.gov* [cited 19 April 2004]).

[18]Cui, J., and Agarwal, R. K., "3D CFD Validation of a Synthetic Jet in Quiescent Air (NASA Langley Workshop Validation: Case 1)," AIAA Paper 2004-2222, 2nd AIAA Flow Control Conference, Portland, OR, 28 June-1 July 2004.

[19] http://www.grc.nasa.gov/WWW/winddocs/user/index.html [19 April 2004].

[20]Bush, R.H., "The Production Flow Solver of the NPARC Alliance," AIAA Paper 88-0935, 1988.

[21]Spalart, P. R., and Allmaras, S. R., "A One-Equation Turbulence Model for Aerodynamic Flows," AIAA Paper 92-0439, 1992.

[22]Menter, F.R., "Zonal Two-Equation k-ω Turbulence Models for Aerodynamic Flows," AIAA Paper 93-2906, 1993.

[23]Mani, M., Ladd, J. A., Cain, A. B., and Bush, R. H., "An Assessment of One- and Two-Equation Turbulence Models for Internal and External Flows," AIAA Paper 97-2010, 1997.

[24]Spalart, P. R., Jou, W. H., Strelets, M., and Allmaras, S. R., "Comments on the Feasibility of LES for Wings, and on a Hybrid RANS/LES Approach," First AFOSR International Conference on DNS/LES, 4-8 August, 1997, Ruston, Louisiana, also in "Advances in DNS/LES," Liu, C. and Liu, Z., eds., Greyden Press, Columbus, OH, 1997.

[25]http://www.grc.nasa.gov/WWW/winddocs/gman/index.html [19 April 2004].

[26]Romer, W. W., and Bush, R. H., "Boundary Condition Procedures for CFD Analyses of Propulsion Systems - The Multi-Zone Problem," AIAA Paper 93-1971, 1993.

[27]http://harpo.wustl.edu/intro.html [19 April 2004].

Parallel Computational Fluid Dynamics – Parallel Computing and Its Applications
J.H. Kwon, A. Ecer, J. Periaux, N. Satofuka and P. Fox (Editors)
2007 Elsevier B.V.

Parallel Performance Assessment of Moving Body Overset Grid Application on PC Cluster

Eugene Kim[a], Jang Hyuk Kwon[a] and Soo Hyung Park[b]

[a]Department of Aerospace Engineering, Korea Advanced Institute of Science and Technology, 373-1 Guseong-dong, Yuseong-gu, Daejeon, 305-701 Republic of Korea

[b]NITRI & Department of Aerospace Engineering, Konkuk University, 1 Hwayang-dong, Gwangjin-Gu, Seoul 143-701, Republic of Korea

Key Words: Moving Body with Relative Motion, Overset Grid, PC cluster, Parallel computing

1. Introduction

The multi-block overset grid method is a powerful technique for high-fidelity computational fluid dynamics(CFD) simulations about complex aerospace configurations[1]. Furthermore, it allows the calculation of moving-body with relative motions, such as applications in the analysis of store separation and many other areas. However, a communication between overlapping grids must be re-established in every time step as the sub-grid is moved. This process of establishing communication between overlapping grids involved in hole cutting and domain connection has been referred to as "grid assembly"[2].

The parallel implementations and performance assessments of the well known chimera codes, Beggar[2] and DCF3D[3] have been accomplished on vendor's machines, such as SGI, IBM, and Cray. In this research, a structured multi-block grid assembly is parallelized by using a static load balancing tied to the flow solver based on the grid size of a decomposed domain. The parallel algorithm is modified for a distributed memory system like a PC cluster, which is more competitive in performance considering the cost. This is why the usage of PC clusters is becoming more widely used in nowadays. To parallelize the grid assembly, a message passing programming model based on the MPI library is implemented using the SPMD(Single Program Multiple Data) paradigm. The parallelized flow solver can be executed in each processor with the static load balancing by pre-processor and the communication load of it is fixed at the initial stage. However, the communication load of grid assembly can not be determined initially and it makes the complexity of chimera algorithms. A coarse-grained communication is optimized with a minimized memory allocation and communication load because the parallel grid assembly can access the decomposed geometry data within other processors only by message passing in the distributed memory system such as the PC cluster.

This parallel grid assembly and flow solver(KFLOW[4]) are tested on the store separation problem from the Eglin wing and pylon configuration[5] in an inviscid flow field. The parallel performance assessment in this problem will be shown in this study.

2. Parallel Implementation of Grid Assembly

The overset grid method is comprised of two major steps, hole cutting and donor cell identification. A zones of interference scheme and a hole-map algorithm are used for the chimera hole cutting. The second step uses the stencil-walk and the gradient searching methods[6]. We will describe the parallel implementation of each step in the following subsections.

2.1. Hole Cutting

Solid walls are used for the hole-cutting surfaces. The cartesian mesh is constructed in cutting boundaries and cutting surfaces are approximately modeled with a collection of cubes that touch the cutting surfaces. These algorithms are relatively simple and effective to discover whether a considering point lies inside of the cutting boundary or not. However, the solid body surface consists of decomposed patches in each processor that is called facets must be a closed surface to make the hole map. The gathering of surface patches in each body is executed by group communication after regenerating a new communicator among the processors related to the same body.

Some processors have multi-blocks for the load balancing. The block numbering in each processor has an independent order as shown in Figure 1.

The flow solver accomplishes its work by the block, however the grid assembly handles grid data by the body. The 'idblk' is a body number that the block constitutes and the 'ipblk' is global-block numbering to connect body indices to independent ordering of multi-block in each processor. Figure 2 is an example of defining newrank as distinct from a MPI_COMM_WORLD by a MPI_COMM_SPLIT library[7]. The 'icolor' is the idblk index in Figure 1. The facets are gathered among the processors with a common new communicator by a MPI_ALLGATHERV library[7] shown as Figure 3.

For the hole cutting, the hole map constructed in each body must be transferred to the processors related to other bodies. However, the direct communication of it causes heavy transfer load because the hole map size is dependent on the grid size. Generally the performance of processors in the PC cluster is better than the network transfer capacity, a Ethernet connection for example. Therefore in this approach, only the facets and the bounds of facets computed in each body are transferred to other body processors. Then the hole map is constructed respectively. We can reduce the execute time because the size of facets is relatively smaller than the total grid size, especially when the background grid is used for the moving body applications.

After construction of the hole map and the hole cutting, a candidate for donor block(CPU rank) of interpolation points is determined by the ipblk index of facet included. For the interpolation points determined by boundary condition inputs, the candidate block will be the first block of body in searching the list input. Then the information of the interpolation points is sent to the candidates. Those are the take rank, coordinates of points and the integer tag to connect send and take ranks.

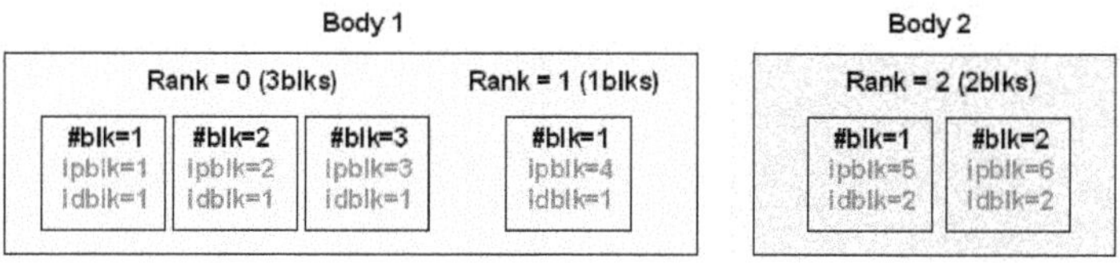

Figure 1. Block data structure of grid assembly (2 bodies and 3CPUs)

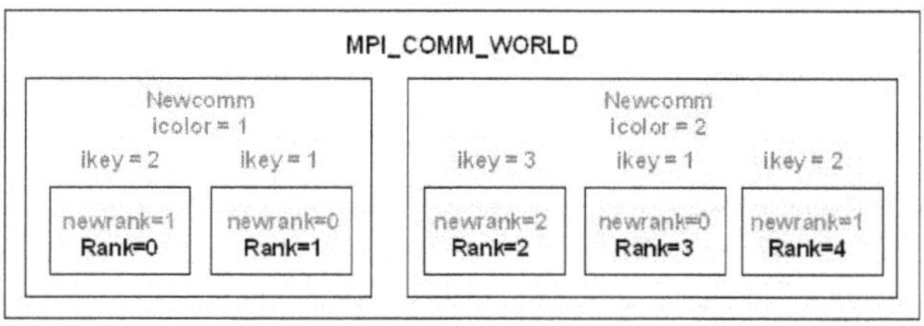

Figure 2. Example of MPI_COMM_SPLIT (2 bodies and 5 CPUs)

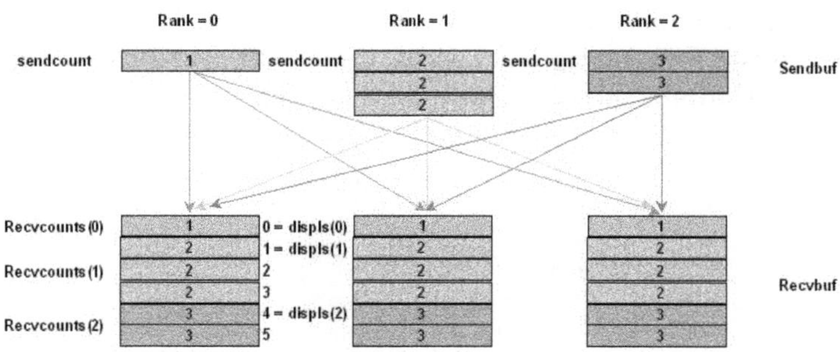

Figure 3. MPI_ALLGATHERV

2.2. Donor Cell Identification

Every processor starts searching donor points by using the information received from the take rank. The two-step search method is used in this procedure as stated before. The stencil walk finds the nearest point to a given interpolation point by comparing the distance between the interpolation point and the donor candidate cells. The gradient search checks whether the interpolation point actually lies inside the cube that is made with the eight donor cells or not. To search through the parallel boundary, we can use coordinates of ghost cell as in Figure 4 and the received and stored facet data of a surface can be used to find the nearest candidate cell when the searching routine meets the solid wall boundary. The information of interpolation points that go over the next block or fail to find a donor is sent to the next candidate block after all processors complete the searching. These procedure is repeated until all interpolation points find donors

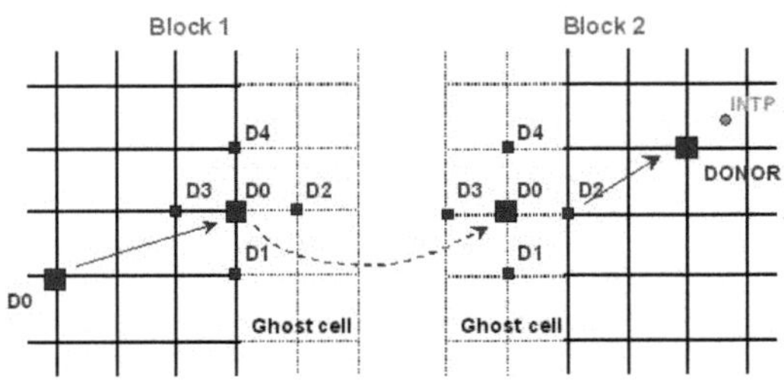

Figure 4. Jumping through parallel boundary

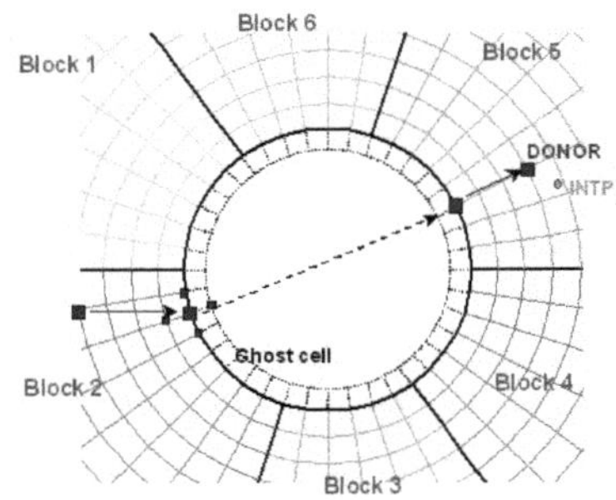

Figure 5. Searching near the solid wall

2.3. Connection between Interpolation and Donor points

During the parallel donor searching, minimized information of interpolation points is communicated between the take and send ranks described before. A take rank doesn't need to recognize the information of donor. Therefore the interpolation and donor data in the take and send ranks respectively are connected by sorting with a tag index. Figure 6 shows this procedure between two processors, the first step is grouping by rank and the second step is sorting by tag index in the same groups. A Heap-sort algorithm is used to sort. It is known as one of the fastest sorting algorithm with *O(nlogn)* operations and has good performance in the worst case.

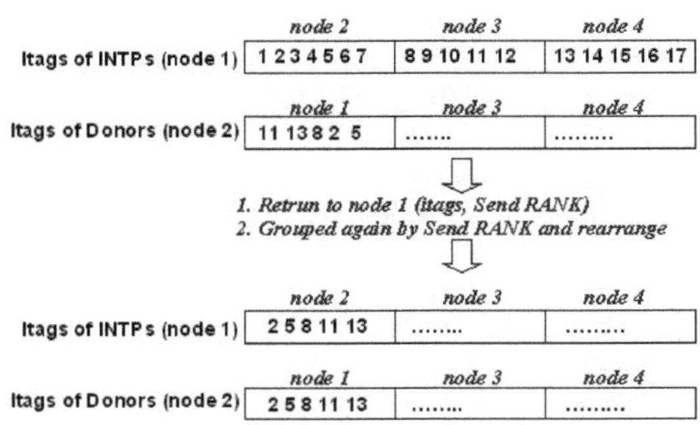

Figure 6. Coupling of interpolation and donor points

Figure 7 is the flow chart of parallel grid assembly. The coarse-grained communication is inserted in three times : construction of hole-map, hole cutting and donor searching routine.

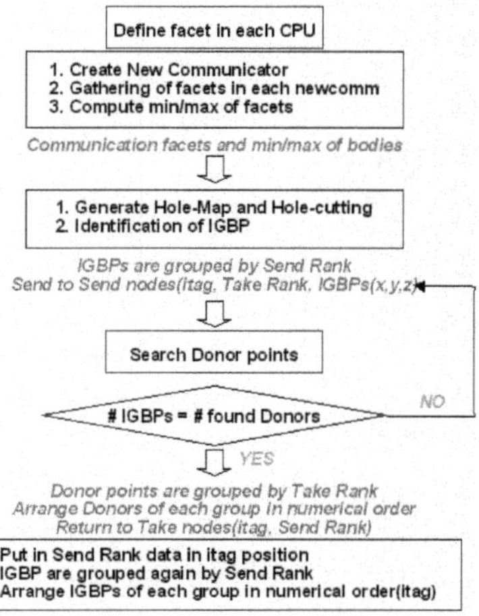

Figure 7. Process of parallel grid assembly

3. Test results

The flow solver is a cell centered finite volume code. The numerical fluxes at cell interface are constructed using the Roe's flux difference splitting method and MUSCL(Monotone Upwind Scheme for Conservation Law) scheme. For the time integration, the diagonalized ADI method is used with the 2nd-order dual time stepping algorithm.

The test problem is the released store from Eglin wing/ pylon configuration in a transonic free stream condition, the Mach number is 0.95. A C-H type grid generated over the wing and an O-O type grid over the pylon are overlapped and used as a major grid system, together with a H-H grid to improve the interpolation quality. The total number of grids is approximately one million. Table 1 is the list of grid system.

A Linux PC cluster with P4-2.6GHz CPU in each node and linked by 100Mbps Ethernet network is used. Serial and parallel computations with 15 and 30 nodes are performed.

Figure 8 shows the constructed overlap grids and the pressure contour at an initial position. The number of IGBPs(Inter-Grid Boundary Points) at this condition is 126,534.

Table 1. Grid system for Eglin wing/pylon/store case

Body	# Block	Type	# node
Wing	1	C-H	129X44X49
Pylon	1	C-O	121X27X26
Store	4	O-H	130X32X25
BG block	1	H-H	105X51X85
2 Extra surfaces for construction of Hole map			

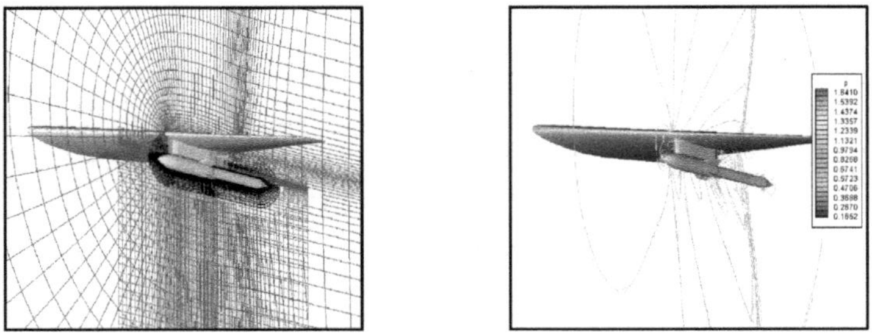

(a) Overlapped grids (b) Pressure distributions

Figure 8. Grids and flow solution in store symmetric section at the initial condition

To simulate the dynamics of store, the unsteady time step is set to 0.00195 second and we simulated up to 0.4 second.

Figure 9 shows the pressure distributions on the store during the separation and the experimental data[5]. The flow solver including the developed grid assembly agrees well with the trend and the magnitude of the CTS data. Figure 10 is the simulated CG location and attitude of store. The results of serial and parallel computations are almost same in this figure.

Tables 2 and 3 show parallel performance of each routine for the grid assembly, the domain connection(the interpolation work time in every time step) and the flow solving. Total speed up ratio in the case of using 30 processors is 17.58. This is about 59% of an ideal parallel performance. The most time consuming parts is the domain connection routine that is just data communication through network.

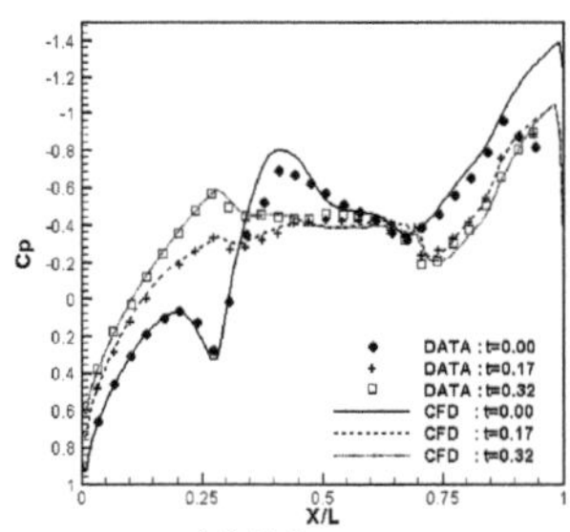

(a) 5 degrees

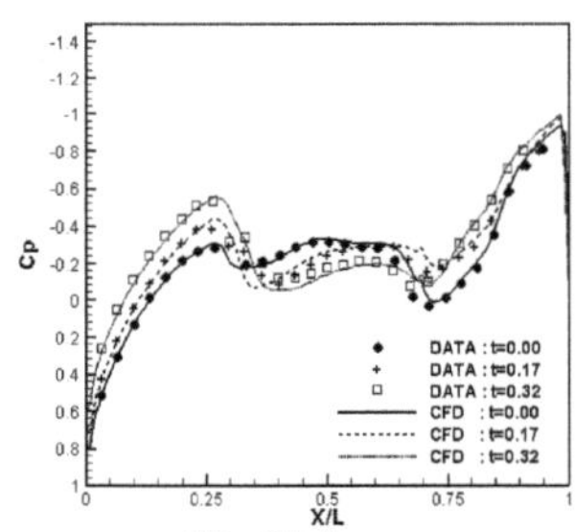

(b) 95 degrees

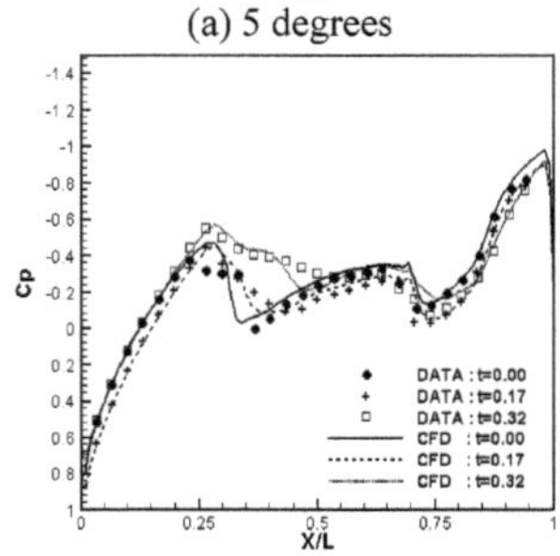

(c) 185 degrees

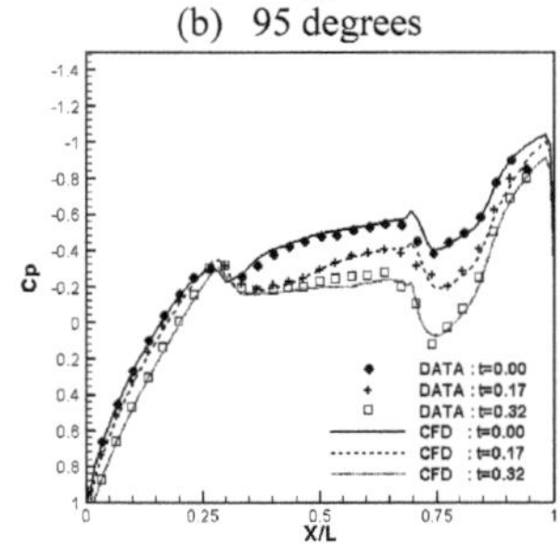

(d) 275 degrees

Figure 9. Sectional Cp distributions on the store surface (various Φ angles)

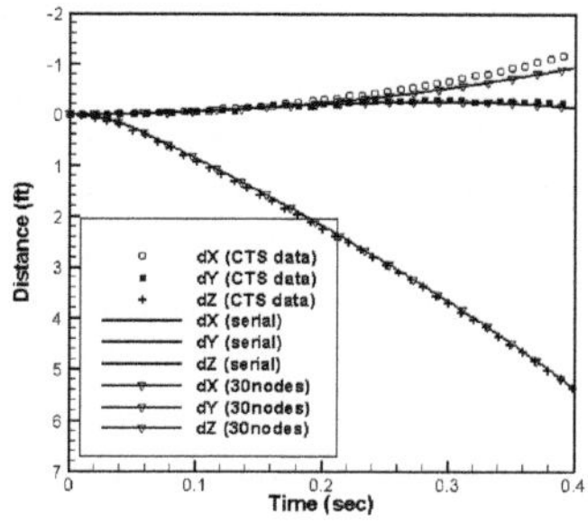

(a) Displacements

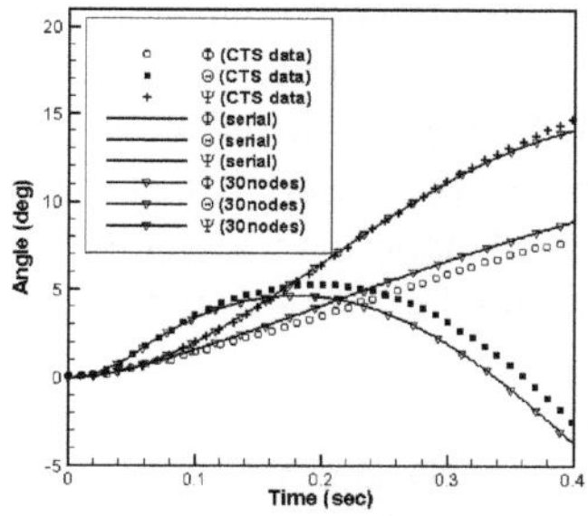

(b) Attitudes

Figure 10. Trajectory of released store (CTS vs CFD : serial and parallel)

Table 2. Parallel performance (wall clock time and CPU time)

Proc.	Grid assembly		Domain connection		Flow solver		Total	
Time(sec)	W time	CPU time	W time	CPU time	W time	CPU time	W time	CPU time
Serial	26754.2	1863.4	3111.9	1587.4	375325.8	315182.8	410459.3	342534.7
15	4781.7	972.2	6231.8	119.5	41524.7	21913.9	53221.1	23534.7
30	4069.5	756.6	4871.3	46.5	14126.5	9213.6	23346.3	10246.8

Table 3. Speed up of each module (fraction of wall clock time)

Proc.	Grid assembly	Domain connection	Flow solver	Total
15	5.596	0.499	9.039	7.712
30	6.574	0.639	26.569	17.581

Table 4. Load imbalance of interpolation and donor points at initial stage (15 CPUs)

Proc.	1	2	3	4	5	6	7	8	9	10	11	12	13	14	15
Grid Size(K)	71	71	71	71	85	77	77	82	77	77	82	104	104	104	104
Body	Wing				Pylon	BG block						Store			
INTP	-	802	1595	-	13655	13461	13652	13314	9715	14796	15122	6192	6192	6195	11846
Donor	308	21128	50813	210	3858	3174	3886	3101	10670	4441	4808	3088	4288	4245	8255

4. Conclusion

The multi-block parallel grid assembly has been developed. The coarse-grained communication is optimized with the minimized memory allocation and communication load without modifications of main algorithms to cut holes and search donors. The flow solver including this parallel grid assembly has been tested on the dynamic problem on a general Linux PC cluster system. The parallel performance of total work is somewhat poor because the network connection is not good enough to reduce the communication time in the domain connection routine. A higher performance network system, for example the Myrinet, can reduce communication time in these routines. However, a larger number of processors is better than the expensive network system so far because the limitation of static load balancing tied to flow solver in grid assembly still remains as in Table 4 and the flow solver routine consumes over 60% of the total work.

REFERENCES

[1] J.L. Steger, F.C. Dougherty and J.A. Benek, "A Chimera Grid Scheme," In Advances in Grid Generation, pp59-69, *ASME FED*-Vol.5, NY, June 1985.

[2] N.C. Prewitt, D.M. Belk and W. Shyy, "Improvements in Parallel Chimera Grid Assembly," *AIAA Journal*, Vol.40, No.3, March 2002, pp.497-500.

[3] A.M. Wissink and R.L. Meakin, "On Parallel Implementations of Dynamic Overset Grid Methods." *SC97: High Performance Networking and Computing*, November 1997.

[4] S.H. Park, Y.S. Kim and J.H. Kwon, "Prediction of Damping Coefficients Using the Unsteady Euler Equations," *Journal of Spacecraft and Rockets*, Vol.40, No.3, 2003, pp.356-362.

[5] L.E. Lijewski and N.E. Suhs, "Time-Accurate Computational Fluid Dynamics Approach to Transonic Store Separation Trajectory Prediction," *Journal of Aircraft*, Vol.31, No.4, 1994, pp. 886-891.

[6] K.W. Cho, J.H. Kwon and S. Lee, "Development of a Fully Systemized Chimera Methodology for Steady/Unsteady Problems," *Journal of Aircraft*, Vol. 36, No.6, 1999, pp. 973-980.

[7] W. Gropp, E, Lusk and A. Skjellum, "Using MPI : Portable Parallel Programming with the Message-Passing Interface," The MIT Press, Cambridge, Massachusetts, London, England.

Parallel Computational Fluid Dynamics – Parallel Computing and Its Applications
J.H. Kwon, A. Ecer, J. Periaux, N. Satofuka and P. Fox (Editors)
2007 Elsevier B.V.

Streamfunction finite element method for Magnetohydrodynamics *

K. S. Kang [a†] and D. E. Keyes[ab ‡]

[a]Brookhaven National Laboratory, Computational Science Center, Building 463, Room 255, Upton NY 11973, USA

[b] Applied Physics & Applied Mathematics, Columbia University, New York, NY 10027, USA

We apply the finite element method to two-dimensional, incompressible MHD, using a streamfunction approach to enforce the divergence-free conditions on the magnetic and velocity fields. This problem was considered by Strauss and Longcope [1] and Kang [2]. In this paper, we solve the problems in magnetic and velocity field coordinates instead of the velocity stream function, magnetic flux, and their derivatives. Considering the multiscale nature of the tilt instability, we study the effect of domain resolution in the tilt instability problem. We use a finite element discretization on unstructured meshes and an implicit scheme. We use the PETSc library with index sets for parallelization. To solve the nonlinear MHD problem, we compare two nonlinear Gauss-Seidel type methods and Newton's method with several time step sizes. We use GMRES in PETSc with multigrid preconditioning to solve the linear subproblems within the nonlinear solvers. We also study the scalability of this program on a cluster.

1. MHD and its streamfunction approach

Magnetohydrodynamics (MHD) describes the macroscopic behavior of electrically conducting fluids in a magnetic field. The fluid motion induces currents, which produce Lorentz body forces on the fluid. Ampere's law relates the currents to the magnetic field. The MHD approximation is that the electric field vanishes in the moving fluid frame, except for possible resistive effects. In this study, we consider finite element methods on an unstructured mesh for two-dimensional, incompressible MHD, using a streamfunction approach to enforce the divergence-free condition on magnetic and velocity fields and an implicit time difference scheme to allow much lager time steps. Strauss and Longcope [1] applied adaptive afinite element method with explicit time difference scheme and

*This manuscript has been authored by Brookhaven Science Associates, LLC under Contract No. DE-AC02-98CH10886 with the U.S. Department of Energy and under Department of Energy cooperative agreement DE-FC02-04ER25595 and National Science Foundation grant CCF-03-52334 to Columbia University.
†email address: `kskang@bnl.gov`
‡email address: `david.keyes@columbia.edu`

Kang applied a finite element method with an implicit time difference scheme with hybrid streamfunction formula [2] to this problem.

The incompressible MHD equations are:

$$\begin{aligned}
&\tfrac{\partial}{\partial t}\vec{B} = \nabla \times (\vec{v} \times \vec{B}),\\
&\tfrac{\partial}{\partial t}\vec{v} = -\vec{v} \cdot \nabla\vec{v} + (\nabla \times \vec{B}) \times \vec{B} + \mu\nabla^2\vec{v},\\
&\nabla \cdot \vec{v} = 0,\\
&\nabla \cdot \vec{B} = 0,
\end{aligned} \tag{1}$$

where $\vec{B}$ is the magnetic field, $\vec{v}$ is the velocity, and μ is the viscosity. To enforce incompressibility, it is common to introduce stream functions: $\vec{v} = \left(\frac{\partial\phi}{\partial y}, -\frac{\partial\phi}{\partial x}\right)$, $\vec{B} = \left(\frac{\partial\psi}{\partial y}, -\frac{\partial\psi}{\partial x}\right)$. By reformulation for symmetric treatment of the fields, in the sense that the source function vorticity Ω and current density C are time advanced, and the potentials ϕ and ψ are obtained at each time step by solving Poisson equations, we obtain

$$\begin{aligned}
&\tfrac{\partial}{\partial t}\Omega + [\Omega, \phi] = [C, \psi] + \mu\nabla^2\Omega,\\
&\tfrac{\partial}{\partial t}C + [C, \phi] = [\Omega, \psi] + 2\left[\tfrac{\partial\phi}{\partial x}, \tfrac{\partial\psi}{\partial x}\right] + 2\left[\tfrac{\partial\phi}{\partial y}, \tfrac{\partial\psi}{\partial y}\right],\\
&\nabla^2\phi = \Omega,\\
&\nabla^2\psi = C,
\end{aligned} \tag{2}$$

where the commutator is defined by $[a, b] = \frac{\partial a}{\partial x}\frac{\partial b}{\partial y} - \frac{\partial a}{\partial y}\frac{\partial b}{\partial x}$.

To solve (2), we have to compute the partial derivatives of potentials. These partial derivatives can be obtained by solutions of linear problems. To do this, we have to introduce four auxiliary variables. Altogether, this requires the solution of eight equations at each step.

We use the hybrid streamfunction approach which was considered in [2] and use the velocity $\vec{v}$ and the magnetic field $\vec{B}$ to reduce the number of equations to solve (2). To this aim, we put $\vec{v} = (v_1, v_2) = \left(\frac{\partial\phi}{\partial y}, -\frac{\partial\phi}{\partial x}\right)$, $\vec{B} = (B_1, B_2) = \left(\frac{\partial\psi}{\partial y}, -\frac{\partial\psi}{\partial x}\right)$ in equation (2) and get the following system:

$$\begin{aligned}
&\tfrac{\partial\Omega}{\partial t} + (v_1, v_2) \cdot \nabla\Omega = (B_1, B_2) \cdot \nabla C + \mu\nabla^2\Omega,\\
&\tfrac{\partial C}{\partial t} + (v_1, v_2) \cdot \nabla C = (B_1, B_2) \cdot \nabla\Omega + 2([v_1, B_1] + [v_2, B_2]),\\
&-\nabla^2 v_1 = -\tfrac{\partial\Omega}{\partial y}, \qquad -\nabla^2 v_2 = \tfrac{\partial\Omega}{\partial x},\\
&-\nabla^2 B_1 = -\tfrac{\partial C}{\partial y}, \qquad -\nabla^2 B_2 = \tfrac{\partial C}{\partial x},\\
&\quad\nabla^2\phi = \Omega, \qquad\qquad \nabla^2\psi = C.
\end{aligned} \tag{3}$$

In the eight equations in (3), the last two equations for potentials need not be solved to advance the solution in each time step. If the potentials are desired at a specific time, they are obtained by solving the last two equations in (3). To solve the Poisson's equations for $\vec{v}$ and $\vec{B}$ in (3), we have to impose boundary conditions that are compatible to boundary conditions of ϕ, ψ, Ω, and C.

2. Finite element discretization

To solve (3), we use the first-order backward difference (Euler) derivative scheme, leading to an implicit scheme that removes the numerically imposed time-step constraint, allowing much larger time steps. This approach is first-order accurate in time and is chosen merely for convenience.

For domain K, let H^1 denote $H^1(K)$, $H^{1,A}$ denote the subset of $H^1(K)$ whose elements satisfy the boundary condition of field A, and $H^{1,A'}$ denote the subspace of $H^1(K)$ whose elements have zero values on the Dirichlet boundary of A for fields $\Omega, v_1, v_2, B_1, B_2$. Multiply the test functions and integrate by parts in each equation and use the appropriate boundary conditions, to derive the variational form of (3) as follows: Find $\vec{X} = (\Omega, C, v_1, v_2, B_1, B_2) \in H^{1,\Omega} \times H^1 \times H^{1,v_1} \times H^{1,v_2} \times H^{1,B_1} \times H^{1,B_2}$ such that

$$\vec{F}^n(\vec{X}, \vec{Y}) = 0 \tag{4}$$

for all $\vec{Y} = (u, w, p_1, p_2, q_1, q_2) \in H^{1,\Omega'} \times H^1 \times H^{1,v_1'} \times H^{1,v_2'} \times H^{1,B_1'} \times H^{1,B_2'}$, where $\vec{F}^n = (F_1^n, F_2^n, F_3, F_4, F_5, F_6)^T$,

$$\begin{aligned}
F_1^n(\vec{X}, \vec{Y}) &= M_t^n(\Omega, u) + (\vec{v} \cdot \nabla\Omega, u) + \mu a(\Omega, u) - (\vec{B} \cdot \nabla C, u),\\
F_2^n(\vec{X}, \vec{Y}) &= M_t^n(C, w) + (\vec{v} \cdot \nabla C, w) - (\vec{B} \cdot \nabla\Omega^n, w) - 2P(\vec{v}, \vec{B}, w),\\
F_3(\vec{X}, \vec{Y}) &= a(v_1, p_1) + \left(\tfrac{\partial\Omega}{\partial y}, p_1\right), \quad F_4(\vec{X}, \vec{Y}) = a(v_2, p_2) - \left(\tfrac{\partial\Omega}{\partial x}, p_2\right),\\
F_5(\vec{X}, \vec{Y}) &= a(B_1, q_1) + \left(\tfrac{\partial C}{\partial y}, q_1\right), \quad F_6(\vec{X}, \vec{Y}) = a(B_2, q_2) - \left(\tfrac{\partial C}{\partial x}, q_2\right),
\end{aligned}$$

$(u, w) = \int_K uw dx$, $M_t^n(u, w) = \frac{1}{\Delta t}(u - u^n, w)$, $a(u, w) = \int_K \nabla u \cdot \nabla w dx$, and $P(\vec{u}, \vec{v}, w) = \int_K [u_1, v_1] w dx + \int_K [u_2, v_2] w dx$.

Let $\mathcal{K}_h$ be a given triangulation of domain K with the maximum diameter h of the element triangles. Let V_h be the continuous piecewise linear finite element space. Let V_h^A, $A = \Omega, v_1, v_2, B_1, B_2$, be the subsets of V_h that satisfy the boundary conditions of A on every boundary point of $\mathcal{K}_h$ and $V_h^{A'}$ be subspaces of V_h and $H^{1,A'}$. Then we can write the discretized MHD problems as follows: For each n, find the solutions $\vec{X}_h^n = (\Omega_h^n, C_h^n, v_{1,h}^n, v_{2,h}^n, B_{1,h}^n, B_{2,h}^n) \in V_h^\Omega \times V_h \times V_h^{v_1} \times V_h^{v_2} \times V_h^{B_1} \times V_h^{B_2}$ on each discretized times n which satisfy

$$F^n(\vec{X}_h^n, \vec{Y}_h) = 0 \tag{5}$$

for all $\vec{Y}_h \in V_h^{\Omega'} \times V_h \times V_h^{v_1'} \times V_h^{v_2'} \times V_h^{B_1'} \times V_h^{B_2'}$.

3. Nonlinear and linear solvers

The problem (5) is a nonlinear problem in the six variables consisting of two time-dependent equations and four Poisson equations. However, if we consider the equations separately, each equation is linear with respect to one variable, i.e., the first two equations are time-dependent problems and the last four equations are linear Poisson problems. From the above observations, we naturally consider a nonlinear Gauss-Seidel iterative solver that solves a linear each equation for each variable in (5) in consecutive order with recent approximate solutions (GS1), or blocked with the first two equations of (5) as one block and the four Poisson equations as a second block(GS2).

The nonlinear Gauss-Seidel iterative method has no convergence guarantee, but converges well in many cases, especially for small time step sizes in time-dependent problems.

Next, we consider the Newton linearization method. Newton's method has, asymptotically, second-order convergence for nonlinear problems and greater scalability with respect to mesh refinement than the nonlinear Gauss-Seidel method, but requires computation of the Jacobian of nonlinear problem which can be complicated.

To solve linear problems in nonlinear solvers, we use Krylov iterative techniques which are suited because they can be preconditioned for efficiency. Among the various Krylov methods, GMRES is selected because it guarantees convergence with nonsymmetric, non-positive definite systems. However, GMRES can be memory intensive and expensive. Restarted GMRES can in principle deal with these limitations; however, it lacks a theory of convergence, and stalling is frequently observed in real applications.

Preconditioning consists in operating on the system matrix J_k, where

$$J_k \delta x_k = -F(x_k), \tag{6}$$

with an operator P_k^{-1} (preconditioner) such that $P_k^{-1} J_k$ is well-conditioned. This is straightforward to see when considering the equivalent linear system:

$$P_k^{-1} J_k \delta x_k = -P_k^{-1} F(x_k). \tag{7}$$

Notice that the system in equation (7) is equivalent to the original system (6) for any nonsingular operator P_k^{-1}. Thus, the choice of P_k^{-1} does not affect the accuracy of the final solution, but crucially determines the rate of convergence of GMRES, and hence the efficiency of the algorithm.

In this study, we use multigrid, which is well known as a successful preconditioner, as well as a scalable solver in unaccelerated form, for many problems. We consider the symmetrized diagonal term of Jacobian as a reduced system, i.e.,

$$J_{S,k} = \frac{1}{2}\left(J_{R,k} + J_{R,k}^T\right), \tag{8}$$

where $J_{R,k}$ is a block diagonal matrix. The reduced system $J_{S,k}$ may be less efficient than $J_{R,k}$ but more numerically stable because it is symmetric and nonsingular.

To implement the finite element solver for two-dimensional, incompressible MHD on parallel machines, we use PETSc library which is well developed for nonlinear PDE problems and easily implements a multigrid preconditioner with GMRES. We use PETSc's index sets for parallelization of our unstructured finite element discretization.

4. Numerical experiments: Tilt Instability

We consider the initial equilibrium state as

$$\psi = \begin{cases} [2/kJ_0(k)]J_1(kr)\frac{y}{r}, & r < 1 \\ (1/r - r)\frac{y}{r}, & r > 1 \end{cases},$$

where J_n is the Bessel function of order n, k is any constant that satisfies $J_1(k) = 0$, and $r = \sqrt{x^2 + y^2}$.

In our numerical experiments, we solve on the finite square domain $K = [-R, R] \times [-R, R]$ with the initial condition of the tilt instability problem from the above initial

(a) $t = 0.0$.

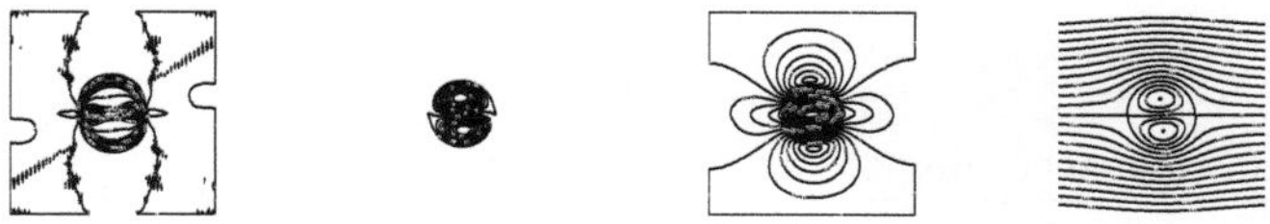

(b) $t = 4.0$.

(c) $t = 6.0$.

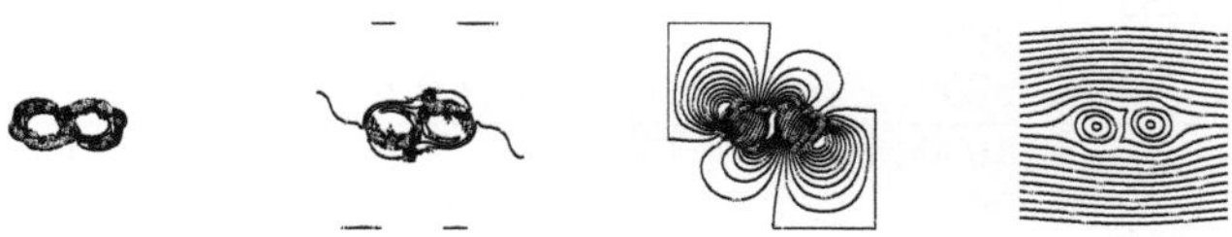

(d) $t = 7.0$.

Figure 1. Contours of Ω, C, ϕ, and ψ at time $t = 0.0, 4.0, 6.0, 7.0$.

equilibrium and perturbation of ϕ (originating from perturbations of velocity) such that

$$\Omega(0) = 0.0, \qquad C(0) = \begin{cases} 19.0272743 J_1(kr) y/r & \text{if } r < 1 \\ 0.0 & \text{if } r > 1 \end{cases},$$

$$\phi(0) = 10^{-3} e^{-(x^2+y^2)}, \quad \psi(0) = \begin{cases} -1.295961618 J_1(kr) y/r & \text{if } r < 1 \\ -(\frac{1}{r} - r) y/r & \text{if } r > 1 \end{cases},$$

where $k = 3.831705970$, and with Dirichlet boundary conditions $\Omega(x, y, t) = 0.0$, $\phi(x, y, t) = 0.0$, and $\psi(x, y, t) = y - \frac{y}{x^2+y^2}$ and Neumann boundary condition for C, i.e., $\frac{\partial C}{\partial n}(x, y, t) = 0.0$. The initial and boundary conditions for velocity $\vec{v}$ and magnetic field $\vec{B}$ are derived from the initial and boundary conditions of Ω, C, ϕ, and ψ. Numerical simulation results are illustrated in Fig. 1.

The tilt instability problem is defined on an unbounded domain. To investigate the effect of size of domains, we compare two methods, one uses streamfunctions ϕ, ψ and its derivatives (standard) and the other uses hybrid streamfunction formulations, i.e., $\vec{v}$ and $\vec{B}$ (hybrid) on several square domains with $R = 2, 3, 3.5$. These numerical simulation results are depicted in Fig. 2; the contours of ψ at $t = 7.0$ and plot the kinetic energies in Fig. 3. The average growth rate γ of kinetic energy is shown in Table 1. These

Table 1
Average growth rate γ of kinetic energy from $t = 0.0$ to $t = 6.0$.

Standard streamfunction formula			Hybrid streamfunction formula		
$R = 2$	$R = 3$	$R = 3.5$	$R = 2$	$R = 3$	$R = 3.5$
2.167	2.152	2.148	1.744	2.102	2.125

Table 2
Average solution time according to the number of processors.

Machine	Solver	DoF/P	Number of processor						
			2	4	8	16	32	64	128
BGC	GS2	2000	133		386		1180		
		8000	425		1205		2263		
	NM	2000	77.6		199		616		
		8000	335		599		1429		
Chee-tah	GS2	1000		37.02		47.90		75.49	
		2000	67.37		76.58		110.7		239.5
		4000		130.1		216.1		311.7	
		8000	255.8		351.0		503.0		
	NM	1000		14.27		17.86		28.37	
		2000	23.08		30.84		41.21		82.09
		4000		50.41		91.43		117.5	
		8000	110.3		149.8		204.1		

numerical simulation results show that the solutions of two formulations are closer when the domain is enlarged, with the standard approach converging from above and hybrid approach converging from below.

For parallel scalability, we measure the simulation time from $t = 0.0$ to $t = 0.5$ with $dt = 0.005$ using the hybrid formulation. We simulate on the BGC (Galaxy Cluster) machine at Brookhaven National Laboratory, which consists of 256 Intel P3 and P4 dual processor nodes running at speeds up to 2.4 GHz with 1 Gbyte of memory per node and the Cheetah 4.5 TF IBM pSeries system at Oak Ridge National Laboratory, which consists of 27 p690 nodes, where each node consists of thirty-two 1.3 GHz Power4 processors, and connected via IBM's Federation interconnect. We report the simulation times according to the number of levels and processors in Table 2 and plot the weak scalability in Fig. 4 and speed up in Fig. 5 on Cheetah. These table and figures show that Newton's method has a better scalability and is also faster than Gauss-Seidel iteration.

REFERENCES

1. Strauss, H. R. and Longcope, D. W.: An Adaptive Finite Element Method for Magnetohydrodynamics., *Journal of Computational Physics*, 147:318–336, 1998.
2. K. S. Kang, New stream function approach method for Magnetohydrodynamics, *Proceedings of 16th International Conference on Domain Decomposition Methods, Springer Lecture Notes in CS & E*, to appear, 2006.

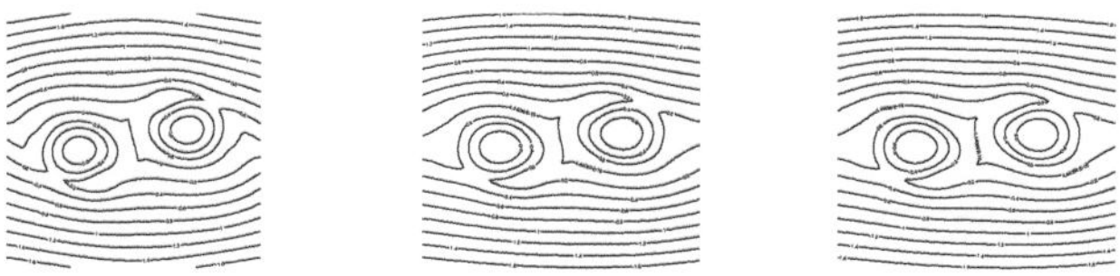

(a) $R = 2.$ (b) $R = 3.$ (c) $R = 3.5$

Standard streamfunction formulation

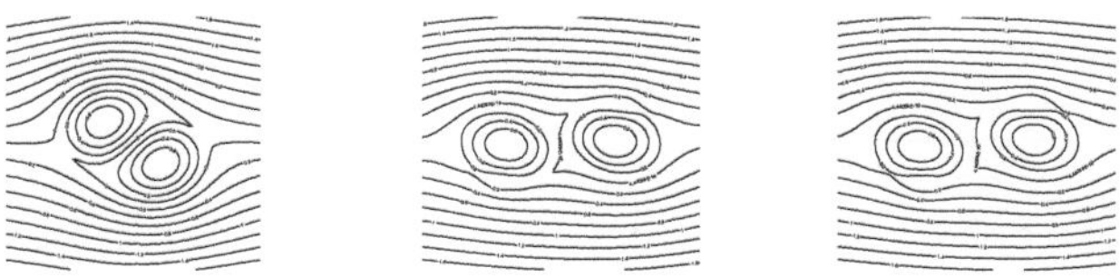

(d) $R = 2.$ (e) $R = 3$ (f) $R = 3.5$

Hybrid streamfunction formulation

Figure 2. Contours of ψ at $t = 7.0$.

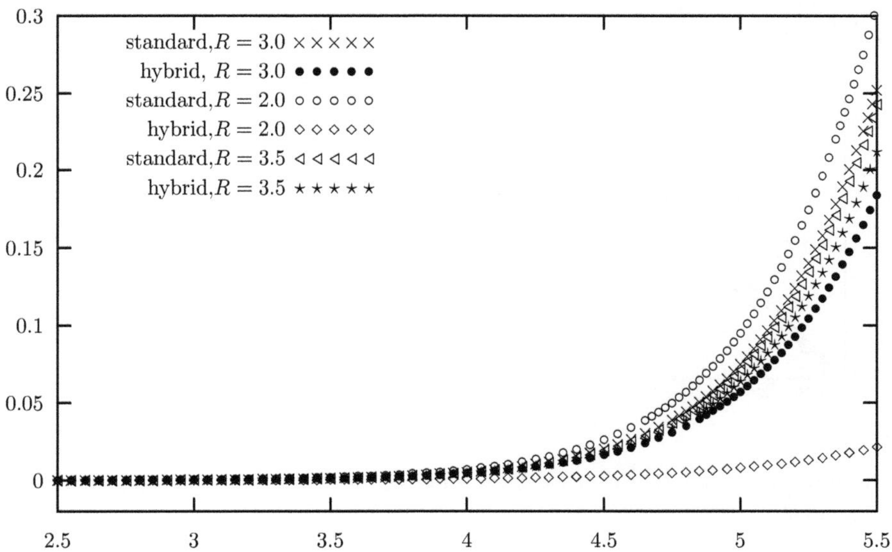

Figure 3. Kinetic Energy, as a function of time.

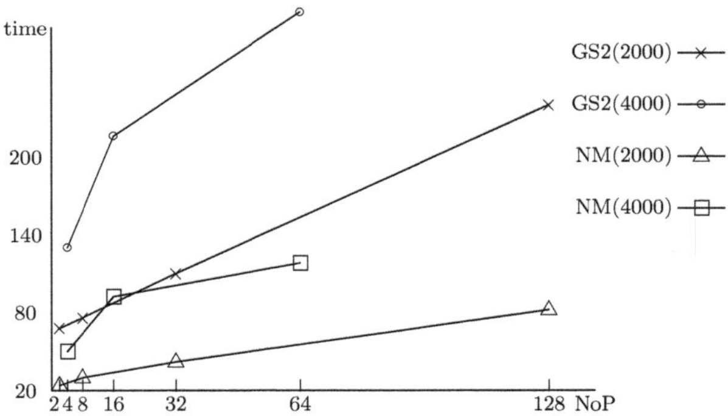

Figure 4. Solution time for a fixed number of degrees of freedom per processor (weak scaling).

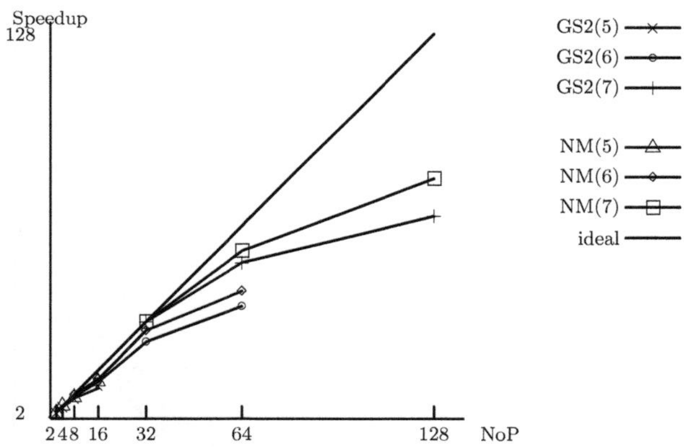

Figure 5. Strong speedup for fixed number of grid levels according to the number of processors.

Parallel Computational Fluid Dynamics – Parallel Computing and Its Applications
J.H. Kwon, A. Ecer, J. Periaux, N. Satofuka and P. Fox (Editors)

Parallelization of Phase-Field Model to Simulate Freezing in High-*Re* Flow—Multiscale Method Implementation

Ying Xu[a*], J. M. McDonough[a] and K. A. Tagavi[a]

[a]Department of Mechanical Engineering, University of Kentucky,
151 RGAN Bldg., Lexington, KY 40506

We implement a heterogeneous two-level multiscale method to solve equations of the phase-field model of solidification in a flow field. Governing equations, boundary and initial conditions on both scales and the detailed algorithm are provided. Freezing in a lid-driven-cavity flow at $Re = 10,000$ has been simulated, and the results in both large-scale and small-scale domains are provided, together with the parallel performance using MPI.

Key Words: Multiscale, Phase-field, Freezing, High-*Re* Flow

1. INTRODUCTION

Phase-field models have been applied to phase transformation problems for decades. Numerous studies have been conducted using phase-field methods on solidification problems, including phase transitions in pure materials and binary alloys, and solidification in an induced flow field. A detailed literature survey can be found in Xu et al. [1], [2].

We note that all previous research on solidification in a flow field involves length and time scales in microns and nanoseconds, respectively, which is not appropriate for studying freezing in a typical flow field of engineering scale. Therefore, the concepts of multiscale methods should be introduced to phase-field models with convection. Even such formulations still have the drawback of being very CPU intensive. Hence, it is necessary to apply parallelization to phase-field model algorithms in order to decrease wall-clock time. The task of this research is to simulate dendrite growth in a high-*Re* flow (in this case lid-driven-cavity flow). We observe that previous solidification with convection problems have been solved under small Reynolds number conditions. We will implement a two-level multiscale method, i.e., a macroscale for flow field and a microscale for dendrite growth, which enables us to solve phase-change problems in a turbulent flow.

We introduce a multiscale method and its implementation in the first part of this paper, followed by the phase-field model with convection. Finally we present the numerical solutions, discuss the approach to parallelization and show the speedups obtained.

*yxu2@uky.edu

2. MULTISCALE METHOD

Many systems in nature involve multiple scales and become interesting because they exhibit different behaviors on different scales. Such systems range from living organisms to geological and geophysical systems, materials and condensed matter systems and even social structures and hierarchies, etc. Therefore two questions become important. What are the rules governing the small scale (microscopic) units in a large (macroscopic) system, and how is the macroscopic behavior influenced by the microscopic behavior? What are the rules governing the large scale behavior, and how does this influence the behavior on the small scale?

The traditional approach for multiscale problems is to obtain, either analytically or empirically, explicit equations for the scale of interest, while eliminating other scales of the problem. Analysis of these systems is conventionally done by construction of approximate solutions using similarity techniques by Barenblatt [3], averaging methods by Lichtenberg and Lieberman [4], renormalization and homogenization theory by Babuska [5], etc. Despite some successes, these approaches also force us to introduce empirical closures for other systems that are not always justified or understood. As a result, the success of resulting phenomenological equations is much less spectacular for a large class of complex systems. Typical examples of such a situation are found in complex fluids, plasticity, fracture dynamics, and important regimes of turbulent flows.

A new approach has emerged in recent years and has quickly attracted attention. The aim of this approach is to model the theoretical input to a coarse-grained model from a more detailed microscopic model, bypassing the necessity of empirical modeling. E and Engquist [6] presented a general framework for designing and analyzing numerical methods that deal with problems that can be divided into two types:

A. A macroscopic description is known but ceases to be valid in a localized region in space and/or time, and where the microscopic description has to be used instead.

B. A macroscopic model may not be explicitly known, or too expensive to obtain, but is known to exist; i.e., there exists a set of macroscopic variables obeying a closed macroscopic model.

In addition, there exists a class of problems, say of Type C, which combines the characteristics of A and B, namely that the macroscopic model is not explicitly known and any typical model ceases to be valid in some regions.

The freezing problem in a flow field can be categorized as type C; i.e., the macroscopic freezing model is not explicitly known, and macroscopic representations cease to be valid on the microscopic level. The reason is that the phase-field-like models are needed to provide details of dendrite structures on microscopic levels. However, these cannot be applied to the macroscopic level unless some techniques, e.g., the homogenization theory, are applied. The so-called heterogeneous multiscale method is implemented in our research because of advantages presented before. In particular, we develop different numerical methods for different physical models at different scales. We make direct use of the microscale model to obtain the information which is required by the macroscale model, e.g., volume fraction of solid phase and energy flux gained or released by the phase-change process. The two-scale phase-field model with flow field will be presented in the next section, followed by details of numerical methods employed in its implementation.

We explain the implementation of a multiscale method via Fig. 1 and Algorithm 1. In Figure 1, we denote a large-scale domain by Ω and a small-scale domain inside one grid cell of Ω by Ω_s. We first solve the equations of fluid motion on a macroscopic level in Ω to obtain u, v and pressure p. We then interpolate the large-scale solutions to every grid point of Ω_s to prepare for the computation on the small scale. Small-scale computation is conducted to simulate ice growth in the flow field during one time step of the large-scale calculation. We assume that the small-scale computation does not affect the velocity field and pressure on the large scale during this time. Phase field ϕ and temperature T on the small scale are computed using Neumann boundary conditions on $\partial\Omega_s$ to approximate an infinite domain. Moreover, velocity and pressure on the small scale are obtained using momentum equations with Neumann boundary conditions which will be provided in Section 3.

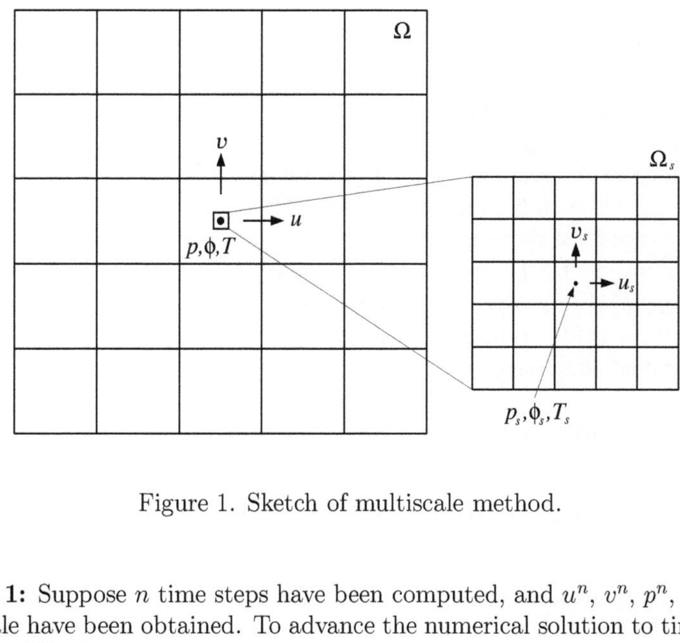

Figure 1. Sketch of multiscale method.

Algorithm 1: Suppose n time steps have been computed, and u^n, v^n, p^n, ϕ^n and T^n for the large scale have been obtained. To advance the numerical solution to time level $n+1$, carry out the following steps.

1. Large-Scale Computation on Ω:
   Solve flow field equations to obtain u^{n+1}, v^{n+1} and p^{n+1};

2. Interpolation from Large Scale to Small Scale:
   Pick a grid cell (i_s, j_s) in Ω such that $\Omega_s \subseteq (i_s, j_s)$ as in Fig. 1. Interpolate large-scale solutions u^n, v^n and p^n to grid points on Ω_s to obtain initial conditions u_s^0, v_s^0, p_s^0 and boundary conditions u_s^b, v_s^b, p_s^b for small-scale computation;

3. Small-Scale Computation on Ω_s:
   Suppose n_s time steps have been computed, and $u_s^{n_s}$, $v_s^{n_s}$, $p_s^{n_s}$, $\phi_s^{n_s}$ and $T_s^{n_s}$ have been obtained. Carry out the following steps to advance to time level n_s+1.

   (a) Solve flow field equations on small scale using well-posed boundary conditions to obtain $u_s^{n_s+1}$, $v_s^{n_s+1}$ and $p_s^{n_s+1}$;

   (b) Solve phase field and energy equations on small scale to obtain $\phi_s^{n_s+1}$ and $T_s^{n_s+1}$;

   (c) If $|\phi_s^{n_s+1} \quad 1| < 10^{\quad 12}$ or $|T_s^{n_s+1} \quad T_m| < 10^{\quad 10}$
      $\Omega_s = 2\Omega_s$
      Interpolate $u_s^{n_s+1}$, $v_s^{n_s+1}$, $p_s^{n_s+1}$, $\phi_s^{n_s+1}$ and $T_s^{n_s+1}$ to new domain Ω_s
      Obtain new boundary conditions u_s^b, v_s^b, p_s^b from u^{n+1}, v^{n+1} and p^{n+1} via interpolation
      goto (a) for next time step of small scale
   else if $n_s+1 =$ total number of time steps for small scale
      $\phi_s^{n+1} = \phi_s^{n_s+1}$, $T_s^{n+1} = T_s^{n_s+1}$
      goto 4;

4. Data transfer from Small Scale to Large Scale:
   To determine ϕ^{n+1} and T^{n+1} of large scale from ϕ_s^{n+1} and T_s^{n+1} of small scale at large domain grid point (i_s, j_s), apply the following procedure:
      Let $\|\phi^{n+1}\| = \|\phi_s^{n+1}\|$ and $\|T^{n+1}\| = \|T_s^{n+1}\|$ over the whole grid cell (i_s, j_s) of the large-scale domain with
      $\phi^{n+1} = \frac{1}{h_x h_y}\int_0^{h_y}\int_0^{h_x}\phi_s^{n+1}dxdy$, $T^{n+1} = \frac{1}{h_x h_y}\int_0^{h_y}\int_0^{h_x}T_s^{n+1}dxdy$
      where $\phi_s^{n+1} = 1$, $T_s^{n+1} = T_{init}$ everywhere outside the small-scale domain Ω_s;
   Assign $u^{n+1} = v^{n+1} = 0$ in solid phase;
   goto 1 for next time step.

Here, h_x and h_y are large-scale spatial step sizes in x and y directions.

3. TWO-SCALE PHASE-FIELD MODEL IN FLOW FIELD

In implementing the heterogeneous multiscale method for the phase-field model with convection, we solve the mass conservation and momentum equations on the macroscopic scale and the whole set of governing equations on the microscopic scale. Therefore, we have

$$u_x + v_y = 0 \tag{1a}$$

$$u_t + (uu)_x + (vu)_y = \frac{p_x}{\rho_0} + \frac{1}{\rho_0}\left[2\left(\mu u_x\right)_x + \left(\mu u_y\right)_y + \left(\mu v_x\right)_y\right] + X_1(\phi)\,, \tag{1b}$$

$$v_t + (uv)_x + (vv)_y = \frac{p_y}{\rho_0} + \frac{1}{\rho_0}\left[\left(\mu v_x\right)_x + 2\left(\mu v_y\right)_y + \left(\mu u_y\right)_x\right] + X_2(\phi) + B(\phi, T)\,, \tag{1c}$$

on the large scale, and

$$u_x + v_y = 0 \tag{2a}$$

$$u_t + (uu)_x + (vu)_y = \frac{p_x}{\rho_0} + \frac{1}{\rho_0}\left[2\left(\mu u_x\right)_x + \left(\mu u_y\right)_y + \left(\mu v_x\right)_y\right] + X_1(\phi)\,, \tag{2b}$$

$$v_t + (uv)_x + (vv)_y = \frac{p_y}{\rho_0} + \frac{1}{\rho_0}\left[\left(\mu v_x\right)_x + 2\left(\mu v_y\right)_y + \left(\mu u_y\right)_x\right] + X_2(\phi) + B(\phi, T)\,, \tag{2c}$$

$$\phi_t + (u\phi)_x + (v\phi)_y = \frac{\epsilon^2}{M}N(\phi) + \frac{\rho_0 L_0 P'(\phi)}{MT_m}(T \quad T_m) \quad \frac{\rho_0 \psi'(\phi) T}{4aM} + \frac{Y(\phi, T)}{M}\,, \tag{2d}$$

$$T_t + (uT)_x + (vT)_y = \alpha \Delta T + \left[\frac{\epsilon^2 N(\phi)}{2\rho_0 c_p} \quad \frac{30 L_0 \psi(\phi)}{c_p}\right]\frac{D\phi}{Dt} + \frac{W(u, v, \phi)}{\rho_0 c_p}\,, \tag{2e}$$

on the small scale. In the above equations, we denote

$$N(\phi) = \xi_1^2 \phi_{xx} + \xi_2^2 \phi_{yy} + \left(2\xi_1 \xi_{1x} + \xi_2 \xi_{1y} + \xi_1 \xi_{2y}\right)\phi_x + \left(\xi_1 \xi_{2x} + \xi_2 \xi_{1x} + 2\xi_2 \xi_{2y}\right)\phi_y + 2\xi_1 \xi_2 \phi_{xy}\,, \tag{3a}$$

$$X_1(\phi) = \frac{\epsilon^2}{\rho_0}\phi_x\left(\xi_1^2 \phi_{xx} + 2\xi_1\xi_2\phi_{xy} + \xi_2^2 \phi_{yy}\right)\,, \tag{3b}$$

$$X_2(\phi) = \frac{\epsilon^2}{\rho_0}\phi_y\left(\xi_1^2 \phi_{xx} + 2\xi_1\xi_2\phi_{xy} + \xi_2^2 \phi_{yy}\right)\,, \tag{3c}$$

$$B(\phi, T) = \frac{\rho(\phi, T) \quad \rho_L}{\rho_0} g\,, \tag{3d}$$

$$Y(\phi, T) = \frac{30(p \quad p_0)}{\rho_0}\psi(\phi)\left[\rho_L \quad \rho_S + \beta\rho_L(T \quad T_m)\right]\,, \tag{3e}$$

$$W(u, v, \phi) = \mu\left[2u_x^2 + \left(u_y + v_x\right)^2 + 2v_y^2\right] + \frac{\epsilon^2}{4}\left(\xi_1^2\phi_x^2 \quad \xi_2^2\phi_y^2\right)\left(v_y \quad u_x\right) \quad \frac{\epsilon^2}{2}\left[v_x\left(\xi_1\xi_2\phi_x^2 + \xi_2^2\phi_x\phi_y\right) + u_y\left(\xi_1^2\phi_x\phi_y + \xi_1\xi_2\phi_y^2\right)\right]\,. \tag{3f}$$

The initial and boundary conditions for a well-posed problem on the large-scale domain $\Omega \equiv [0, L] \times [0, L]$ follow:

Initial conditions

$$u = v = p \equiv 0 \qquad \text{in } \Omega\,, \tag{4a}$$

Boundary conditions

$$\begin{aligned} u &= 0 && \text{on } \partial\Omega \backslash \{y = L\}\,, \\ u &= U && \text{on } \{y = L\}\,, \\ v &= 0 && \text{on } \partial\Omega\,, \\ \frac{\partial p}{\partial n} &= 0 && \text{on } \partial\Omega\,, \end{aligned} \tag{4b}$$

while conditions on a small-scale domain $\Omega_s \equiv [0, L_s] \times [0, L_s]$ are:

Initial conditions

$$\begin{aligned}
&\phi = 0 && \text{in } \Omega_0 \,, \\
&\phi = 1 && \text{in } \Omega_s \backslash \Omega_0 \,, \\
&T = T_m && \text{in } \Omega_0 \,, \\
&T = T_{init} < T_m && \text{in } \Omega_s \backslash \Omega_0 \,,
\end{aligned} \tag{5a}$$

Boundary conditions

$$\begin{aligned}
&u_x = \quad v_y \,; \quad v_x = u_y + \omega && \text{on } \{x = 0\} \cup \{x = L_s\} \,, \\
&u_y = v_x \quad \omega \,; \quad v_y = \quad u_x && \text{on } \{y = 0\} \cup \{y = L_s\} \,, \\
&\frac{\partial \phi}{\partial n} = \frac{\partial T}{\partial n} = \frac{\partial p}{\partial n} = 0 && \text{on } \partial \Omega_s \,,
\end{aligned} \tag{5b}$$

where Ω_0 is the domain of a small seed which causes the onset of freezing. In the above boundary conditions, vorticity ω on boundary $\partial \Omega_s$ is obtained by solving the vorticity transport equation

$$\omega_t = \quad u\omega_x \quad v\omega_y + \frac{\mu(\phi)}{\rho_0} \Delta \omega \qquad \text{only on} \quad \partial \Omega_s \,.$$

Here, velocity boundary conditions are employed to satisfy the mass conservation law, and to avoid the effect of the boundary condition to the flow field, especially when moving boundary problems are involved, as shown by Ziaei et al. [7].

For simplicity, it is reasonable to assume that small-scale velocity and pressure are "frozen" within one time step of the large-scale calculation. Hence, velocity and pressure in Eq. (2) are unchanged during the small-scale computation. Therefore, the governing equations (2) on the small-scale domain $\Omega_s \equiv [0, L_s] \times [0, L_s]$ are reduced to

$$\phi_t + (u\phi)_x + (v\phi)_y = \frac{\epsilon^2}{M} N(\phi) + \frac{\rho_0 L_0 P'(\phi)}{M T_m} (T \quad T_m) \quad \frac{\rho_0 \psi'(\phi) T}{4aM} + \frac{Y(\phi, T)}{M} \,, \tag{6a}$$

$$T_t + (uT)_x + (vT)_y = \alpha \Delta T + \left[\frac{\epsilon^2 N(\phi)}{2\rho_0 c_p} \quad \frac{30 L_0 \psi(\phi)}{c_p} \right] \frac{D\phi}{Dt} + \frac{W(u, v, \phi)}{\rho_0 c_p} \,, \tag{6b}$$

with u, v and p independent of time. Initial conditions for the above equations are given by Eq. (5a), and homogeneous Neumann boundary conditions for ϕ and T as in Eq. (5b) are applied. We emphasize here that the physical meaning of all notation in this section can be found in [1].

4. NUMERICAL METHODS AND RESULTS

The governing equations (1) and (2) are nonlinear parabolic equations in conserved form. We apply Leray projection to the momentum equations to provide pressure-velocity coupling and use a Shuman filter [8] to mollify the solutions. Since time-splitting methods are efficient for solving multi-dimensional problems by decomposing them into sequences of 1-D problems, a δ-form Douglas & Gunn [9] procedure is applied to the current model. Quasilinearization of ϕ is constructed by Fréchet–Taylor expansion in "δ-form" (Ames [10]).

We prescribe the large-scale domain as $\Omega = [0, L] \times [0, L] \equiv (0, 0.1m) \times (0, 0.1m)$, and the small-scale domain $\Omega_s = l \times l \equiv 40\mu m \times 40\mu m$. The Reynolds number is determined by the velocity of the lid on the top via $Re = UL/\nu_L$, with ν_L being viscosity of water at $T = T_m$. The initial conditions for small-scale computations are $\phi_0 = 0$, $T_0 = 233K < T_m$, the melting temperature, in Ω_0, where Ω_0 is a square-shaped seed with side length $l_c = 1.2\mu m$ in the center of the domain. Six fold anisotropy is introduced for ice crystal growth. The spatial and time step sizes for large-scale computation are $\Delta x = \Delta y = 10^{-3}m$, $\Delta t = 10^{-3}s$, respectively, and those for small-scale are $\Delta x = \Delta y = 10^{-7}m$, $\Delta t = 4\times10^{-9}s$.

Figure 2 displays streamlines for large-scale velocity and the velocity vector field for the small-scale at the bottom center, where the small-scale computation is introduced at $t = 100s$ and at the point $(0.05m, 5 \times 10^{-4}m)$. Here, we consider the simplified case, i.e., u, v and p are independent of time for small-scale computation. We observe in Fig. 2 that the flow field affects the ice crystal growth tremendously. It increases the ice growth rate in the streamwise direction, and decreases the rate in the opposite direction. The differential growth rate is caused by the temperature gradient induced by the fluid flow. The temperature gradient at each main branch is essentially the same without a flow field, but it is one order of magnitude greater in the streamwise direction than in the opposite one with existence of the flow field.

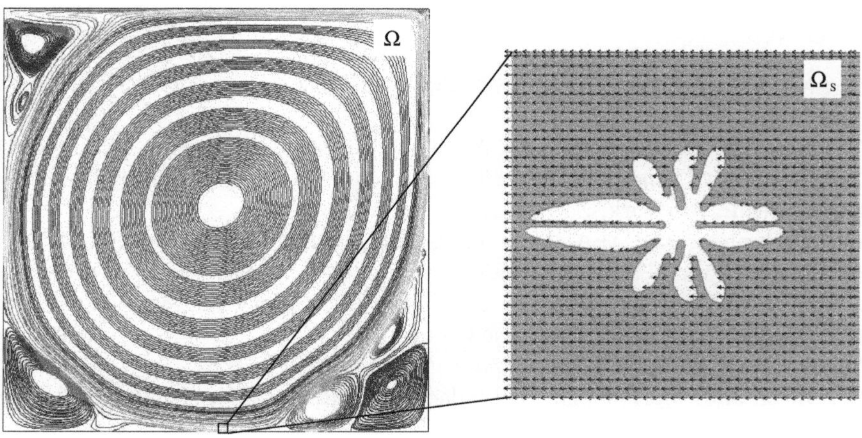

Figure 2. Left: streamlines of lid-driven-cavity flow at $t = 100s$ with $Re = 10,000$; Right: velocity vector field around ice crystal for small-scale computation at bottom center in large-scale domain.

5. APPROACH TO PARALLELIZATION AND RESULTS

Parallelization of the numerical solution procedure is based on the MPI paradigm using the HP Fortran 90 HP-UX compiler and executed on the HP SuperDome at the University of Kentucky Computing Center. We parallelize the loops in the Douglas & Gunn time splitting and line SOR, which is equivalent to distributing the computations on a 2-D domain line by line to each processor. The maximum number of threads available on the HP SuperDome is 64, and in the current study each processor is used to compute one part of the whole domain (number of lines of the domain divided by number of processors). We foresee that this decomposition increases communication time as number of processors is increased. As a result, the speed-up performance will be sublinear.

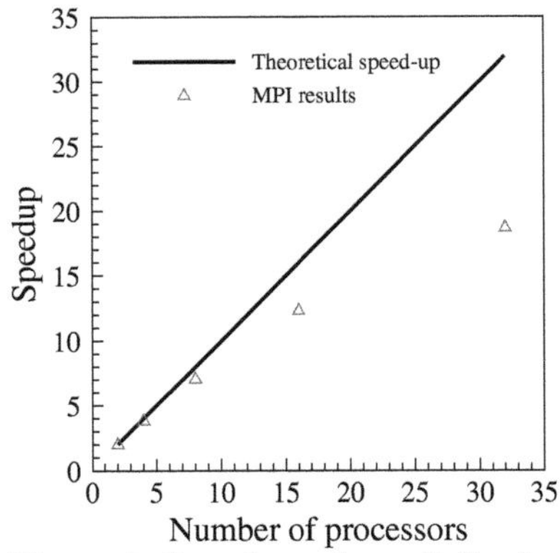

Figure 3. Speedup of parallelized phase-field model with convection

To study speed-up of parallelization, different numbers n of processors ($n = 2, 4, 8, 16, 32$ are used to execute the algorithm. Figure 3 shows the speed-up factor versus the number of processors. It shows that, as the number of processors increases, the speed-up factor increases sub-linearly, suggesting that the demonstrated speedups are not extremely good for $n \geq 32$.

REFERENCES

1. Ying Xu, J.M. McDonough and K.A. Tagavi, In *Proceedings of the Parallel CFD 2005*, Maryland, USA, May 2005.
2. Ying Xu, J.M. McDonough and K.A. Tagavi, In Press, *J. Comput. Phys.* (2006).
3. G.I. Barenblatt, *Similarity, Self-Similarity, and Intermediate Asymptotics*, Consultants Bureau, New York, NY, 1979.
4. A.J. Lichtenberg and M.A. Lieberman, *Regular and Chaotic Dynamics*, Springer-Verlag, New York, NY, second edition, 1992.
5. I. Babuska, In *Numerical Solution of Partial Differential Equations, III*, B. Hubbard (eds.), page 89, Academic Press, New York, 1976.
6. Weinan E and Bjorn Engquist, *Comm. Math. Sci.* 1 (2003) 87.
7. A.N. Ziaei, J.M. McDonough, H. Emdad and A.R. Keshavarzi, Submitted to *Int. J. Numer. Meth. Fluids* (2006).
8. T. Yang and J.M. McDonough, *Journal of Discrete and Continuous Dynamical System*, supplement volume (2003) 951.
9. J. Douglas Jr. and J.E. Gunn, *Numer. Math.* 6 (1964) 428.
10. W.F. Ames, *Numerical Methods for Partial Differential Equations*, Academic Press, New York, NY, 1977.

Parallel Computational Fluid Dynamics – Parallel Computing and Its Applications
J.H. Kwon, A. Ecer, J. Periaux, N. Satofuka and P. Fox (Editors)
2007 Elsevier B.V.

Parallel Property of Pressure Equation Solver with Variable Order Multigrid Method for Incompressible Turbulent Flow Simulations

Hidetoshi NISHIDA[a]* and Toshiyuki MIYANO[b]

[a]Department of Mechanical and System Engineering, Kyoto Institute of Technology, Matsugasaki, Sakyo-ku, Kyoto 606-8585, Japan

[b]Toyota Industries Corporation, Toyoda-cho 2-1, Kariya 448-8671, Japan

In this work, the parallel property of pressure equation solver with variable order multigrid method is presented. For improving the parallel efficiency, the restriction of processor elements (PEs) on coarser multigrid level is considered. The results show that the parallel property with restriction of PEs can be about 10% higher than one without restriction of PEs. The present approach is applied to the simulation of turbulent channel flow. By using the restriction of PEs on coarser multigrid level, the parallel efficiency can be improved in comparison with the usual without restriction of PEs.

1. INTRODUCTION

The incompressible flow simulations are usually based on the incompressible Navier-Stokes equations. In the incompressible Navier-Stokes equations, we have to solve not only momentum equations but also elliptic partial differential equation (PDE) for the pressure, stream function and so on. The elliptic PDE solvers consume the large part of total computational time, because we have to obtain the converged solution of this elliptic PDE at every time step. Then, for the incompressible flow simulations, especially the large-scale simulations, the efficient elliptic PDE solver is very important key technique.

In the parallel computations, the parallel performance of elliptic PDE solver is not usually high in comparison with the momentum equation solver. When the parallel efficiency of elliptic PDE solver is 90% on 2 processor elements (PE)s, that is, the speedup based on 1PE is 1.8, the speedup on 128PEs is about 61 times of 1PE. This shows that we use only the half of platform capability. On the other hand, the momentum equation solver shows almost theoretical speedup [1], [2]. Therefore, it is very urgent problem to improve the parallel efficiency of elliptic PDE solver.

In this paper, the parallel property of elliptic PDE solver, i.e., the pressure equation solver, with variable order multigrid method [3] is presented. Also, the improvement of parallel efficiency is proposed. The present elliptic PDE solver is applied to the direct numerical simulation (DNS) of 3D turbulent channel flows. The message passing interface

*This work was supported in part by a Grant-in-Aid for Scientific Research (16560146) from the Japan Society for the Promotion of Science. We would like to thank Japan Atomic Energy Agency (JAEA) for providing a parallel platform.

(MPI) library is applied to make the computational codes. These MPI codes are implemented on PRIME POWER system with SPARC 64V (1.3GHz) processors at Japan Atomic Energy Agency (JAEA).

2. NUMERICAL METHOD

The incompressible Navier-Stokes equations in the Cartesian coordinates can be written by

$$\frac{\partial u_i}{\partial x_i} = 0, \tag{1}$$

$$\frac{\partial u_i}{\partial t} + u_j \frac{\partial u_i}{\partial x_j} = -\frac{\partial p}{\partial x_i} + \nu \frac{\partial^2 u_i}{\partial {x_i}^2}, \tag{2}$$

where u_i $(i = 1, 2, 3)$ denotes the velocity, p the pressure and ν the kinematic viscosity. The pressure equation can be formally written by

$$\frac{\partial^2 p}{\partial x_i \partial x_i} = \nabla^2 p = f, \tag{3}$$

where f is the source term.

2.1. Variable Order Method of Lines

The solution procedure of the incompressible Navier-Stokes equations (1) and (2) is based on the fractional step approach on the collocated grid system.

In the method of lines approach, the spatial derivatives are discretized by the appropriate scheme, so that the partial differential equations (PDEs) in space and time are reduced to the system of ordinary differential equations (ODEs) in time. The resulting ODEs are integrated by the Runge-Kutta type time integration scheme.

In the spatial discretization, the convective terms are approximated by the variable order proper convective scheme [2], because of the consistency of the discrete continuity equation, the conservation property, and the variable order of spatial accuracy. This scheme is the extension of the proper convective scheme proposed by Morinishi [4] to the variable order. The variable order proper convective scheme can be described by

$$u_j \frac{\partial u_i}{\partial x_j}\Big|_{\mathbf{x}} = \sum_{\ell=1}^{M/2} c_{\ell'} \overline{\overline{u_j}^{x_j} \frac{\delta_{\ell'} u_i}{\delta_{\ell'} x_j}}^{\ell' x_j}\Big|_{\mathbf{x}}, \quad (\ell' = 2\ell - 1), \tag{4}$$

where M denotes the order of spatial accuracy, and the operators in eq.(4) are defined by

$$\overline{f}^{\ell' x_j}\Big|_{\mathbf{x}} = \frac{1}{2}(f_{\mathbf{x}_j+\ell'/2} + f_{\mathbf{x}_j-\ell'/2}), \tag{5}$$

$$\overline{u_j}^{x_j}\Big|_{\mathbf{x}_j \pm \ell'/2} = \sum_{m=1}^{M/2} \frac{c_{m'}}{2}[u_j|_{\mathbf{x}_j \pm \{\ell'+m'\}/2} + u_j|_{\mathbf{x}_j \pm \{\ell'-m'\}/2}], \tag{6}$$

$$\frac{\delta_{\ell'} u_i}{\delta_{\ell'} x_j}\Big|_{\mathbf{x}_j \pm \ell'/2} = \frac{\pm 1}{\ell' \Delta x_j}(u_i|_{\mathbf{x}_j \pm \ell'} - u_i|_{\mathbf{x}_j}), \tag{7}$$

where $m' = 2m - 1$. In this technique, the arbitrary order of spatial accuracy can be obtained automatically by changing only one parameter M. The coefficients $c_{\ell'}$ and $c_{m'}$ are the weighting coefficients and Δx_j denotes the grid spacing in the x_j direction.

On the other hand, the diffusion terms are discretized by the modified differential quadrature (MDQ) method [5] as

$$\frac{\partial^2 u_i}{\partial x_i{}^2}|_{\mathbf{x}} = \sum_{m=-M/2}^{M/2} \Phi_m{}''(\mathbf{x}) u_i|_{\mathbf{x}_i+m}, \tag{8}$$

where $\Phi_m{}''(\mathbf{x})$ is the second derivative of the function $\Phi_m(\mathbf{x})$ defined by

$$\Phi_m(\mathbf{x}) = \frac{\Pi(\mathbf{x})}{(\mathbf{x} - \mathbf{x}_{i+m})\Pi'(\mathbf{x}_{i+m})}, \tag{9}$$

$$\Pi(\mathbf{x}) = (\mathbf{x} - \mathbf{x}_{i-M/2}) \cdots (\mathbf{x} - \mathbf{x}_i) \cdots (\mathbf{x} - \mathbf{x}_{i+M/2}). \tag{10}$$

The coefficients of the variable order proper convective scheme, $c_{\ell'}$, can be computed automatically by using the MDQ coefficients. Then, the incompressible Navier-Stokes equations are reduced to the system of ODEs in time. This system of ODEs is integrated by the Runge-Kutta type scheme.

2.2. Variable Order Multigrid Method

The pressure equation described by

$$\frac{\partial^2 p^{n+1}}{\partial x_i^2} = \frac{1}{\alpha \Delta t} \frac{\partial \overline{u_i{}^*}^{x_i}}{\partial x_i}, \tag{11}$$

is solved by the variable order multigrid method [3]. In eq.(11), $u_i{}^*$ denotes the fractional step velocity, α is the parameter determined by the time integration scheme. The overbar denotes the interpolation from the collocated location to the staggered location. In the variable order multigrid method, the unsteady term is added to the pressure equation. Then, the pressure (elliptic) equation is transformed to the parabolic equation in space and pseudo-time, τ.

$$\frac{\partial p^{n+1}}{\partial \tau} = \frac{\partial^2 p^{n+1}}{\partial x_i^2} - \frac{1}{\alpha \Delta t} \frac{\partial \overline{u_i{}^*}^{x_i}}{\partial x_i}. \tag{12}$$

Equation (12) can be solved by the variable order method of lines. The spatial derivatives are discretized by the aforementioned MDQ method, so that eq.(12) is reduced to the system of ODEs in pseudo-time,

$$\frac{d\overrightarrow{p^{n+1}}}{d\tau} = \overrightarrow{L}(\overrightarrow{p^{n+1}}) \tag{13}$$

This system of ODEs in pseudo-time is integrated by the rational Runge-Kutta (RRK) scheme [6], because of its wider stability region. The RRK scheme can be written by

$$p^{m+1} = p^m + \frac{2\overrightarrow{g_1}(\overrightarrow{g_1}, \overrightarrow{g_3}) - \overrightarrow{g_3}(\overrightarrow{g_1}, \overrightarrow{g_1})}{(\overrightarrow{g_3}, \overrightarrow{g_3})}, \tag{14}$$

$$\vec{g_1} = \Delta\tau \vec{L}(\vec{p^m}), \quad \vec{g_2} = \Delta\tau \vec{L}(\vec{p^m} + c_2 \vec{g_1}), \quad \vec{g_3} = b_1 \vec{g_1} + b_2 \vec{g_2}, \tag{15}$$

where $\Delta\tau$ and m are the pseudo-time step and level, and the operators such as $(\vec{g_1}, \vec{g_3})$ denote inner product of vectors $\vec{g_1}$ and $\vec{g_3}$. Also the coefficients b_1, b_2, and c_2 satisfy the relations $b_1 + b_2 = 1, \ c_2 = -1/2$.

In addition, the multigrid technique [7] is incorporated into the method in order to accerelate the convergence. Then, the same order of spatial accuracy as the momentum equations can be specified.

3. MULTIGRID PROPERTY OF PRESSURE EQUATION SOLVER

First, the 2D case with the source term $f(x, y) = -5cos(x)cos(2y)$ and the Neumann boundary conditions $\partial p/\partial x_i = 0$ are considered. In order to compare the results, the point Jacobi (PJ) method and checkerboard SOR (CB-SOR) method are considered as the relaxation scheme. Figure 1 shows the computational region and analytic solution.

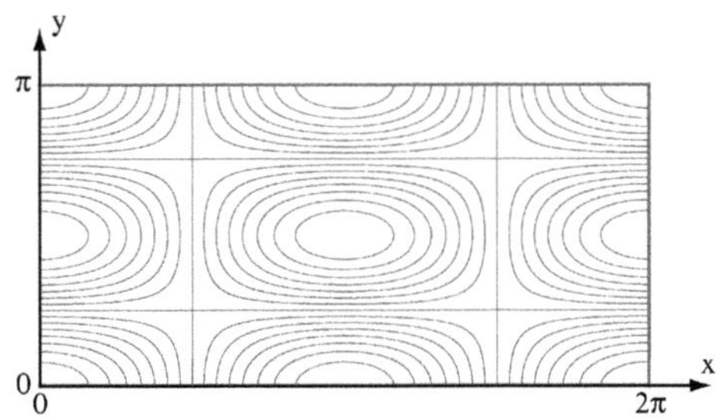

Figure 1. Computational domain and analytic solution of 2D problem.

3.1. Multigrid Property

Figure 2 shows the work unit until convergence. The comparison of computational time and work unit is shown in Table 1. The computation is executed with the second order of spatial accuracy. The present three relaxation schemes indicate the mulatigrid convergence. In these relaxation schemes, the RRK scheme gives the good multigrid property with respect to the work unit and computational time. In order to check the influence of spatial accuracy, the comparison of work unit is shown in Fig.3. In both relaxation schemes, the multigrid convergence can be obtained and the RRK scheme has the independent property of spatial accuracy.

3.2. Parallel Property

In order to implememt to a parallel platform, the domain decomposition technique is used. In this work, 16 sub-domains are assigned to 16 PEs. Figure 4 shows the parallel efficiency defined by

$$Efficiency = \frac{T_{single}}{N \cdot T_{parallel}} \times 100 \ (\%) \tag{16}$$

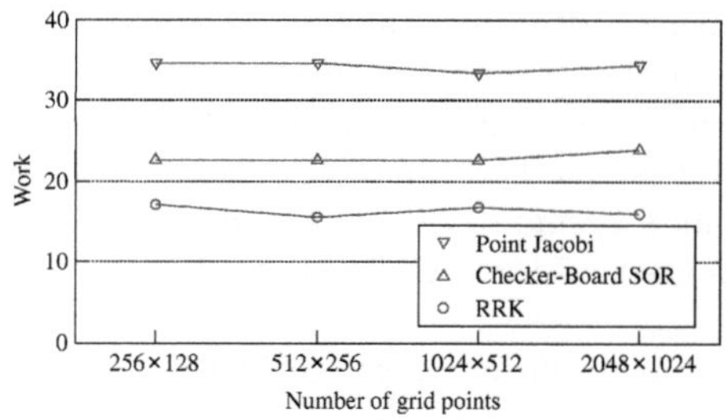

Figure 2. Comparison of work unit until convergence.

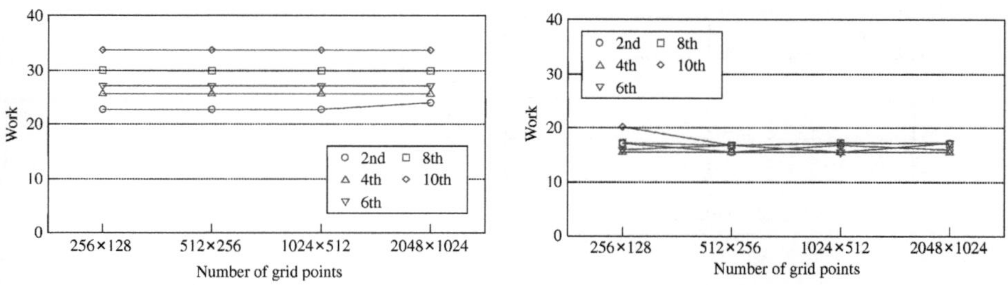

(a) chekerboard SOR method.

(b) RRK scheme.

Figure 3. Multigrid property of spatial accuracy.

Table 1
Work unit and computational time.

work	256 × 128	512 × 256	1024 × 512	2048 × 1024
PJ	34.629	34.629	33.371	34.390
CB-SOR	22.655	22.666	22.666	23.917
RRK	17.143	15.583	16.833	16.000
time (sec)	256 × 128	512 × 256	1024 × 512	2048 × 1024
PJ	0.348	1.448	5.770	23.332
CB-SOR	0.238	0.847	3.650	15.218
RRK	0.255	0.943	4.405	16.448

where N denotes number of PEs, T_{single} and $T_{parallel}$ are the CPU time on single PE and N PEs, respectively. In both relaxation schemes, as the order of spatial accuracy becomes higher, the parallel efficiency becomes higher. The RRK scheme gives the higher efficiency than the checkerboard SOR method. In 2D multigrid method, from fine grid to coarse grid, the operation counts become 1/4, but the data transfer becomes 1/2. Then, the parallel efficiency on coarser grid will be lower, so that the total parallel efficiency is not too high.

In order to improve parallel efficiency, we consider the restriction of PEs on coarser grid level. Figure 5 shows the parallel efficiency with the restriction of PEs on coarser than 64×32 grid level. In Fig.5, version 1 and version 2 denote the parallel efficiency without and with restriction of PEs on coarser grid level, respectively. It is clear that the parallel efficiency of version 2 can be about 10% higher than version 1.

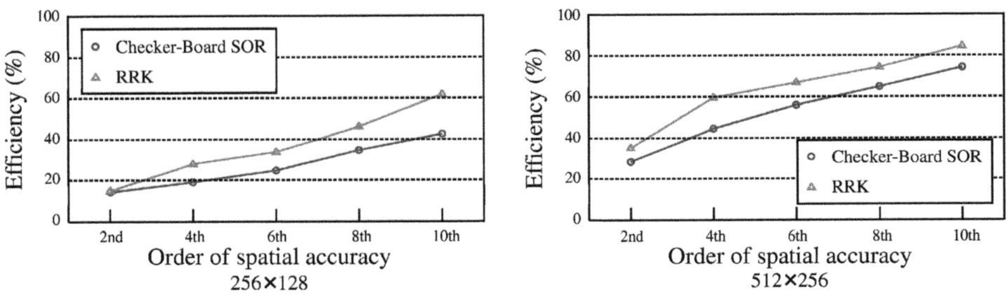

Figure 4. Parallel efficiency.

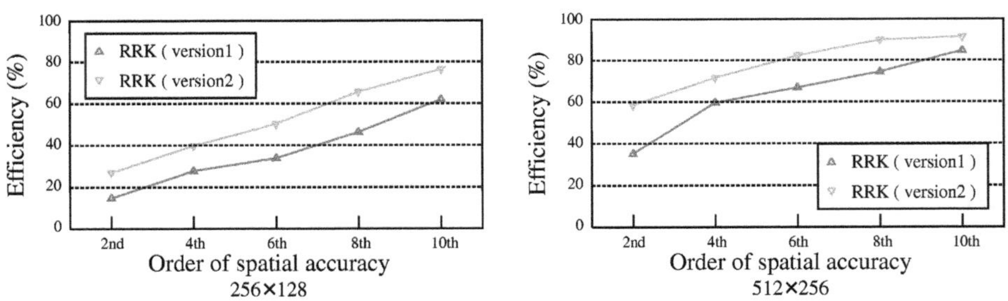

Figure 5. Improved parallel efficiency.

4. TURBULENT CHANNEL FLOW SIMULATION

The DNS of 3D turbulent channel flows is performed by the variable order method of lines. The order of spatial accuracy is the 2nd order and the number of grid points are 32 × 64 × 32 and 64 × 64 × 64 in the x, y and z directions. The numerical results with Reynolds number $Re_\tau = 150$ are compared with the reference database of Kuroda and Kasagi [8] obtained by the spectral method. Figures 6 shows the mean streamwise velocity, Reynolds shear stress profiles, velocity fluctuation, and the iso-surface of second invariant of velocity gradient tensor, respectively. The present DNS results are in very good agreement with the reference spectral solution.

Table 2 shows the parallel efficiency of 2nd order multigrid method in the DNS of 3D turbulent channel flow. The parallelization is executed by the domain decomposition with 8 sub-domains. As the number of grid points becomes larger, the parallel efficiency becomes higher. The version 2 with restriction of PEs shows the higher parallel efficiency than the usual without restriction. In comparison with the checkerboard SOR method, the RRK scheme has the higher parallel efficiency not only in version 1 but also in version 2.

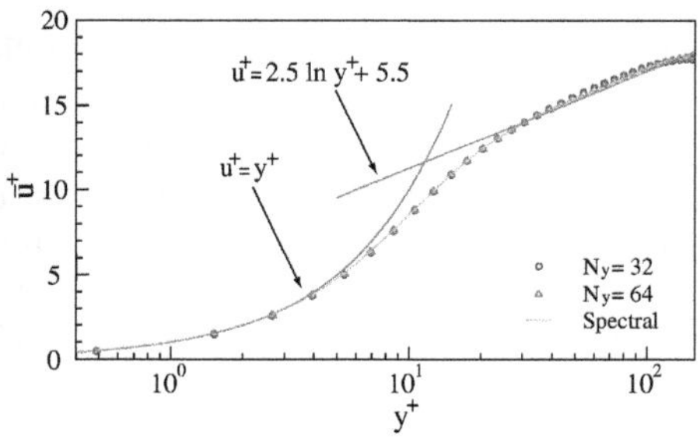

(a) Mean streamwise velocity.

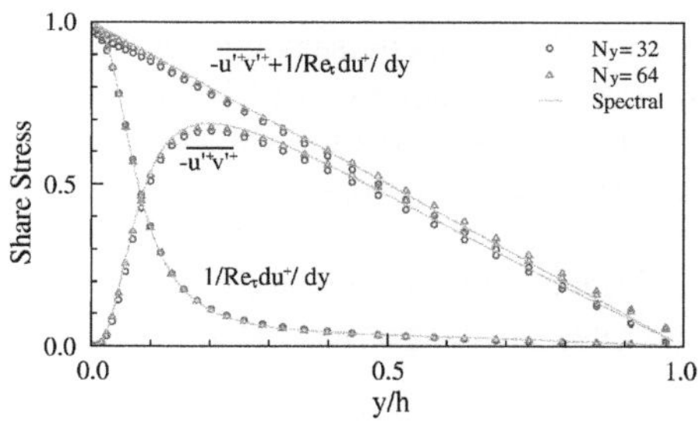

(b) Reynolds shear stress.

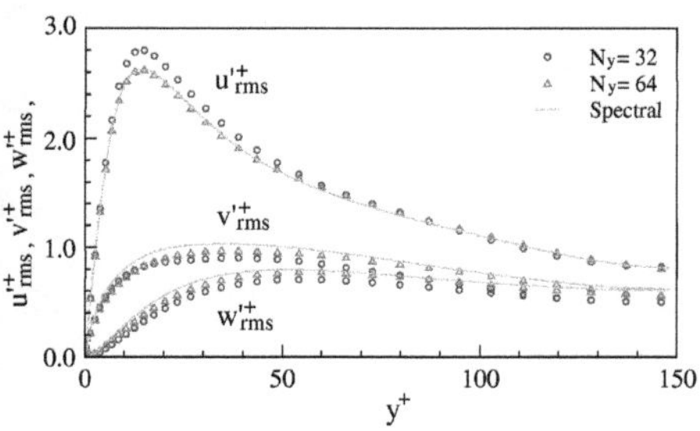

(c) Velocity fluctuation.

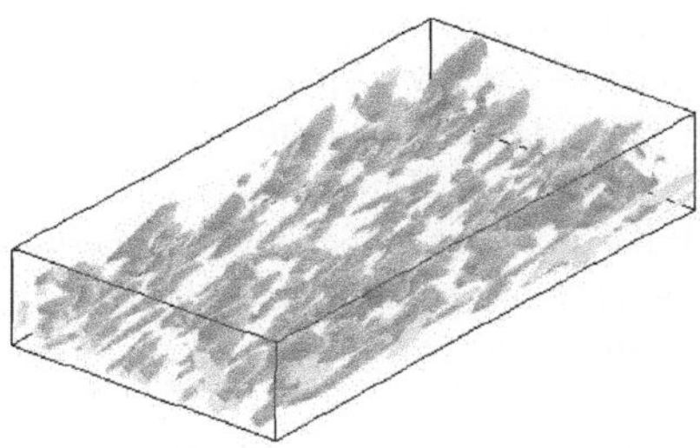

(d) Iso-surface of second invariant of velocity gradient tensor.

Figure 6. Turbulent channel flow.

5. CONCLUDING REMARKS

In this work, the parallel property of pressure equation solver with variable order multigrid method is presented. For improving the parallel efficiency, the restriction of processor

Table 2
Parallel efficiency for 3D turbulent channel flow.

	Checkerboard SOR		RRK	
$32 \times 64 \times 32$	time (sec/step)	efficiency (%)	time (sec/step)	efficiency (%)
single	0.711	-	1.113	-
parallel(version1)	0.147	60.459	0.207	67.210
parallel(version2)	0.143	62.150	0.188	74.003
$64 \times 64 \times 64$	time (sec/step)	efficiency (%)	time (sec/step)	efficiency (%)
single	3.711	-	7.731	-
parallel(version1)	0.628	73.865	1.242	77.812
parallel(version2)	0.582	79.704	1.120	86.307

elements (PEs) on coarser multigrid level is considered. The results show that the parallel property with restriction of PEs can be about 10% higher than one without restriction of PEs. The present pressure equation solver is applied to the simulation of turbulent channel flow. The incompressible Navier-Stokes equations are solved by the variable order method of lines. By using the restriction of PEs on coarser multigrid level, the parallel efficiency can be improved in comparison with the usual without restriction of PEs. Therefore, the present approach is very hopeful for parallel computation of incompressible flows.

REFERENCES

1. H. Nishida, H. Yoshioka and M. Hatta, Parallel Computational Fluid Dynamics - Multidisciplinary Applications, Elsevier, (2005) 121.
2. H. Nishida, S. Nakai and N. Satofuka, Parallel Computational Fluid Dynamics - New Frontiers and Multi-Disciplinary Applications, Elsevier, (2003) 321.
3. H. Nishida and N. Satofuka, Memoirs of the Faculty of Engineering and Design, Kyoto Institute of Technology 36 (1987) 24.
4. Y. Morinishi, T.S. Lund, O.V. Vasilyev, P. Moin, J. Comput Phys. 143 (1998) 90.
5. N. Satofuka, K. Morinishi, NASA TM-81339 (1982).
6. A. Wambecq, Computing 20 (1978) 333.
7. A. Brandt, Mathematics of Computation 31-138 (1977) 333.
8. N. Kasagi, Y. Tomita and A. Kuroda, Trans. ASME, J. Heat Transf. 114 (1977) 598.

Parallel Computational Fluid Dynamics – Parallel Computing and Its Applications
J.H. Kwon, A. Ecer, J. Periaux, N. Satofuka and P. Fox (Editors)

Parallel Numerical Simulation of Shear Coaxial LOX/GH_2 Jet Flame in Rocket Engine Combustor

S. Matsuyama,[a] J. Shinjo,[a] Y. Mizobuchi[a] and S. Ogawa[a]

[a]*Institute of Aerospace Technology, Japan Aerospace Exploration Agency, 7-44-1 Jindaiji-higashi, Chofu, Tokyo 182-8522 Japan*

Keywords: Combustion; Liquid Rocket engine

1. INTRODUCTION

Liquid rocket engines are used in a number of launch vehicles worldwide. In these rocket engines, e.g. LE-7A engine of H-2A, Space Shuttle Main Engine (SSME) and Vulcain engine of ARIANE 5, liquid oxygen (LOX) and gaseous hydrogen (GH_2) are injected through coaxial injectors (a schematic is shown in Fig.1). Although the injector design is a critical component of the combustion device, the processes associated with LOX/GH_2 injection, such as atomization, mixing and combustion, are still not well understood. The methodology of injector design, therefore, mostly relies on a time-consuming and costly trial-and-error. In order to reduce the development cost for future or improved rocket engines, a large part of such trial-and-error phase must be eliminated.

A lot of works has been conducted both numerically [1-5] and experimentally [6-11] on the LOX/GH_2 injection and combustion, which provided an improved understanding of the processes in liquid rocket engine combustion chambers. Despite these efforts, many basic phenomena, such as turbulent mixing and combustion, is still far from being understood to a satisfactory level.

For recent years, numerical simulation has been becoming a powerful tool in combustion research. Due to the rapid improvement of computer resources, a time-dependent three-dimensional simulation of actual size flame with detailed chemistry and transport models has been enabled. Our research group succeeded in capturing a turbulent hydrogen jet lifted flame using the direct numerical simulation (DNS) approach [12,13]. In the study, important and interesting features of the flame have been revealed by analysis of simulated flowfield data. We believe that this kind of huge and detailed simulation will give us a good understanding also for the case of LOX/GH_2 jet flame in rocket engines.

In the present study, we will aim to clarify the detail structure of a shear coaxial LOX/GH_2 jet flame in rocket engine combustor by means of the detailed numerical simulation. In this paper, a preliminary result by an axisymmetric numerical simulation with detailed chemistry and fine resolution mesh is shown for a single shear coaxial injector element which follows the experiment by Mayer and Tamura [6]. By the analysis of the simulated flame, fundamental features of LOX/GH_2 jet flame are explored. To reduce the required computing time accompanying with the time-dependent simulation with fine resolution mesh and detailed chemistry, our combustion simulation code is fully parallelized. The parallel performance exceeds 50 GFLOPS using 83 processors on the Central Numerical Simulation System (CeNSS) installed at Institute of Aerospace Technology (IAT), Japan Aerospace Exploration Agency (JAXA).

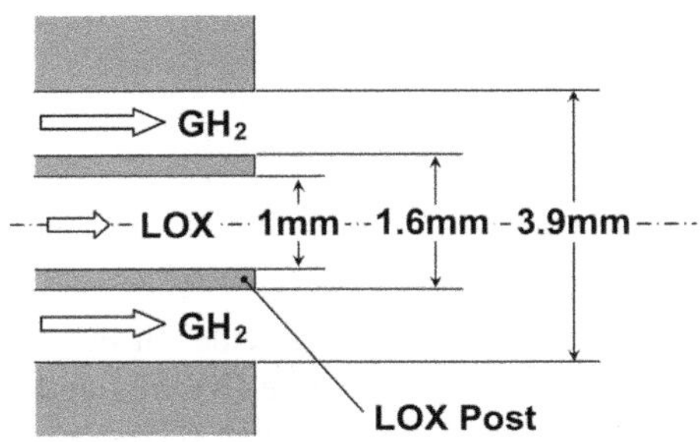

Figure 1 A schematic of shear coaxial injector.

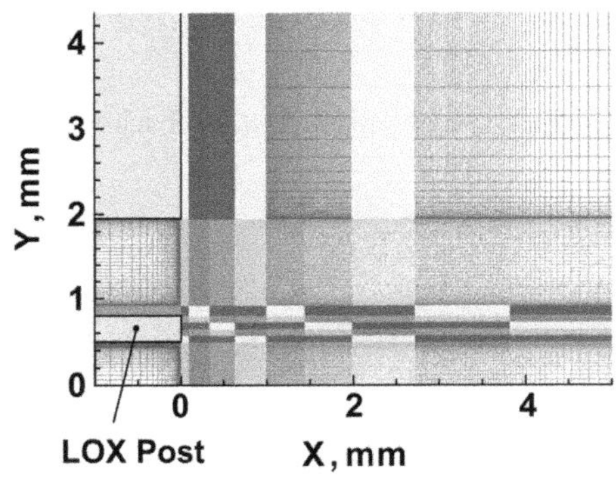

Figure 2 Computational mesh near the LOX post tip.

Table 1 LOX/GH_2 injection conditions

Chamber pressure	10 MPa
Oxygen injection velocity	30 m/sec
Oxygen injection temperature	100 K
Hydrogen injection velocity	300 m/sec
Hydrogen injection temperature	300 K

2. FLOW CONDITIONS

The present simulation is conducted for a single shear coaxial injector element which follows the experiment by Mayer and Tamura [6]. The dimensions of the LOX/GH_2 injector are shown in Fig. 1. The injection conditions are summarized in Table 1. In Fig.2, the computational mesh employed in the present simulation is shown for the region near the injector exit. The computational domains consist of the LOX injector, the hydrogen annulus, and the combustion chamber. The diameter and length of combustion chamber are 40 and 400 mm, respectively. The overall mesh size is 571×401 points along the axial and radial directions, respectively. To resolve a thin reaction layer and a shear layer, grid points are clustered in the wake of LOX injector post. 161 grid points are used to cover the LOX post thickness of 0.3 mm in the radial direction. The grid spacing at the LOX post wall is 1μm. In the axial direction, grid points are distributed with the stretching factor of 1.02.

The boundary conditions are as follows: The velocity distributions at injector inlets are assumed to be a fully developed turbulent pipe flow velocity profiles. At the exit boundary, the non-reflection condition is imposed [14]. The LOX injector post and combustion chamber walls are assumed to be no slip and isothermal wall. The wall temperatures of LOX injector post and combustion chamber are assumed to be 500 and 300 K, respectively.

3. NUMERICAL METHOD

The governing equations are the Navier-Stokes equations in an axisymmetric form. Eight chemical species (H_2, O_2, OH, H_2O, H, O, H_2O_2 and HO_2) are assumed and 18 reactions model by Petersen and Hanson [15] is employed. The numerical flux function is given by the AUSM-DV scheme [16] with the 2nd order MUSCL interpolation. The viscous terms are evaluated with 2nd order difference formulae. The time integration method is 1st order Euler explicit method.

The equation of state employed in the present simulation is Soave-Redlich-Kwong equation of state (SRK EoS) which is a cubic equation of state for predicting pressure-volume-temperature behavior of dense fluid at high pressure environment [17]. The nonideality of thermodynamic properties under high pressure conditions is expressed by departure functions [17]. The speed of sound is also derived from SRK EoS. The viscosity and thermal conductivity are calculated by the method based on the extended corresponding state principle by Ely and Hanley [18,19].

Parallel computation is implemented by the domain decomposition strategy, and carried out using 83 processors on the CeNSS installed at IAT, JAXA. The entire computational domain is decomposed into 83 domains (Fig.2). Boundary values are exchanged between neighboring domains by the Message Passing Interface (MPI). The total performance of parallel computation exceeds 50 GFLOPS.

4. RESULTS

In the present study, simulation is conducted for about 1msec, and a stable flame is successfully obtained through the simulation. Figure 3 shows the instantaneous contours of temperature at 0.5msec. The stoichiometric line (solid black line) is also shown in the temperature contours. The flame is attached to the LOX post tip, and the hot product gas completely separates the hydrogen and oxygen streams. This result is consistent with the experimental study by Mayer and Tamura [6], and results of LES by Oefelein and Yang [1-3]. The shear layer shed from the outer rim of the LOX post tip becomes unstable and generates a series of vortices. These vortices interact and coalescence with their neighboring vortices in the hydrogen stream side while convecting downstream. Developed vortices impinge directly onto the flame, and the thickness of the flame becomes very thin at the flow downstream region. As indicated by the study of Juniper et al. [20], a hydrogen/oxygen flame is quite resistant to flow straining. Also in our simulation, no local extinction is observed when a strong straining due to vortices occurs.

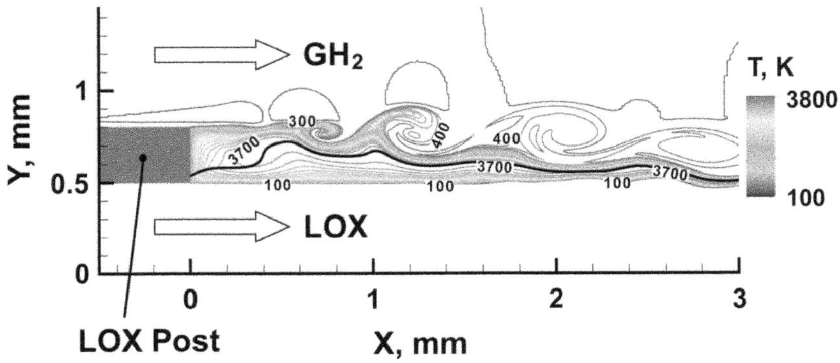

Figure 3 Instantaneous temperature contours at 0.5msec. The stoichiometric line (solid black line) is also shown.

In Fig.4, instantaneous temperature contours and stream lines near the LOX injector post tip at 0.5 msec are shown. Near the LOX injector post, a recirculation zone is present as reported by many researchers [1-3,6]. Due to these recirculation vortices, a stationary combustion is sustained, and the maximum temperature within the recirculation zone reaches about 3750K, which is almost identical to the adiabatic flame temperature of hydrogen/oxygen at stoichiometric condition.

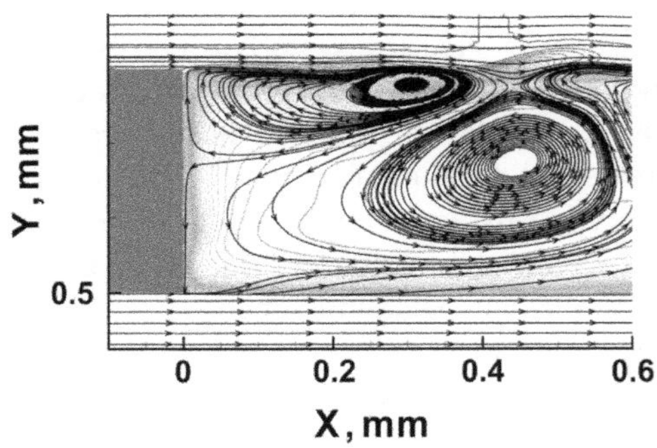

Figure 4 Instantaneous temperature contours and stream lines near the LOX injector post tip at 0.5 msec.

Figure 5 shows the corresponding contours of H_2, O_2, OH, and H_2O mass fraction. The stoichiometric line is shown also in the figure by a solid black line. The observed feature of flame obtained in the present simulation is almost identical to that of a diffusion flame. The layer of OH radical distributions, which is indicative of the flame location, almost attaches to the LOX Post tip, and the hydrogen (fuel) stream and the oxygen (oxidant) stream are separated. In the region where the flame is strained by the developed vortices, the thickness of layer of OH radicals becomes very thin. At the hydrogen stream side, unburned hydrogen is entrained and mixed with the hot products by developed vortices. On the contrary, the oxygen stream is less influenced by the developed vortices and its surface is relatively smooth. At just behind the LOX post tip, a recirculation zone exists, and the combustion product gases partially recirculate back.

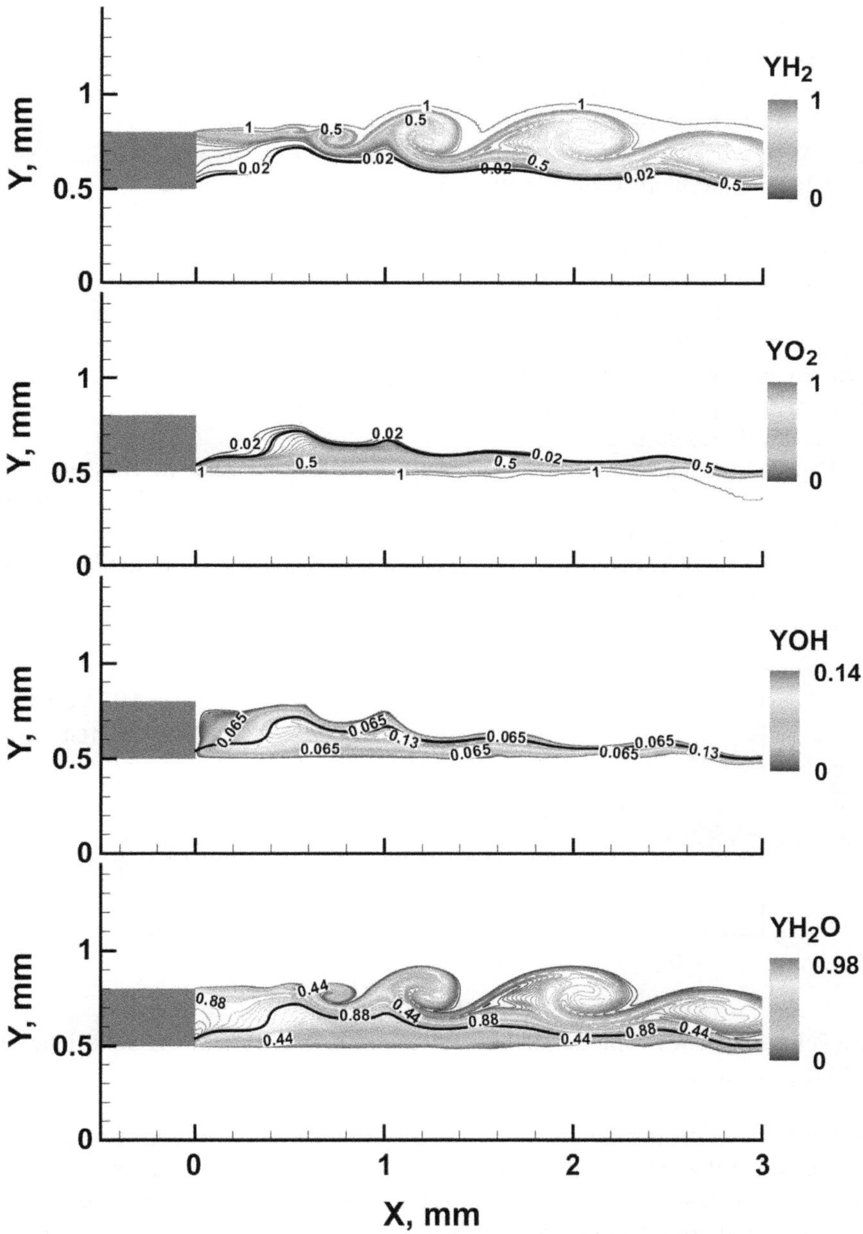

Figure 5 Instantaneous contours of H_2, O_2, OH, and H_2O mass fraction at 0.5msec. The solid black line shows the stoichiometric line.

The local flame structures at the axial location of x=2mm is shown in Fig. 6. It is apparent from these profiles that the local flame structure at the flow down stream region is analogous to a diffusion flame. The maximum flame temperature is about 3700K. At this location, the flame is strained by the developed eddy, and the reaction layer is very thin. As shown in the figure, the thickness of OH distribution (corresponds to reaction layer) is about 0.06 mm. Although no significant mixing occurs at the oxygen stream side, hydrogen stream is mixed with high temperature product by the developed vortices.

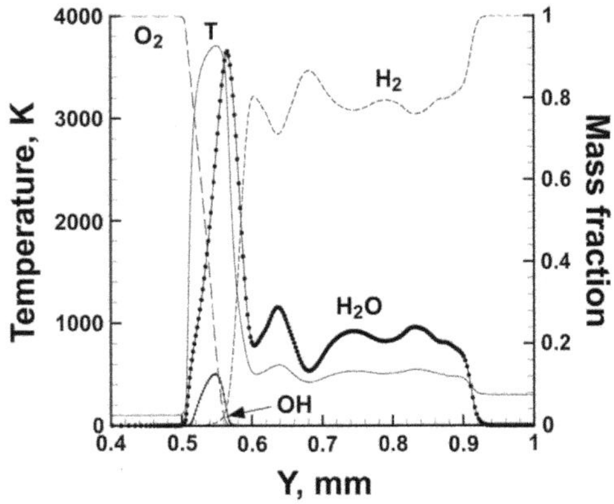

Figure 6 Local flame structure at axial location of x=2mm.

In Fig. 7, instantaneous contours of heat release rate are shown. The heat release layer almost aligns with the stoichiometric line. There is intense heat release at the location where the flame is strained by the developed vortices. The magnitude of heat release rate at the flow downstream region is grater than that at the recirculation zone near the LOX post tip by a factor of 10 to 100.

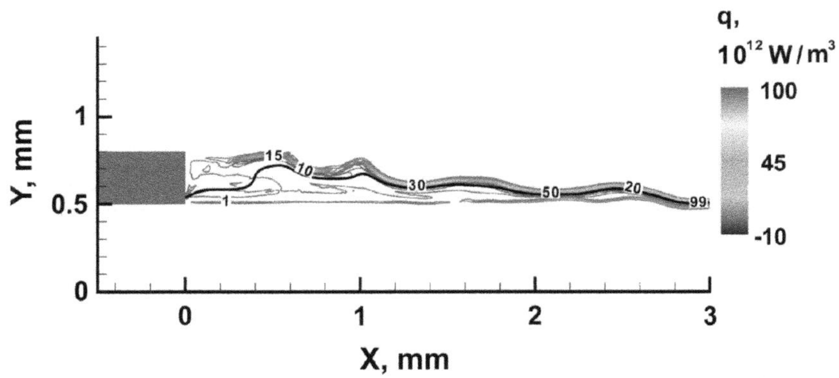

Figure 7 Instantaneous contours of heat release rate at 0.5msec. The solid black line shows the stoichiometric line.

In Fig.8, the vortical structure of flowfield is summarized. The instantaneous vorticity contours and the power spectral densities of radial velocity oscillation at three different axial locations along the mixing layer are shown. Probe locations where temporal signals are stored are also indicated in the figure. In the present simulation, the dominant frequencies are 450, 147 and 97.7 kHz at each probe location, respectively. As the flow convecting downstream, the dominant frequency decreases to around 100 kHz.

The corresponding Strouhal numbers are 0.45, 0.147 and 0.097 at each probe location, respectively. We note that the Strouhal number is defined based on the LOX Post thickness (0.3 mm) and the mean inlet velocity of hydrogen stream (300 m/sec). According to the two-dimensional simulation conducted for the backward facing step flow [21], the Strouhal number of dominant oscillation decreases toward the flow downstream and reaches almost constant value of O(0.1). The result of present simulation, therefore, is consistent with the tendency observed in the backward facing step flow.

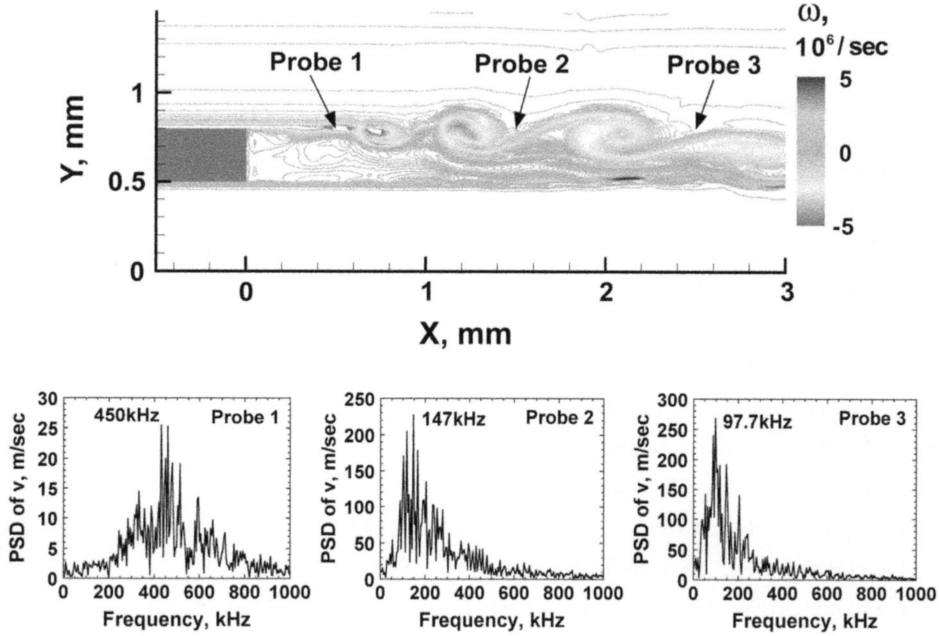

Figure 8 Instantaneous vorticity contours at 0.5msec, and frequency spectra of radial velocity oscillation at three different axial locations along the mixing layer.

5. SUMMARY

An axisymmetric simulation with detailed chemistry and fine resolution mesh is conducted for the LOX/GH_2 jet flame in rocket engine combustor. A preliminary result is shown for a single shear coaxial injector element. The fundamental features of the LOX/GH_2 coaxial jet flame is explored by the analysis of simulated flame. The major results obtained are summarized below.

1. The flame is attached to the LOX post tip, and the hot product gas separates the hydrogen and oxygen streams. The observed feature of flame is almost identical to that of a diffusion flame.

2. Near the LOX post tip, the recirculation zone exists, and the combustion products reciruclate back. The flame holding mechanism may be attributed to these recirculating combustion products. Further study is needed to clarify the detail of flame holding mechanism.
3. Intense heat release is observed at the location where the flame is strained by vortices. The thickness of the strained flame is very thin, and less than 0.1 mm.
4. Vortical structure shed from the outer rim of the LOX injector post tip is analogous to the backward facing step flow. The tendency of the dominant frequency of oscillation and the Strouhal number is consistent with that of the backward facing step flow.

REFERENCES

[1] Oefelein, J. C., and Yang, V., *J Propul. Power*, 14 (1998) 843-857.

[2] Oefelein, J. C., AIAA Paper 2003-0479.

[3] Oefelein, J. C., *Proc. of Combust. Inst.*, 30 (2005) 2929-2937.

[4] Zong, N., et al., *Phys. Fluids*, 16 (2004) 4248-4261.

[5] Zong, N. and Yang, V., *Proc. of Combust. Inst.*, 31 (2006).

[6] Mayer, W., and Tamura, H., *J Propul. Power*, 12 (1996) 1137-1147.

[7] Mayer, M., et al., *J Propul. Power*,16 (2000) 823-828.

[8] Mayer, W., et al., *J Propul. Power*, 17 (2001) 794-799.

[9] Candel, S., et al., *J Propul. Power*, 14 (1998) 826-834.

[10] Singla, G., et al., *Combust. Flame*, 144 (2006) 151-169.

[11] Kendrick, D., et al., *Combust. Flame*, 118 (1999) 327-339.

[12] Mizobuchi, Y., et al., *Proc. of Combust. Inst.*, 29 (2002) 2009-2015.

[13] Mizobuchi, Y., et al., *Proc. of Combust. Inst.*, 30 (2005) 611-619.

[14] Poinsot, T. J., and Lele, S. K., *J Comput. Phys.*, 101 (1992) 104-129.

[15] Petersen, E. L., and Hanson, R. K., *J Propul. Power*, 15 (1999) 591-600.

[16] Wada, Y. and Liou, M. S., NASA TM-106452, 1994.

[17] Polling, B. E., and Prausnitz, J. M., and O'Connell, J. P., *The Properties of Gases and Liquids*, McGraw-Hill, 5th ed., 2001.

[18] Ely, J. F., and Hanley, H. J. M., *Ind. Eng. Chem. Fundam.*, 20 (1981) 323-332.

[19] Ely, J. F., and Hanley, H. J. M., *Ind. Eng. Chem. Fundam.*, 22 (1981) 90-97.

[20] Juniper, M., et al., *Combust. Flame*, 135 (2003) 87-96.

[21] Wee, D., et al., *Phys. Fluids*, 16 (2004) 3361-3373.

Parallel Computational Fluid Dynamics – Parallel Computing and Its Applications
J.H. Kwon, A. Ecer, J. Periaux, N. Satofuka and P. Fox (Editors)

Construction of Numerical Wind Tunnel on the e-Science Infrastructure

Jin-ho Kim[a], Jae Wan Ahn[a], Chongam Kim[a], Yoonhee Kim[b] and Kum Won Cho[c]

[a]*School of Mechanical and Aerospace Engineering, Seoul National University, Seoul 151-742, Korea*

[b]*Department of Computer Science, Sookmyung Women's University, Seoul 140-742, Korea*

[c]*Department of Supercomputing Applications, Korea Institute of Science and Technology Information(KISTI), Daejeon 305-333, Korea*

Key Words: Grid computing, e-Science, Numerical Wind Tunnel

1. INTRODUCTION

Aerospace engineering is system-integrated research field. In order to design an aircraft, engineers in various disciplines such as aerodynamics, structure, thrust, control, etc. have to collaborate and integrate their research results. In addition, as the aerospace engineering objects become more complex and enormous, the necessity of collaboration among the engineers in different branches also increases. By the way, there are some drawbacks of the aerospace engineering research system. As for numerical simulation, the scale of aerospace engineering objects basically requires vast amount of high-performance computing resources. With regard to experiments, scarcity of large-scale wind tunnel causes researchers to spend much time on obtaining equipments and performing experiments. And, geographical dispersion of research institutes and inadequacy of collaboration infrastructure occasionally brings about repeated investment on the same equipments.

Thus, construction of the collaborative research system is inevitable. In that system, researchers will be able to execute large-scale computations automatically, share numerical and experimental data and perform remote monitoring/discussion. That system should provide the integrated research platform of multiple disciplines in aerospace field.

The present research developed an integrated research system of numerical and experimental scientists by adopting the next-generation computing technologies like Grid & e-Science. Grid[1][2] technology enables services for sharing the resources of an interconnected worldwide network of tens of thousands of computers and storage devices. e-Science[3][4][5] is an extended concept of the Grid. e-Science refers to the large-scale science that will increasingly be carried out through distributed global collaborations enabled by the Internet. With these technologies, scientists can generate, analyse, share and discuss their insight, experiments and computation results in an effective manner.

The product of current research is named e-AIRS(e-Aerospace Integrated Systems) and e-AIRS system is provided in the form of a portal service(Figure 1). e-AIRS portal supports both CFD simulation service and remote wind tunnel experiment service. Portal frames are developed by using GridSphere[6] and basic ideas of service architecture in Ref. 7 are referenced.

2. e-AIRS Portal

The e-AIRS portal is a problem solving environment composed of various portlets which are developed by the GridSphere[8]. The GridSphere portlet framwork provides a solution to construct an open-source web portal[9]. The GridSphere supports standard portlets, and these can be extended or added to in the form of new portlets. The portlets are implemented in Java and can be modified easily.

The main goal of the e-AIRS is to establish the powerful and user-friendly research/collaboration environment to aerospace engineers. In this point of view, the e-AIRS portal should provide the main interface through which all of computational and experimental services can be accessed. This portal hides the complexity of the system from the end-users. In addition, the portal also provides an interface to the processes required in the CFD computation(generation of the CFD mesh system, definition of the physical boundary conditions and inflow conditions, choice of the type of CFD solver,

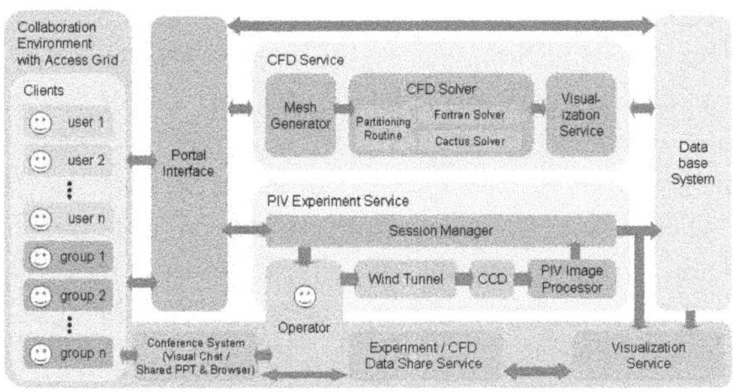

Fig. 1. the architecture of the e-AIRS portal. The e-AIRS portal is the main interface enables users to access internal software and hardware.

execution of calculation, monitoring of the convergence, visualization of temporary and final results of computation, and so on) and the remote wind tunnel experiment(request of experiment, management of requested experiment, and display of the experiment information). Users can access to all of e-AIRS software and hardware resources via this portal interface. The brief synopsis of the e-AIRS is depicted in the Fig. 1.

3. Computational Simulation Service

The computational simulation service consists of three main components: mesh generation service, CFD solver service, and monitoring & visualization service. These three components are implemented as the three sections of the computational simulation portlet. This portlet is built upon reusable Java Server Pages(JSP). The computational simulation portlet provides services of selecting data files and computational resources, job submissions to remote machines, and file transfer between mesh generation and CFD simulation. All instructions concerning mesh generation, CFD calculation, and visualization are prescribed interactively using a graphical user interface.
The computational simulation service presents the mesh generator(e-AIRSmesh) and the visualization application(e-AIRSview) at necessary stages during numerical calculation. This service also enables numerical analysis with high performance numerical tools. The e-AIRSmesh and the e-AIRSview are developed in Java applet forms. The Java applet makes it feasible to work without installing any stand-alone program. Moreover, users can check the progress of calculation and the result files through the portal interface.

3.1. Mesh Generation Service

The e-AIRS portal provides a mesh generation service supported by the e-AIRSmesh Java applet. The mesh system is the set of discrete cells dividing the flow field.

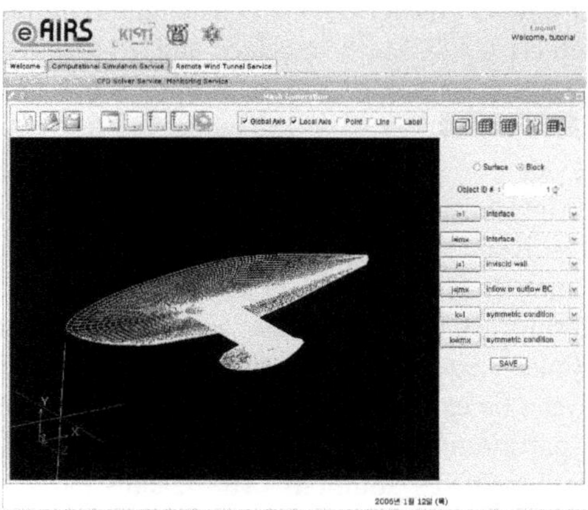

Fig. 2. the interface of the e-AIRSmesh mesh generator. The graphic icons enable a user to make a mesh system by intuition.

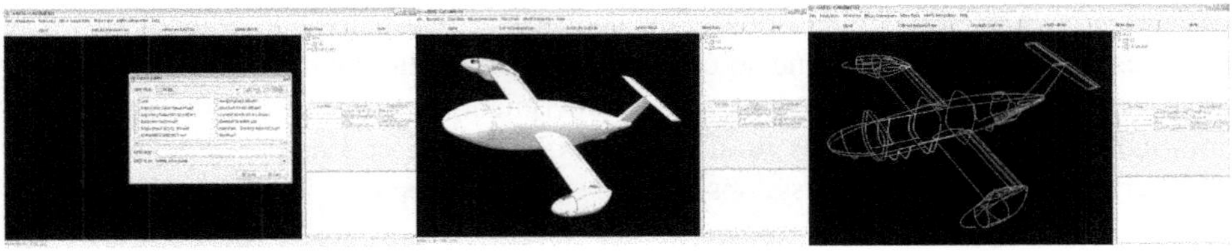

Fig. 3. The interface of the CAD2Mesh and procedure of extraction of indexd line from CAD data(VRML file).

CFD simulation process[10,11] is to change the physical values of gas or air by calculating the flux between neighboring spatial cells. The e-AIRSmesh has an convenient interface for creation of new geometry, making of a mesh system, and configuration of boundary conditions. This interface brings together most of the portal process technologies just in one environment. To support parallel calculations of CFD simulations, the e-AIRSmesh alse provides multi-block mesh generation. The approach adopted for parallel mesh generation is geometrical partitioning of the calculation domain. The Fig. 2 shows the e-AIRSmesh interface.

The generation of a mesh is the most difficult phase of the CFD processes and is anticipated to be the most time-consuming part for new users. In order to make mesh generation job more convenient, the e-AIRSmesh suggests some default mesh templates of NACA 4-digit airfoils.

Not only the mesh generation but also setting of boundary conditions is provided in the e-AIRSmesh service. If a user selects any condition for a specific boundary, the e-AIRSmesh hides other boundaries to avoid confusing. In addition, mouse control functions(zoom-in/out, shift, rotation, and so on) and various display options make the e-AIRSmesh convenient and user-friendly software, too.

Additionally, mesh generation service offers CAD2Mesh module. CAD2Mesh enables researchers to extract indexed lines from CAD data and to build more complicated mesh easily. The Fig. 3 depicts CAD2Mesh interface and a process of extraction of indexed lines.

3.2. CFD Solver Service

Fig. 4 shows the user interface of the CFD solver service. Under the interface, there are two kinds of CFD solvers: Fortran-based CFD solver(developed and validated in Aerodynamic Simulation & Design Lab., Seoul National University) and Cactus-based CFD solver. The former contains accurate and numerical schemes to solve Euler equations, Navier-Stokes equations containing the turbulence model[10,11]. The Fortran-based solver can be executed on any parallel environment because it includes a generalized domain-partitioning algorithm. The Cactus-based CFD solver can also produce Euler or Navier-Stokes solutions. The Cactus-based solver is prepared as a prerequisite work for future extension of the e-AIRS project considering workflow environment.

The execution of the CFD solver requires several flow parameters such as Mach number, Reynolds number, Inflow temperature, and so on. These parameters created by user's input are written on a flow condition file. With this flow condition file and the prepared mesh data file, the parallel CFD calculation can be executed.

1. The solver divides the mesh data into multiple partitions and transfers these partitions to distributed computing resources of the e-AIRS.
2. Execution parallel solver at each computing resource.
3. Exchange related boundary data among subdomains.
4. When the solver has converged, the result data are collected and combined.

Upon the completion of a job, the server collects the outputs of the tasks, aggregates those data, and stores the final result file in the data storage for the user.

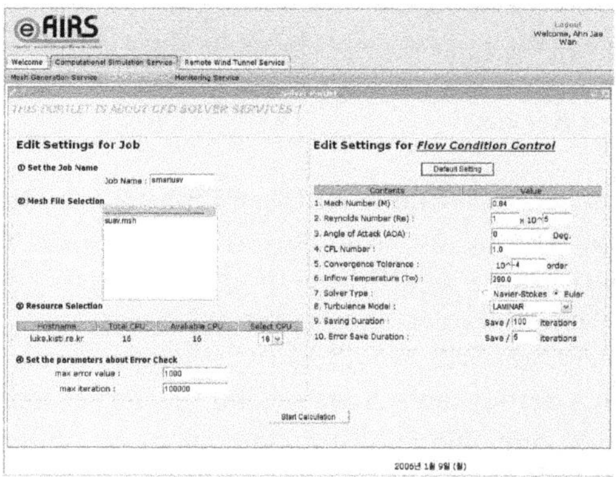

Fig. 4. The portal GUI of the e-AIRS CFD solver service. The default flow parameters are suggested automatically when the 'Default Setting' button is clicked.

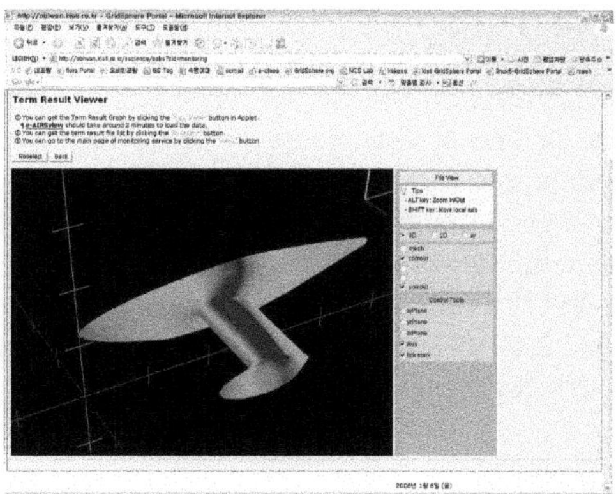

Fig. 5. The e-AIRSview interface. The CFD result data can be visualized on the e-AIRS portal.

3.3. Monitoring and Visualization Service

A user can monitor the latest status of his calculation. The user also can see the convergence history graph. With this information and temporary results the user can interrupt a calculation if he decides the calculation is wrong. Basically, a job submission using CFD solver service returns a unique job ID which can be used for enquiry about the job status. If the job is submitted to a server, it is in the 'PREPARE' state while sitting in the queue waiting to be executed. In case of normal completion the job status is 'DONE', otherwise the job status is one of 'INRPT', 'ERR', and 'FAILED'.

When the computation finishes on remote computing machine, all output data are transferred to a storage server. In this stage, the location of these result files is recorded in a database system again. Thus for each calculation, input parameters, location of output data with some extra information such as owner information, execution date are recorded. The user can download and visualize the result data through the portal. The e-AIRSview is used for the visualization of the temporary and final result data. Fig. 5 is an example of the visualization of the smart UAV by the e-AIRSview.

4. Remote Experiment Service

The remote experiment service of the e-AIRS performs a wind tunnel test through the portal interface without visiting the institute possesses a wind tunnel. The clients of wind tunnel service can use the e-AIRS portal as a communicational interface with a experiment operator who really performs experiments and manages the wind tunnel. The remote experiment service provides PIV(Particle Image Velocimetry) type measurements[12] for a aerodynamic model. The PIV experiment outcomes velocity vector field of gas flow. For more skillful experiment, the transonic wind tunnel of Korea Aerospace Research Institute(KARI)[13] is used for the remote service. In the first year of the e-AIRS project, the smart UAV[13] model has been tested.

The remote experiment service consists of three services: the experiment request service, the experiment managing service, and the experiment information service. A client can request an experiment through the experiment request service. Then the operator of the wind tunnel checks new requested experiments on the experiment managing service. This managing service can offer information of requested experiment such as Reynolds number, angle of attack, data form, test area on the aerodynamic model, and so on to the operator. Then the operator can carry out adequate experiments and upload result data files including particle image through web UI. A client user can browse and check the status of his experiment through the experiment information service. The states are 'NEW', 'ON-GOING', and 'FINISHED'. This information service also shows various images with which a user is able to see the result image files conveniently. Fig. 6 shows the interfaces of these services.

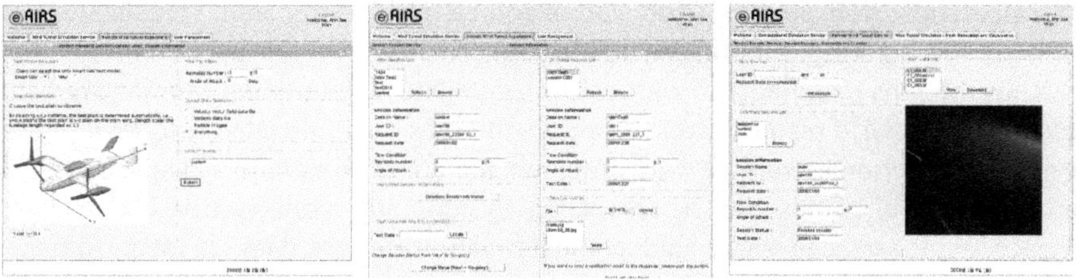

Fig. 6. The remote experiment service interfaces: The experiment request service(left), the experiment managing service(middle), and the experiment information service(right).

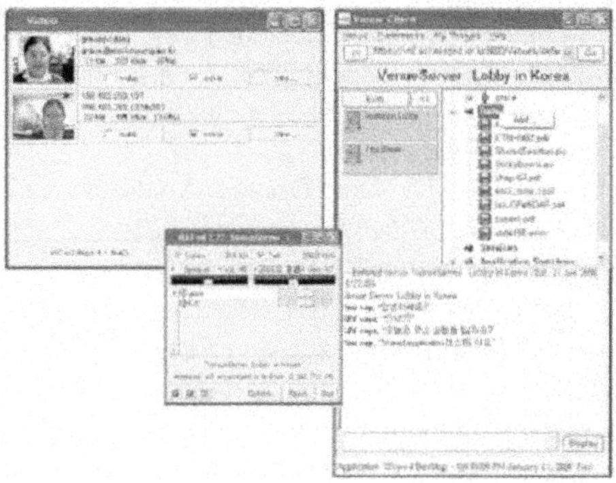

Fig. 7. The snapshots of video and audio collaboration service using AG venue client

5. Collaboration Environment

The collaborative research teams require collaborative communication system. The Access Grid[14] technology is adopted in the e-AIRS project as the collaborative communication system. The Access Grid Toolkit(AGTk) was installed in KARI, SNU, and Sookmyung Univ. on each AG node hardware. Not only as a communication system but also as a research data sharing system, the AG presents powerful sharing activities. The experiment operator, client users can discuss their computational and experimental data through shared applications such as 'shared PPT', 'shared browser', 'shared PDF', 'shared desktop'[15,16]. The AG client software which is more convenient and powerful than the original AGTk venue client is under construction.

6. Conclusions

The e-AIRS which is still under development, is suggested an easy-to-use research environment. Through the portal services, the non-experts can produce their own research outputs without the knowledge of the Grid system architecture. The Grid portal with many service modules and user-friendly UI makes the research environment more convenient. In the e-AIRS project, main modules of service framework for sequential

CFD calculation, job steering, job and resource monitoring and visualization were presented. Job steering submits an application to resource for its execution, that is, dispatches the jobs to suitable resources and aggregates the outputs. Monitoring handles access control to Grid resources and monitors the jobs over allocated resource.

The remote experiment service was prepared for client researchers who don't possess their wind tunnel. With the Access Grid system and collaborative activities, the aerospace engineers can share and compare their research data. In near future, this collaborative research trend will make a role of extending research topics.

References

1. I. Foster, C. Kesselman, S. Tuecke, (2001). "The Anatomy of the Grid: Enabling Scalable Virtual Organizations" *International J. Supercomputer Applications*
2. I. Foster, C. Kesselman, J. Nick, S. Tuecke, (2002). The Physiology of the Grid: An Open Grid Services Architecture for Distributed Systems Integration, Open Grid Service Infrastructure. *W.G., Global Grid Forum*
3. T. Hey, A.E. Trefethen, (2003). The Data Deluge: An e-Science Perspective. In *Grid Computing – Making the Global Infrastructure a Reality*
4. D. Roure, N. Jennings, N. Shadbolt, (2001). Research Agenda for the Future Semantic Grid: A Future e-Science Infrastructure. In F. Berman, G. Fox and A.J.G. Hey (Eds): *Grid Computing - Making the Global Infrastructure a Reality*, pp.437-470
5. L. Chen, N.R. Shadbolt, F. Tao, C. Puleston, C. Goble, S.J. Cox, (2003). "Exploiting Semantics for e-Science on the Semantic Grid", *Proceedings of Web Intelligence (WI2003) workshop on Knowledge Grid and Grid Intelligence*, pp.122-132
6. J. Novotny, M. Russell, O. Wehrens, (2004). "GridSphere: A Portal Framework for Building Collaborations", *Concurrency - Practice and Experience*, Vol.16, No.5, pp.503-513
7. G. von Laszewski, I. Foster, J. Gawor, P. Lane, N. Rehn, M. Russell, (2001). Designing Grid-based Problem Solving Environments and Portals. *Proceedings of the 34th Hawaiian International Conference on System Science*
8. http://www.gridsphere.org/gridsphere/gridsphere
9. http://developers.sun.com/prodtech/portalserver/reference/techart/jsr168/pb_whitepaper.pdf
10. http://www.cfd-online.com/
11. Charles Hirsch: Numerical Computation of Internal and External Flows. John Wiley & Sons. Inc. (1992)
12. http://www.dantecdynamics.com/piv/princip/index.html
13. http://www.kari.re.kr/
14. http://www.accessgrid.org/
15. http://portal.accessgrid.org/
16. http://www.unix.mcs.anl.gov/fl/research/accessgrid/documentation/SHARED_APPLICAT-ION_MANUAL/ProgrammersManual_SharedApplicationsHTML.htm

Parallel Computational Fluid Dynamics – Parallel Computing and Its Applications
J.H. Kwon, A. Ecer, J. Periaux, N. Satofuka and P. Fox (Editors)

Efficient distribution of a parallel job across different Grid sites

E. Yilmaz, R.U. Payli, H.U. Akay, and A. Ecer
Computational Fluid Dynamics Laboratory, Department of Mechanical Engineering
Indiana University-Purdue University Indianapolis (IUPUI)
Indianapolis, Indiana 46202 USA, http://www.engr.iupui.edu/cfdlab

Key Words: TeraGrid, Distributed Computing, Parallel Computing, Load Balancing, Graph Partitioning.

Abstract

Cross-cluster load distribution of a parallel job has been studied in this paper. Two TeraGrid sites was pooled as a single resource to run 128 and 512 mesh blocks of a CFD problem having 18M mesh elements. Blocks were evenly distributed across the Grid clusters. Random block distribution was compared with two different distributions of graph partitioning. Interface size and block size were considered as weighting factors to group blocks between the Grid sites. Communication between Grid sites is major bottleneck though there are minor differences in timing when considering interface size and block size as weighting factors. As number of mesh blocks increase, communication dependency across the Grid sites increases.

1 Introduction

With the extended infrastructure investments in recent years, Grid computing continues to challenge the way scientific computing has been done up until a decade ago. Tera-scale computational capacity such as TeraGrid [1], which is an NSF-supported grid computing environment established by combining computational resources of six geographically different centers in a single pool is now available to researchers for grand challenging problems. In near future, Peta-scale computing infrastructure is expected to be available.

Along with the Tera or Peta-scale investments, computational researchers need more advanced tools which can facilitate using full capacity of these environments. One such tool for the researchers in the parallel computational fluid dynamics field is MPICH-G2, which is the Grid enabled implementation of the Message Passing Interface (MPI) [2]. This utility is capable of starting a parallel job across Grid sites with some predetermined CFD data block distribution. The fact that CFD data blocks are highly interconnected requires efficient distribution of the blocks such that communication between the Grid sites for the particular parallel job would be at minimum. While the blocks communicate at high speed within the same Grid site, they communicate with whatever bandwidth is provided between the Grid sites. Therefore, the bottleneck is the inter-connection between the Grid sites. As a result, the CFD code developers and researchers need to optimize this communication by efficiently allocating the blocks among the Grid sites.

In the past, we have developed Dynamic Load Balancing (DLB) tools [3, 4] which can facilitate the CFD block distribution among available computing resources, mostly clusters of processors. DLB is very useful in environments where processors loads are changing dynamically. However, if the computing resource is dedicated for a specific job, then a static distribution would be sufficient for this job. In TeraGrid sites, resource schedulers handle internal distributions of parallel tasks or chunks of dispersed sub-tasks across Grid sites. Therefore, a static distribution of parallel CFD blocks between Grid sites would be sufficient provided that this distribution is optimum or efficient.

We had experience with running jobs in TeraGrid independently at different sites [5, 6]. In this paper, we will demonstrate efficient CFD block distribution across Grid sites such as in TeraGrid. The blocks will be grouped together to ensure optimum distribution of the blocks, hence minimum communication between the Grid sites. We apply Metis graph partitioning libraries [7] for the distribution of the CFD blocks across the clusters.

2 Block distribution across Grid sites

A parallel CFD job executes on partitioned data blocks called mesh blocks. As the number of the partitioned mesh blocks increases, dependencies of one block to other increases, hence the communication overhead. When using multiple sites, a user should consider the balanced load and communication between the Grid sites so that overall time for the job would be lowest achievable. If users are unaware of which block communicate to which, then the opportunity of using different clusters or Grid sites will be obsolete.

There are several paths to distribute blocks across the Grid sites. In our studies, we adopted graph partitioning feature of the Metis package [7]. Metis offers three different options to decide on the weights, lower the weights more efficient the distribution, to be used as criteria to distribute given blocks across partitions of a given graph. For future references in our case, all of the Grid sites compose the graph itself and each Grid site construct partitions of this graph.

In partitioning the graph, Metis offers to use either of the communication volume between pairs of mesh blocks or mesh block size. It is also possible to use combination of block-to-block communication and block size as *weights* for partitioning of the graph. In our studies, random distribution and *weights on edge* are used for comparison purposes. The following criteria for weights of each block can be considered:

a) Random distributions of mesh blocks: consecutive mesh block is assigned to the Grid site following the current site.
b) *Weights on edges*: communication volume, which is interface size between mesh blocks, is used as weight.
c) *Weights on vertices*: mesh block size is used as weight.
d) *Weights on edges and vertices*: communication volume between blocks and mesh block size are used as *weights*.

Based on these weight criteria, the graph is partitioned for optimum cut into as many as number of Grid sites to be used. Figure 1 shows a simple four mesh blocks and their inter-connects distributed between two Grid sites with edges used as *weights*.

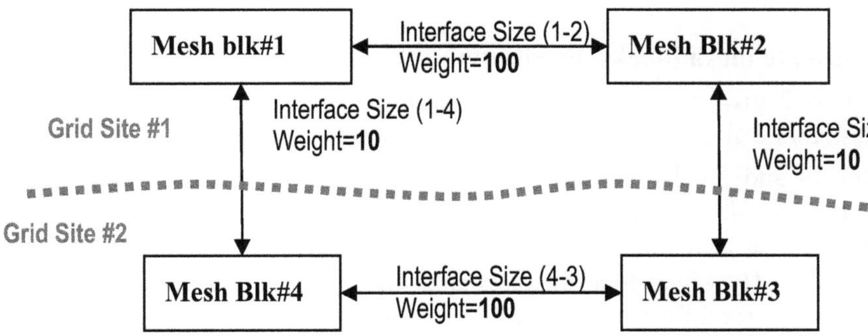

Figure 1: Four mesh blocks distributed between two Grid sites based on communication volume.

Given parallel mesh blocks, a program was written to generate graph data, to be used in Metis graph partitioner. The program extracts parallel mesh information, such as neighbor block numbers and interface size between the blocks as well as block size. Then, using the mesh information it prepares Metis input file based on the four criteria given above. Metis, *pMetis* library, creates block distribution information for given block connectivity and number of Grid sites. After that, this block distribution information is used to rearrange original block numbering so that when submitted to the scheduler of the Grid sites blocks are assigned to right Grid sites. Even block distribution is assumed at this time. For the case where process distribution file option is available for MPICHG2, then users should replace this step with generation of new process distribution file. Finally, job submission to the Grid is done. At this time, some of the intermediate processes are manual as we are more focused on efficient distribution part of the whole problem.

3 Test Case

3.1 Problem Definition

In this study, we performed computational experiments for 128 and 512 mesh blocks between two Grid sites. Mesh has about 18 million tetrahedral elements. Partitioned mesh blocks of this problem for 512 blocks at airplane surface is given in Figure 2. The airplane geometry is known as DLR-F6 wing-body configuration. The flow solver to be used is PACER3D [8], a parallel unstructured mesh flow solver.

Figure 2: DLR F6 wing body with 512 CFD mesh blocks and 18 million elements.

3.2 Multi-site mesh blocks distribution

Tables 1 to 2 give some cumulative quantities for different numbers of Grid sites (2, 4, and 8 sites) for 128 and 512 mesh blocks, respectively.

Tables 1a and 1b shows interface size across the clusters for 128 blocks as number of Grid sites increases. For the distribution based on *weights on edge*, as the number of grid sites increased total number of neighbor increases almost exponentially. However, total interface size changes almost linearly. The same behavior is observed for the random distribution for 128 blocks as well.

Tables 2a and 2b give change of the same quantities as given in Table 1 for 512 blocks. Although the interface size for the distribution based on the *weights on edge* shows similar linear behavior, the random distribution has highest interface size for two grid sites.

Total number of interface and interface size distributed over the grid sites are given in Table 3 for four grid sites by the distribution based on *weights on edge*. Interface size varies about 1/5 ratio among the Grid sites. This makes wait time for the communications between grid sites very high. Figures 3 and 4 show how mesh blocks are distributed across four Grid sites for random and *weights on edge* distributions, respectively. Figure 4 shows that mesh blocks are agglomerated well to form a single interface front to reduce dependency.

Table 1a: Distribution based on *weights on edge* for 128 blocks.

# of Grid Sites	Total # of Neigh. Site	Total # of cross-site block interface	Total interface size of the blocks across clusters
2	2	144	78,730
4	10	330	212,930
8	42	560	344,428

Table 1b: Random distribution for 128 blocks.

# of Grid Sites	Total # of Neigh. Site	Total # of cross-site block interface	Total interface size of the blocks across clusters
2	2	752	621,552
4	12	1,082	882,208
8	56	1,254	1,001,708

Table 2a: Distribution based on *weights on edge* for 512 blocks.

# of Grid Sites	Total # of Neigh. Site	Total # of cross-site block interface	Total interface size of the blocks across clusters
2	2	2,360	736,316
4	12	4,151	1,237,229
8	56	4,892	1,483,427

Table 2b: Random distribution for 512 blocks.

# of Grid Sites	Total # of Neigh. Site	Total # of cross-site block interface	Total interface size of the blocks across clusters
2	2	3,020	1,654,208
4	12	4,523	1,378,346
8	56	5,263	1,603,964

Table 3a: Number of interfaces for four Grid sites.

	Site 1	Site 2	Site 3	Site 4
Site 1	0	55	0	19
Site 2	55	0	18	35
Site 3	0	18	0	38
Site 4	19	35	38	0

Table 3b: Total interface size across clusters for four Grid sites.

	Site 1	Site 2	Site 3	Site 4
Site 1	0	39,645	0	12,926
Site 2	39,645	0	8,583	19,392
Site 3	0	8,583	0	25,919
Site 4	12,926	19,392	25,919	0

3.3 Computational experiment for two TeraGrid Sites combined

Actual computation of the test case has been performed by using two TeraGrid sites (SDSC and NCSA) for the two types of block distributions: random and *weights on edge*. Average elapsed and interface (communication) time per time step for 128 blocks is given in Figure 5. Timing for the distribution based on the *weights on edge* (interface size) demonstrates about 8% benefit compared to *weights on vertices* (block size). However, random distribution is about 90% more costly than other two methods which use Metis graph partitioning. An important observation is in comparison of elapsed time with communication time. Communication time is about 90% of the total elapsed time for all distribution methods.

Figure 6 shows timing for 512 blocks. As number of partitions increase the gap in timing between *random* distribution and *weights on edge* narrows down to 75% from 90% for 128 blocks. Contrary to the timing results of 128 blocks, distribution with *weights on edge* is about 6% more costly than the one in *weights on vertices*. Communication also takes longer in this case, about 98% of the total elapsed time.

Figure 3: Mesh blocks by random distribution across four Grid sites.

Figure 4: Mesh blocks by distribution based on *weights on edge* across four Grid sites.

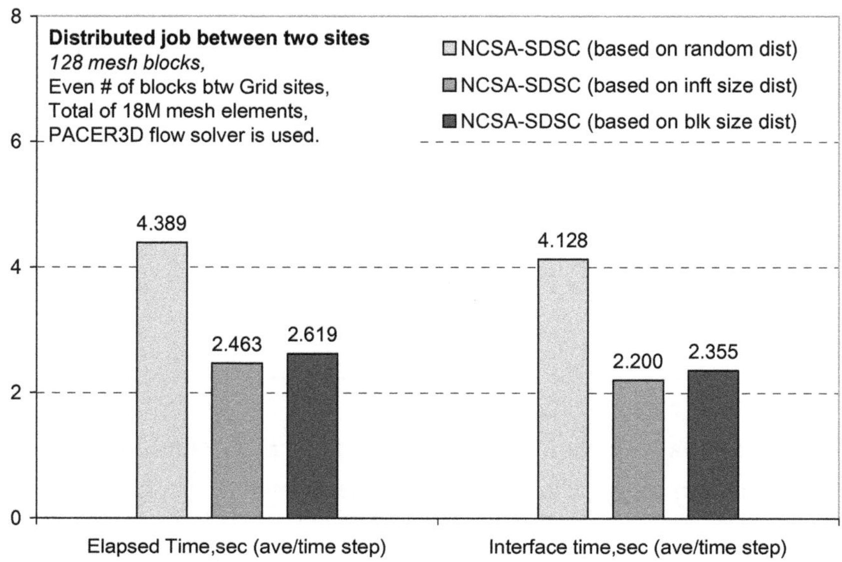

Figure 5: Elapsed time for 128 blocks between two TeraGrid sites.

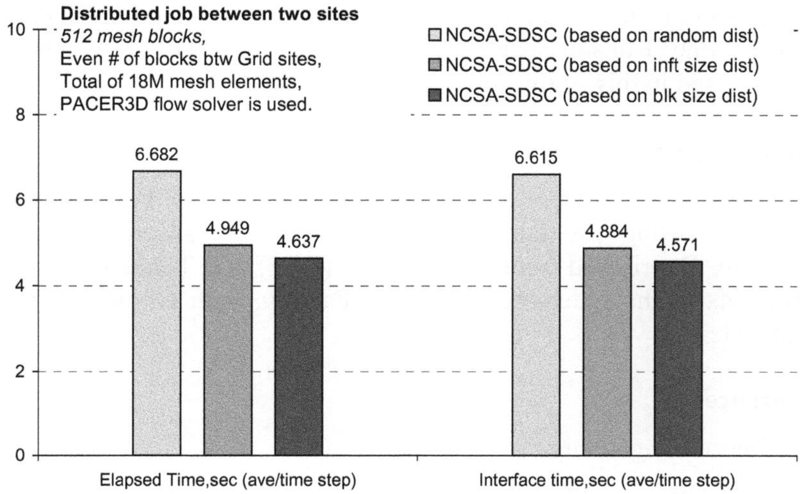

Figure 6: Elapsed time for 512 blocks between two TeraGrid sites.

4 Conclusions

Current study for parallel computing application between Grid sites reveals three conclusions. *First*, using a graph partitioning based block distribution between grid sites gives lower communication time compared to the random block distribution. This can be inferred by just looking at total interface size of the mesh blocks distributed between Grid sites or actual time measurement which also includes communication latency and wait time. *Second*, as number of mesh partitions increase the elapsed time difference between the random and graph partition based mesh block distributions reduces. Due to increased number of neighbors, wait time between the mesh blocks across the Grid sites would be dominating the total elapsed time. *Third*, increasing number of the grid sites would be even more costly compared to two Grid sites due to the simple fact that inter-dependency would result in more wait time. The interconnect speed within the nodes of the Grid sites and network speed between the Grid sites are not comparable as first one is much faster than latter. Although, this conclusion is based on only change of the interface size between the Grid sites, actual time measurement would reveal the effect of latency and wait time as well. As a final outcome from this research study, one can conclude that bottleneck is communication speed and algorithm between Grid sites for a parallel job. It plays an important role for overall parallel efficiency. However, there is still room to improve communication latency and wait time across the clusters for a given Grid infrastructure. With the existing approaches, the parallel computing over the Grid sites would be more beneficial if data size and memory requirement of the parallel job would be large enough not to fit into one Grid site, such as couple of 100 millions of mesh elements for the problem used in this study.

5. Acknowledgement

TeraGrid access was supported by the National Science Foundation (NSF) under the following programs: Partnerships for Advanced Computational Infrastructure, Distributed Terascale Facility (DTF) and Terascale Extensions: Enhancements to the Extensible Terascale Facility, with Grant Number: TG-CTS050003T.

6. References

1. *http://www.teragrid.org.*
2. N.T. Karonis, B. Toomen, and I. Foster, "MPICH-G2: A Grid-Enabled Implementation of the Message Passing Interface," *J. Parallel and Distributed Computing*, vol. 63, no. 5, 2003, pp. 551-563.
3. A. Ecer, H.U. Akay, W.B. Kemle, H. Wang, D. Ercoskun, and E.J. Hall, "Parallel Computation of Fluid Dynamics Problems," *Computer Methods in Applied Mechanics and Engineering*, 112, 1994, pp. 91-108.
4. S. Chien, Y. Wang, A. Ecer, and H.U. Akay, "Grid Scheduler with Dynamic Load Balancing for Parallel CFD," *Parallel CFD 2003,* Edited by B. Chetverushkin, et al., Elsevier Science, pp. 259-266, 2004.
5. R.U. Payli, H.U. Akay, A. Baddi, E. Yilmaz, A. Ecer, and E. Oktay, "CFD Applications on TeraGrid," *Proceedings of Parallel CFD 2005*, Edited by A. Deane, et al., Elsevier Science, 2006, pp. 141-148.
6. R.U. Payli, E. Yilmaz, H.U. Akay, and A. Ecer, "From Visualization to Simulation: Large-scale Parallel CFD Applications on The Teragrid," *TeraGrid Conference: Advancing Scientific Discovery*, Indianapolis, IN, June 12-15, 2006.
7. G. Karypis and V. Kumar, "Multilevel Algorithms for Multi-constraint Graph Partitioning," *Technical Report TR 98-019*, Department of Computer Science, University of Minnesota, 1998.
8. E. Yilmaz, M.S. Kavsaoglu, H.U. Akay, and I.S. Akmandor, "Cell-vertex Based Parallel and Adaptive Explicit 3D Flow Solution on Unstructured Grids," *International Journal of Computational Fluid Dynamics*, Vol. 14, 2001, pp. 271-286.

Parallel Computational Fluid Dynamics – Parallel Computing and Its Applications
J.H. Kwon, A. Ecer, J. Periaux, N. Satofuka and P. Fox (Editors)

New cooperative parallel strategy for massive CFD computations of aerodynamic data on multiprocessors

S. Peigin[a], B.Epstein[b]

[a] Israel Aircraft Industries, Israel

[b]The Academic College of Tel-Aviv Yaffo, Israel

1. ABSTRACT

A new parallel cooperative strategy for iterative multiparametric numerical processes is proposed. The idea of the proposed technology consists in employing the information exchange between the members of the same parametric family. With this end in view, the information on already computed (or partially computed) solutions is distributed in the multilevel way which results in an essential speed-up of the whole parallel process.

Based on this general approach, a new efficient strategy of parallel cooperative CFD computations is implemented which allows the practical construction of large-size aerodynamic data-bases by means of high-accuracy Navier-Stokes simulations.

The results show that the proposed technology essentially reduces CPU time needed for large-scale aerodynamic computations.

2. INTRODUCTION

In practical aerodynamics, massive computations of aerodynamic data, especially the computations of aerodynamic forces and moments, are required.

Aerodynamic data-bases of this kind include drag/lift polars at different free-stream Mach numbers and Reynolds numbers, Mach drag rise curves which are indicative of MDD (Mach Drag Divergence), lift vs. angle of attack graphs, C_M (pitching moment) vs. C_L (lift coefficient) and other moment/angle of attack/lift/drag combinations.

An usual way of attaining the above aerodynamic information is to perform massive CFD computations since the wind-tunnel tests are not able of supplying a quick response to the needs of the aircraft industry. Besides, the wind-tunnel tests are prohibitively expensive (in terms of both money and time), especially in the stage of the configuration design where a considerable amount of different configurations must be tested as fast as possible.

However even the computer-aided construction of parametrically dense aerodynamic data-bases is highly time-consuming (though of course it is much cheaper than the corresponding wind-tunnel tests).

It is especially true where geometrically complicated configurations are treated such as a complete aircraft or even simpler wing-body configurations.

For example, a single high-accuracy simulation of a complete aircraft may take a number of days (or even weeks for a high-accuracy computation) on presently available serial computers. To attain a comprehensive aerodynamic data-base, hundreds (or even thousands) of such simulations are needed.

Two approaches are presently in use in industrial environment intended to diminish the overall computation time.

With the first approach, the computations are performed by means of low-accuracy CFD codes which are based on incomplete gas-dynamic models (such as panel codes). This may give reasonably accurate results for lift evaluations in the subsonic flow regime but fails in more complicated flight conditions even in the case of lift/angle of attack computations.

The computation of drag and pitching moment in a flow regime where considerable viscous effects are present and/or in the presence of shock waves requires high-accuracy full Navier-Stokes simulations which (as already explained above) constitute a highly time-consuming task.

A more comprehensive approach is to make use of massive parallelization of CFD codes on distributed memory clusters. This may essentially decrease the computation time as the CFD parallelization allows for concurrent computation of a number of flight points. Still the overall computing time remains high in the following three practically crucial cases:

- Construction of large-size aerodynamic data-bases
- Numerical testing of concurrent configurations
- Multi-objective aerodynamic design

For example, in the last case, the situation is especially troublesome since the process of the computer-aided design requires testing of tens of thousands of flight points.

The present work makes use of usually neglected information which may be extracted from already computed (or partially computed) flow-fields. A sophisticated use of this (off-line and on-line) information may essentially reduce the number of iterations needed for convergence thus dramatically diminishing the overall time of multiple aerodynamic simulations.

The presented results show that the proposed technology may result in the parallel efficiency which is formally higher than 100%. This is of course due to the information exchange, infeasible in the case of serial runs.

Note, that though the proposed technology is especially efficient for the computation of large-size aerodynamic data-bases, it can be successfully employed for the computation of even relatively small data sets, such as a single lift/drag polar.

3. NEW COOPERATIVE STRATEGY FOR MASSIVE PARALLEL ITERATIVE COMPUTATIONS

In general, parallelization of any computational algorithm leads to distribution of total computational work among participating parallel processors and organization of message-passing between the processors. The crucial question raised by the parallel implementation of iterative algorithms is the efficiency of the data exchange between different processors.

It is widely accepted that the optimal strategy to reach the main goal of parallelization (to maximize parallel efficiency of the algorithm), is to minimize the CPU time related to the process of data transfer. Based on this point of view, we take into account the above CPU time as a loss only.

From a more general (more "philosophical") point of view, the incidence of the data transfer between processors it twofold. On one hand this results in a loss (increasing the total CPU time). But on the other hand it also results in a gain, since individual parallel processes gain additional information, thus improving the rate of convergence of the iterative algorithm, or providing a more accurate solution. Consequently, at least theoretically, the gain may exceed the loss. Of course, the most challenging question is how to recognize which information should be transferred in order that the gain surpasses the loss.

Apparently there exists no general answer to this question. In this report, the problem is being solved in the context of massive iterative computations for the aerodynamic analysis and design.

The first example is the optimization of aerodynamic shapes by means of Genetic Algorithms (GAs). In this case, it is necessary to accelerate the optimum search. It appeared that this goal may be achieved through exchanging information related to the best individuals of parallel populations which essentially improves the convergence rate of GA iterative process.

The second example is the CFD analysis intended to the computations of large-size aerodynamic data. The computer aided aerodynamic flow analysis is performed by numerical solution of flow equations (which represent a non-linear Partial Differential Equations (PDE) system). The equations are usually solved by an iterative numerical algorithm which may involve (for high accuracy solutions) hundreds, and sometimes thousands, of iterations. Thus the convergence acceleration may result in essential gain in the overall time performance of the algorithm.

It was found that in this case, the relevant exchange information consists in the (possibly partial) flow-field data related to a limited subset of the considered solutions. It is important to note that the dimension of this subfamily is essentially lower than the overall dimension of the whole parametric family.

4. IMPLEMENTATION OF THE COOPERATIVE STRATEGY FOR PARALLEL CFD COMPUTATIONS

4.1. Parallel GAs search for the aerodynamic shape optimization

The basic idea behind Genetic Algorithms is to mathematically imitate the evolution process of nature. They are semi-stochastic optimization methods that are conveniently presented using the metaphor of natural evolution: a randomly initialized population of individuals (set of points of the search space at hand) evolves following a crude parody of the Darwinian principle of the survival of the fittest. The GA algorithms are based on the evaluation of a set of solutions, called population. The suitability of an individual is determined by the value of objective (fitness) function.

A new population of individuals is generated using variations of "genetic" operators: selection, crossover and mutation. Parents are chosen by selection, and new offsprings are produced with crossover and mutation. All these operations include randomness. The success of the optimization process is improved by the principle of elitism, where the best individuals of the current generation are included in the next generation. The main point is that the probability of survival of new individuals depends on their fitness: the best are kept with a high probability, the worst are rapidly discarded.

In the optimization of aerodynamic shapes by means of GAs [1] the population represents a set of aerodynamic shapes.

In order to improve the optimal search and to accelerate the convergence to the optimum shape, the above cooperative strategy was implemented in the following way. The crucial question is what kind of information is to be exchanged. Fortunately, for GAs the situation is favourable to giving a positive constructive answer to this general question: by exchanging information related to the best individual of parallel populations, one can expect the convergence rate of GA iterative process to improve.

The parallel asynchronous GA algorithm involved the following main stages:

a) The initial population (based on random search) is independently obtained on each processor of the group.

b) Non-blocking data acquisition from other processors of the group. The received individuals are included in the current population. If the requested message has not arrived this step is skipped.

c) With a preassigned value of the exchange step L, the information related to the best individual in the generation is broadcast to the other processors of the group.

4.2. Parallel cooperative strategy for massive CFD computations of aerodynamic data-bases

First, it is necessary to explain why a straightforward CFD parallelization on distributed memory clusters is not sufficiently efficient where large-scale aerodynamic computations are performed.

Consider a parallelized CFD code aimed to solve the full Navier-Stokes equations. The full-scale computation of a complicated aircraft configuration is usually infeasible on available serial processors.

The code similar to NES [2] which exhibits the high parallel efficiency runs about 6 hours on 106 processors (see [3] - [4]). This means that an average aerodynamic user which is rarely eligible to more than 100-200 processors may get only 1-2 flight point computations in a time. A more economic way to perform these computations is to employ fewer processors (30-40) in a time but then, of course, the overall CPU time greatly increases.

Note that in the case of a popular commercial code FLUENT, the peak parallel performance is achieved (on a typical platform of HP J6000 PA 8600,550) on already 16 processors.

The computation time is almost proportional to the number of iterations (sometimes measured in Multigrid cycles). Due to the complexity of the problem involved, the number of these iterations (cycles) may be as great as 200-400 (depending on the flight conditions).

We propose a parallel cooperative technology which essentially decreases the number of cycles needed to get a sufficiently accurate solution. This is done by performing the computations in an order which allows making use of already performed computations on the one hand, and, by exchanging the information in the course of parallel runs in a sophisticated way, on the other hand.

The idea of the parallel algorithm may be illustrated as follows. Consider, for example, the problem of computing a number of lift/ drag polars for different Mach numbers. For example, 10 polars are needed, each of them containing 20 different angles of attack.

In the first stage, a very limited number of full-size Navier-Stokes calculations are performed. The number of these calculations may amount to 10-15 (of a total of 200 lift/drag curves). The flight points chosen parse the parametric space of Mach/ angle of attack by covering it in an approximately uniform way. (Of course, this may change according to specific flight conditions since harder flight points necessitate a denser covering).

Already in this stage, completed CFD runs may be employed as a starting point for the runs in queue. This can be done by initializing the initial flow-field by an available information taken from a different (than the current) flight point (in contrast to a standard computation which starts from an appropriate free-stream solution).

In this stage, a flexible amount of processors may be employed starting from a cluster of 15-20 processors. On the other hand, the computational volume of the stage makes also efficient the use of large-size clusters including up to 1000 processors.

The next stage of computations deals with the rest of the tested flight points. The specific implementation depends on the number of parallel processors available. The flight points are distributed among the processors in such a way that the points which are close (in the metric of the parametric space of flight conditions) to those computed in the first stage, start immediately after the completion of the first stage.

If necessary, the second wave of computations start originating from the already existing solutions, the third group employs the solutions of the previous two groups and so on. The approach allows to essentially decrease the number of iterations (cycles) needed to get the converged solution and thus to decrease the overall computation time.

An additional important source of increasing the parallel efficiency and thus diminishing the overall CPU time is the inter-level data exchange. Specifically this means that the current solution is interpolated with more advanced solutions at the flight points close to the current one. For example, this may be done by weighting the solutions U_1 and U_2: $U_{new} = a \cdot U_1 + (1-a) \cdot U_2$, where the parameter $0 < a < 1$, in a way minimizing the current density residual of the Navier-Stokes equations.

This approach involves an insignificant amount of additional computations and a certain amount of additional message passing. The tests show that the approach can be highly efficient provided the frequency of exchange is tuned to the parallel configuration.

5. ANALYSIS OF RESULTS

The above described cooperative parallelization strategy based on the public domain PVM software package was implemented on a cluster of MIMD multiprocessors consisting of several hundreds nodes. Each node has 2 processors, 2GB RAM memory, 512KB Level 2 Cache memory and full duplex 100Mbps ETHERNET interface.

5.1. Acceleration of GAs optimal search

The results presented in this section are related to the optimization of RAE2822 airfoil at fixed values of free-stream Mach and Reynolds at $C_L = 0.745$, $M = 0.75$.

The flight conditions are those of a high transonic regime. At the considered values of target C_L, the initial flow is characterized by strong shock/boundary layer interaction.

It appeared that a parallelization of GA algorithms can essentially improve accuracy and robustness of the optimization search.

In fig. 1, a comparison between serial and parallel GAs (in terms of convergence rate of optimum search) is given. Lines 1,2 and lines 3,4 correspond to two different search domains; the exchange step $L = 30$. It can be assessed from fig. 1 that the exchange of information related to the best individuals coming from each evolution process (each processor) essentially improved the convergence of the algorithm.

Thus we observe that the present approach of parallel cooperative implementation of a genetic algorithm provides an exemplary case of a situation where the gain due to exchange of information among processors improves the convergence rate of GA and surpasses the loss in CPU time related to the data exchange itself.

5.2. Massive parallel computations of aerodynamic data-bases

The results of this section are related to massive computations of aerodynamic data for a sample wing-body configuration. This configuration is typical of high-subsonic transport aircraft. It includes a cranked wing with both forward and backward sweep. General view of the configuration is given in fig. 2.

The CFD computations were performed by a multiblock parallel full Navier-Stokes code NES [3]. The computational grids contained about 500000 grid points united in three Multigrid levels, each of the levels included 16 blocks.

We constructed the aerodynamic data bases corresponding to all the possible combinations of the following flight conditions:

- $M = 0.60$, $M = 0.61$, $M = 0.62$, $M = 0.63$, $M = 0.64$, $M = 0.65$, $M = 0.66$, $M = 0.67$, $M = 0.68$, $M = 0.69$, $M = 0.70$
- $\alpha = -2.0°$, $\alpha = -1.5°$, $\alpha = -1.0°$, $\alpha = -0.5°$, $\alpha = 0.0°$, $\alpha = 0.5°$, $\alpha = 1.0°$, $\alpha = 1.5°$, $\alpha = 2.0°$, $\alpha = 2.5°$, $\alpha = 3.0°$, $\alpha = 3.5°$, $\alpha = 4.0°$
- $Re = 3 \cdot 10^6$, $Re = 5 \cdot 10^6$,

A total of 286 flight points were computed.

In fig. 3, drag polars C_L vs. C_D are depicted at $M = 0.60$ and $Re = 3 \cdot 10^6$ which show results achieved by different computation strategies. The conventional strategy (solid line) consisted in the computation at all the angles of attack, listed above. In this case, each converged solution required 60 Multigrid cycles on the fine level. Thus achieved results serve us as the exact reference. Two versions of the cooperative parallel strategy were applied here (the dotted line and crosses, respectively). In the first case, 30 Multigrid cycles were performed at $\alpha = -0.5°$, $\alpha = 1.5°$ and $\alpha = 3.0°$. In the next stage, the information on the flow-fields, obtained in the first stage was transferred to the processors responsible for the computation of the remaining angles of attack. Then, at each angle of attack, 30 Multigrid cycles were performed.

In the second case, the first stage was reduced to the 30 cycles computation at only one angle of attack $\alpha = 0.0°$. In the next stage, the information on this flow-field, was sent to the processors responsible for the computation of the remaining angles of attack. Then, once again, at each angle of attack, 30 Multigrid cycles were performed.

Since the computational cost of the interstage data transfer is negligible, the overall computational volume needed for the construction of the whole data-base (a complete drag polar in this case) can be estimated in terms of Multigrid (MG) cycles. The conventional approach necessitated 780 MG cycles while the first and the second versions of the cooperative strategy required 480 and 420 MG cycles, respectively.

It may be be assessed from fig. 3 that the results achieved by the three computations are virtually identical.

In fig. 4, drag polars C_L vs. C_D are depicted at $M = 0.60$ and $Re = 5 \cdot 10^6$ which show results achieved by different computation strategies.

The conventional strategy (solid line) consisted in the computation at all the angles of attack, listed above, at the target value of Reynolds number. In this case, each converged solution required 60 Multigrid cycles on the fine level. The present comparison is aimed to estimate the cooperative strategy in the case where the information intended for the interstage data transfer, is computed at a different than the target, combination of Reynolds number and angle of attack.

The following cooperative parallel strategy was applied in this case (represented by the dotted line). In the beginning, 30 Multigrid cycles were performed at $\alpha = 0.0°$, $Re = 3 \cdot 10^6$. In the next stage, the information on the just obtained flow-field, was sent to the processors responsible for the computation of the remaining angles of attack. Then at each angle of attack, 30 Multigrid cycles were performed at the corresponding target flight conditions.

For the considered example, the conventional approach required 780 MG cycles while the computation by means of the cooperative strategy amounted to 420 MG cycles.

Similar to the previous example, it is clearly seen from fig. 4 that the results are very close one to another.

In the next example, a challenging case of the free-stream Mach variation, was considered. That is, the corresponding cooperative strategy consists in the computation of a drag/lift polar by using the information coming from the flow-field computed not only at a different angle of attack but also for a different than the target free-stream Mach value. The corresponding results are given in fig. 5, where drag polars are depicted at $M = 0.62$ and $Re = 3 \cdot 10^6$.

In this figure, the solid line corresponds to the computation at all the angles of attack, listed above, at the target value of Mach number (conventional strategy). Similar to the previous examples, each converged solution required 60 Multigrid cycles on the fine level.

In the first version of the cooperative strategy (dashed-dotted line), 30 Multigrid cycles were performed at $\alpha = 0.0°$, $M = 0.60$, $Re = 3 \cdot 10^6$. In the next stage, the information on the just obtained flow-field, was sent to the processors responsible for the computation of the remaining angles of attack. Then at each angle of attack, 30 Multigrid cycles were performed at the corresponding target Mach value $M = 0.62$.

The second version of the cooperative strategy (dashed line) differs from the first one only in the number of Multigrid cycles in the second stage of computations: 40 instead of 30.

For the considered example, the conventional approach required 780 MG cycles while the computation by means of the cooperative strategy amounted to 420 and 550 MG cycles in the first and the second versions of the cooperative strategy, respectively. It is clear that in this challenging case, the 30 cycles version does not provide a sufficient accuracy, but the 40 cycles version provides highly satisfactory results.

In the final example, another challenging case of the transonic free-stream Mach variation, was considered. Similar to the previous example, the cooperative strategy consists in the computation of a drag/lift polar by using the information coming from the flow-field computed for a different than the target free-stream Mach value. The corresponding results are given in fig. 6, where drag polars corresponding to the conventional and cooperative approach (solid and dashed lines, respectively) are shown ($M = 0.70$ and $Re = 3 \cdot 10^6$).

With the conventional approach each converged solution required 60 Multigrid cycles on the fine level. With the cooperative strategy 30 Multigrid cycles were performed at $\alpha = 0.0°$, $M = 0.69$, $Re = 3 \cdot 10^6$. In the next stage, the information on the just obtained flow-field, was sent to the processors responsible for the computation of the remaining angles of attack. Then at each angle of attack, 30 Multigrid cycles were performed at the corresponding target Mach value $M = 0.70$.

In this example, the conventional approach required 780 MG cycles while the computation by means of the cooperative strategy amounted to 420 MG cycles. It is clear that in this challenging case, already the 30 cycles version provides accurate results. This may be attributed to a closer than in the previous example proximity of the transferred flow-field (at $M = 0.69$) to the flow-field at the target Mach value $M = 0.70$.

REFERENCES

1. Epstein, B., and Peigin, S., "A Robust Hybrid GA/ROM Approach to Multiobjective Constrained Optimization in Aerodynamics" *AIAA Journal*, V.42, No. 8, pp.1572–1581, 2004.
2. Epstein, B., Rubin, T., and Seror, S., "Accurate Multiblock Navier-Stokes Solver for Complex Aerodynamic Configurations", *AIAA Journal*, V.41, No.4, pp.582–584, 2003.
3. Peigin, S., Epstein, B., Rubin, T., and Seror, S., "Parallel Large Scale High Accuracy Navier-Stokes Computations on Distributed Memory Clusters", *The Journal of Supercomputing*, Vol. 27, No.1, pp.49–68, 2003.
4. Peigin, S., Epstein, B., and Gali, S., "Multilevel Parallelization Strategy for Optimization of Aerodynamic Shapes", in *Parallel Computational Fluid Dynamics, New Frontiers and Multi-disciplinary Applications*, Elsevier, 2004.

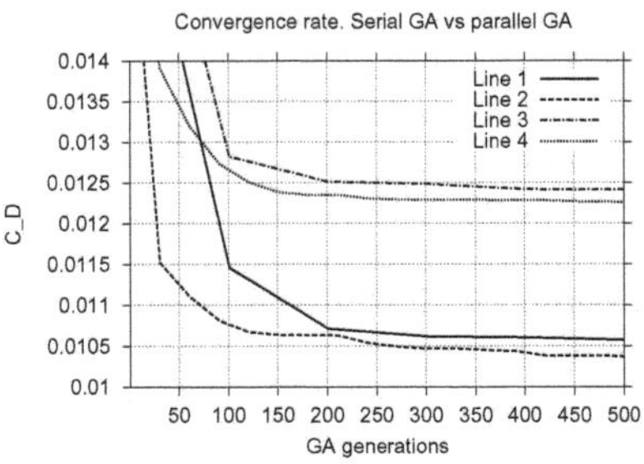

Figure 1. Convergence rate of GAs optimum search. Serial GA (lines 1,3) vs parallel GA (lines 2,4).

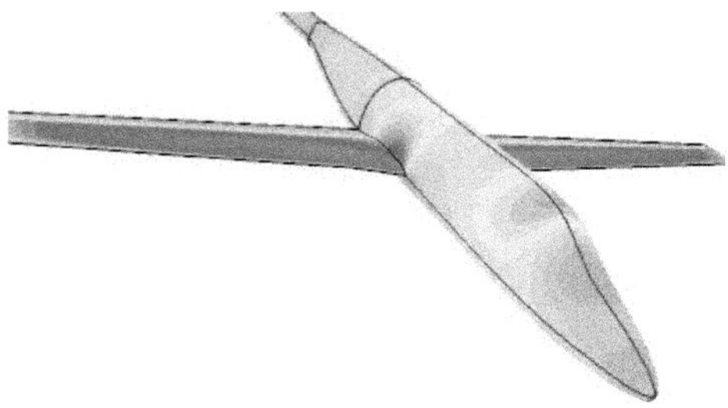

Figure 2. General view of wing-body configuration.

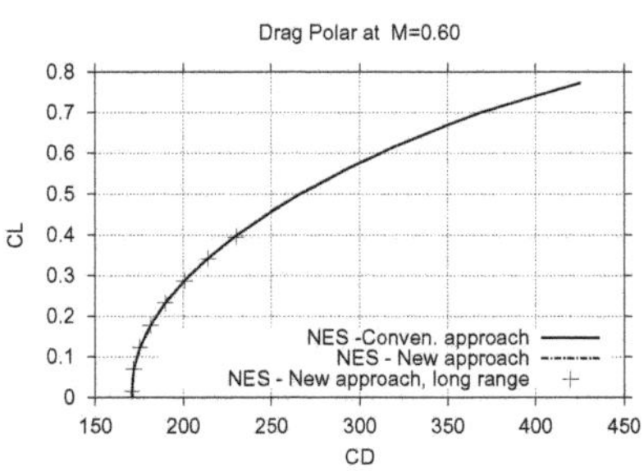

Figure 3. Comparison of drag polars at $M = 0.60$, $Re = 3\dot{1}0^6$. Conventional approach vs. cooperative parallel computation.

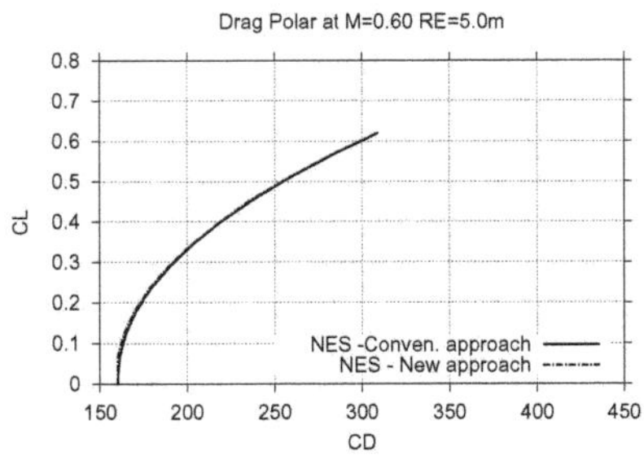

Figure 4. Comparison of drag polars at $M = 0.60$, $Re = 5\dot{1}0^6$. Conventional approach vs. cooperative parallel computation.

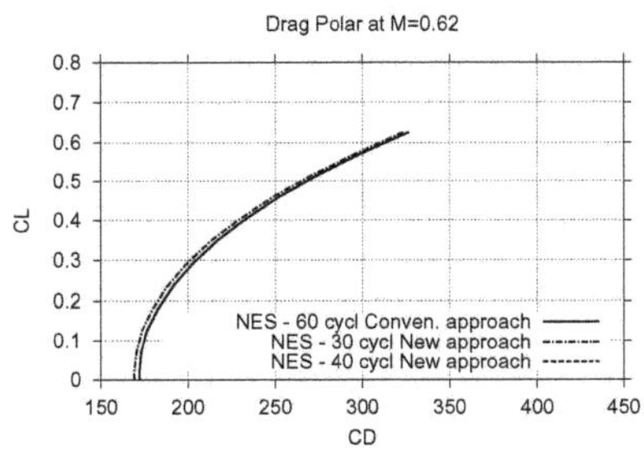

Figure 5. Comparison of drag polars at $M = 0.62$, $Re = 3\dot{1}0^6$. Conventional approach vs. cooperative parallel computation.

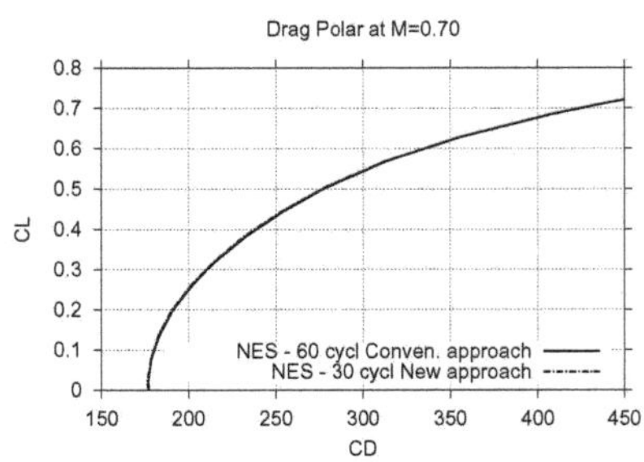

Figure 6. Comparison of drag polars at $M = 0.70$, $Re = 3\dot{1}0^6$. Conventional approach vs. cooperative parallel computation.

Parallel Computational Fluid Dynamics – Parallel Computing and Its Applications
J.H. Kwon, A. Ecer, J. Periaux, N. Satofuka and P. Fox (Editors)

Performance Analysis of Fault Tolerant Algorithms for the Heat Equation in Three Space Dimensions

H. Ltaief[a], M. Garbey[a*] and E. Gabriel[a]

[a]Dept of Computer Science, University of Houston, Houston, TX 77204, USA

With the emerging of new massively parallel systems such as the *BlueGene* with tens of thousands of processors, the Mean Time Between Failures is considerably decreasing. Therefore, the reliability of the machine becomes a very important issue especially for time consuming simulations such as Computation Fluid Dynamics applications. This paper focuses on the performance analysis of a fault tolerant algorithm for the time integration of parabolic three dimensional problem. The method was first presented at *PARCFD05* on the two dimensionnal case. Based on numerical explicit schemes and uncoordinated checkpointing, these algorithms allow to retrieve efficiently the lost data on the failed process(es) by computation, avoiding the expensive roll-back operations required in most of standard Checkpoint/Restart schemes.

1. Introduction

Nowadays, a single failing node or processor on a large High Performance Computing system does not imply, that the entire machine has to go down. Typically, the parallel application utilizing this node has to abort, all other applications on the machine are not affected by the hardware failure. The reason that the parallel application, which utilized the failed processor, has to abort is mainly because the most widespread parallel programming paradigm MPI [1], is not capable of handling process failures. Several approaches how to overcome this problem have been proposed, most of them relying on some forms of Checkpoint/Restart schemes (C/R) [2,3]. However, these schemes have inherent performances and scalability limitations.

The objective of this paper is to extend the fault tolerant algorithms first introduced in [4,5] to higher dimension, based on numerical explicit schemes and uncoordinated checkpointing, for the time integration of parabolic problems. We will also present a detailed performance analysis. First of all, our method allows the application to recover from process failures and to retrieve mathematically the lost data on the failed process(es) avoiding the roll-back operation required in C/R schemes. While for low time consuming applications requiring small amount of resources, aborting after a failure does not really matter, for critical applications running for several weeks on hundreds of processors, the application shutdown appears really to be a major concern.

Then, while it is fairly easy to recover numerically from a failure with a relaxation

*Research reported here was partially supported by Award 0305405 from the National Science Foundation.

scheme applied to an elliptic problem, the problem is far more challenging with the time integration of a parabolic problem. Indeed, the integration back in time is a very ill-posed problem. Further time integration of unsteady problem may run for very long time and are more subject to process failures.

In this paper, we concentrate on the three dimensional heat transfer problem that is a representative test case of the main difficulty and present a new explicit recovery technique which avoids the roll-back operation and is numerically efficient.

The paper is organized as follows: section 2 defines briefly our model problem. In section 3, after recalling in few words the major points of our two fault tolerant algorithms, we give an overview of the checkpointing procedure settings and compare both techniques. Section 4 presents some performance results for the recovery operation and draws a complete performance analysis. Finally, section 5 summarizes the results of this paper and highlights our ongoing work.

2. Definition of the Application Benchmark

We construct an algorithmic solution of the problem in the context of domain decomposition and distributed computing. The model application used throughout the paper is the three dimensional heat equation that is given by

$$\frac{\partial u}{\partial t} = \Delta u + F(x,y,z,t), \tag{1}$$
$$(x,y,z,t) \in \Omega \times (0,T), \quad u_{|\partial\Omega} = g(x,y,z), \; u(x,y,z,0) = u_o(x,y,z).$$

We suppose that the time integration is done by a first order implicit Euler scheme,

$$\frac{U^{n+1} - U^n}{dt} = \Delta U^{n+1} + F(x,y,z,t^{n+1}), \tag{2}$$

and that Ω is partitioned into N subdomains Ω_p, $p = 1..N$.

The work presented in the paper is based on the Fault Tolerant Message Passing Interface (FT-MPI) [6,7] framework developed at the University of Tennessee. While FT-MPI is capable of surviving the simultaneous failing of $n-1$ processes in an n processes job, it remains up to the application developer to recover the user data, since FT-MPI does not perform any (transparent) checkpointing of user-level data items.

We remind in the next section the main characteristics of the two parallel fault tolerant algorithms used to solve the three dimensionnal reconstruction problem based on checkpointed data.

3. Fault Tolerant Approaches

The main purpose of these algorithms is to avoid the expensive roll-back operation to the last consistent distributed checkpoint, loosing all the subsequent work and adding a significant overhead for applications running on thousands of processors due to coordinated checkpoints.

3.1. Forward Implicit Method

For the first approach, the application process p has to store every K time steps its current solution $U_p^{n(p)}$. Additionally, the artificial boundary conditions $I_p^m = \Omega_p \cap \Omega_{p+1}$

have to be stored for all time steps $m < M$ since the last checkpoint. The solution U_p^M can then be reconstructed on the failed process with the forward time integration (2). While it is fairly easy to implement the algorithm, this technique requires to save the boundary conditions at each time step besides the local solution. We will see that this procedure can increase significantly the communication traffic and therefore, can be very expensive.

3.2. Backward Explicit Method

The second approach does not require the storage of the boundary conditions of each subdomain at each time step, and allows to retrieve the interface data by computation instead. There are three main steps.

Firstly, the neighbors of the failed process(es) compute backward in time from the available data on main memory using the explicit formula provided by (2):

$$\begin{aligned} U_{i,j,k}^{n} &= U_{i,j,k}^{n+1} - dt\,\Big(\frac{U_{i+1,j,k}^{n+1} - 2U_{i,j,k}^{n+1} + U_{i-1,j,k}^{n+1}}{hx^2} + \frac{U_{i,j+1,k}^{n+1} - 2U_{i,j,k}^{n+1} + U_{i,j-1,k}^{n+1}}{hy^2} \\ &+ \frac{U_{i,j,k+1}^{n+1} - 2U_{i,j,k}^{n+1} + U_{i,j,k-1}^{n+1}}{hz^2}\Big) - F_{i,j,k}^{n+1}. \end{aligned} \tag{3}$$

The existence of the solution is granted by the forward integration in time. However, since the numerical procedure is unstable, the precision deteriorates rapidly in time and blows up after few time steps. Furthermore, one is restricted to a cone of dependence.

To stabilize the scheme, one may add a hyperbolic regularization such as the telegraph equation that is a perturbation of the heat equation. Further details regarding this stabilization procedure can be found in [4,8].

Then, to construct the solution outside the cone of dependencies and therefore to determine the solution at the subdomain interface, we used a standard procedure in inverse heat problem, the so-called space marching method [9].

Finally, the neighbors of the failed process(es) provide the computed boundary solutions. The freshly restarted process(es) can rebuild their lost data by applying locally (2). Figure 1 illustrates the numerical accuracy of the overall reconstruction scheme for $\Omega = [0,1] \times [0,1] \times [0,1]$, $dt = 0.5 \times h$, $K = 5$ and F such that the exact analytical solution is $cos(q_1\,(x+y+z))(sin(q_2\,t)+\frac{1}{2}cos(q_2\,t))$, $q_1 = 2.35, q_2 = 1.37$. As expected, the maximum of the L2 norm of the error on the domain decreases when we increase the global size. Moreover, the overall order achieved by the method is 2.

The next section describes the checkpointing procedure settings.

3.3. Implementation Details

The checkpointing infrastructure implemented in the 3D Heat equation is using two groups of processes: a solver group composed by processes which will only solve the problem itself and a spare group of processes whose main function is to store the local data from solver processes. The main objective is to minimize as much as possible the checkpointing overhead. Therefore, on the one hand, we use non blocking checkpointing to introduce overlap areas between the time of computation and the time spent in communication. On the other hand, we combine it with uncoordinated checkpointing to unload the interconnect network.

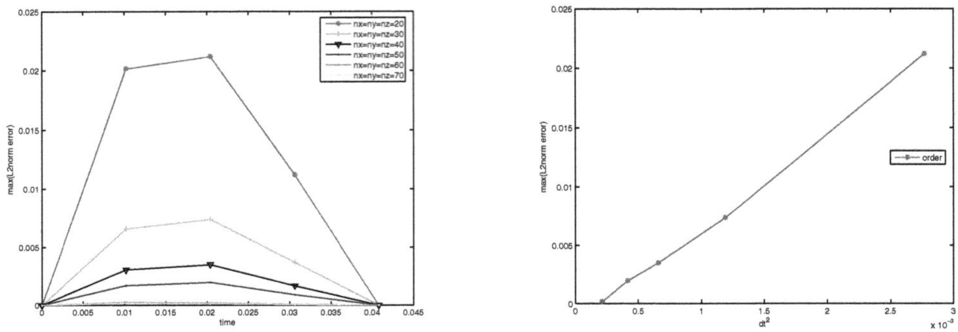

Figure 1. Numerical accuracy of the overall reconstruction scheme in three dimensions.

The communication between the solver processes and the checkpointing processes is handled by an inter-communicator. Since it does not make sense to have as many checkpointing processes as solver processes, the number of spare processes is equal to the number of solver processes in the x-direction (px). Thus, while the solver processes are arranged in the 2-D cartesian topology, the checkpoint processes are forming a 1-D cartesian topology. Figure 2 gives a geometrical representation of the two groups with a local

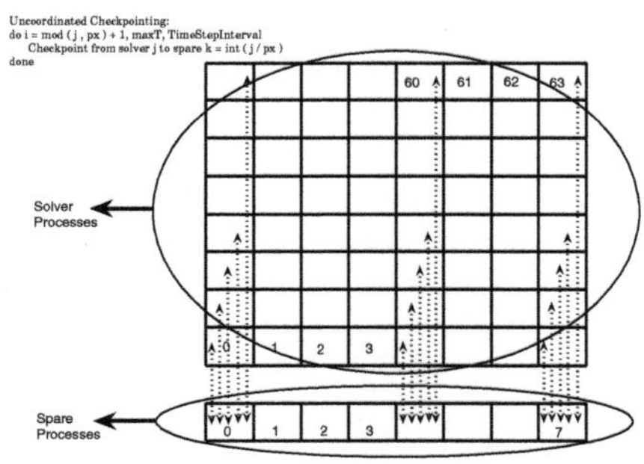

Figure 2. Uncoordinated Checkpointing Settings

numbering. This figure shows furthermore, how the local checkpointing is applied for a configuration of 64 solver processes and 8 spare processes. Each spare process is in charge of storing the data of a column of the 2D processor mesh.

In the following section, we would like to compare the communication and checkpointing overhead required by the two reconstruction methods described up to now.

3.4. Comparison Results

The machine used for these experiments is a twenty Dual Opteron 1.6GHz with 2MB of cache and 4GB of RAM each with a Gigabit Ethernet interconnect. For the performance comparison between both methods 3.1 and 3.2, we have evaluated the three-dimensional version of (1) with three different problem sizes per processor ($10 \times 10 \times 50$, $15 \times 15 \times 76$ and $18 \times 18 \times 98$) on four different processor configurations (9, 16, 25 and 36 processes). The processes are arranged in a regular two dimensional mesh. Each column of the processor-mesh has a separate checkpoint processor assigned to it. The data each checkpoint process receives is stored in their main memory, avoiding therefore expensive disk I/O operations.

The results are displayed in figures (3-4-5). The horizontal coordinate gives the checkpointing interval in number of time steps between each checkpoint, while the verticale coordinate shows the overhead compared to the same code and problem size without any checkpointing. As expected, the overhead is decreasing with increasing distance between two subsequent checkpoints. However, saving the artificial boundary conditions at each time step slows down the code execution significantly. While interface conditions are one dimension lower than the solution itself, the additional message passing is interrupting the application work and data flow. For example, saving the local solution each 10 time steps add 6% overhead with 36 processes and a large problem size while saving additionnaly the boundary conditions engender 26% overhead for the same configuration. This confirms our preliminary results in [10].

In the next section, we have simulated process failures and we have maesured the time needed for the recovery and the reconstruction steps using the explicit numerical methods where only the local solution has to be checkpointed.

4. Performance Recovery Results

We have randomly killed processes and we show some timing measurements on how long the application takes to recover and to compute the lost solution by successively applying the Explicit Backward in time scheme followed by the Space Marching algorithm. We have set the backup interval to 10 time steps. Here, we do only need to checkpoint the local solution on each process.

Figures (6-8-10) present the overhead in percentage on the overall time of execution when different numbers of simulteanously failures happen compare to the standard code without any checkpointing procedure. As a consequence, we can see that the overhead induced by the failures is fairly low especially when running with large data set and high process numbers. For example, with 36 solver processes (plus 6 spare processes) and large problem size, killing randomly 4 processes produces an overhead around 8%. Since we are limited by the amount of the resources available in the cluster, our performance could have been better. Indeed, some of the 4 freshly restarted processes are running on dual nodes which are already busy with 2 processes. This slows down the application and can be avoided simply by running on larger clusters.

Figures (7-9-11) draw the evolution of the recovery time in seconds from small to large data set. The recovery time corresponds to the time needed to rebuild the lost data. As a matter of fact, these results prove that the explicit numerical methods (Explicit Backward scheme associated with the Space Marching algorithm) are very fast and very cheap in

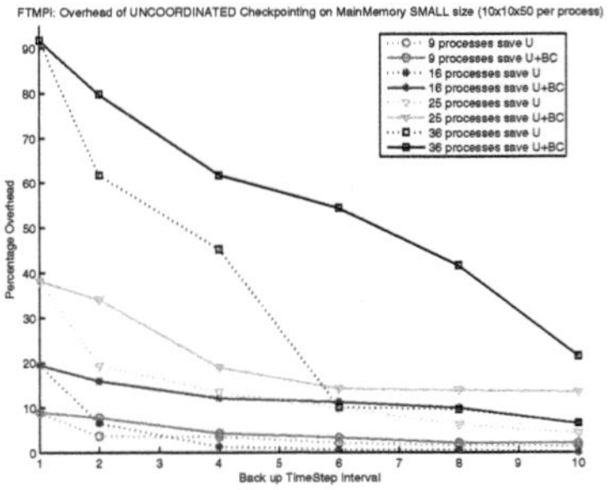

Figure 3. Uncoordinated Checkpointing Overhead with SMALL sizes.

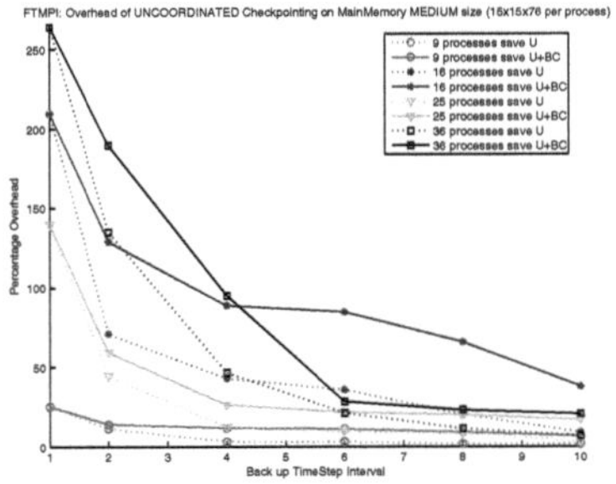

Figure 4. Uncoordinated Checkpointing Overhead with MEDIUM sizes.

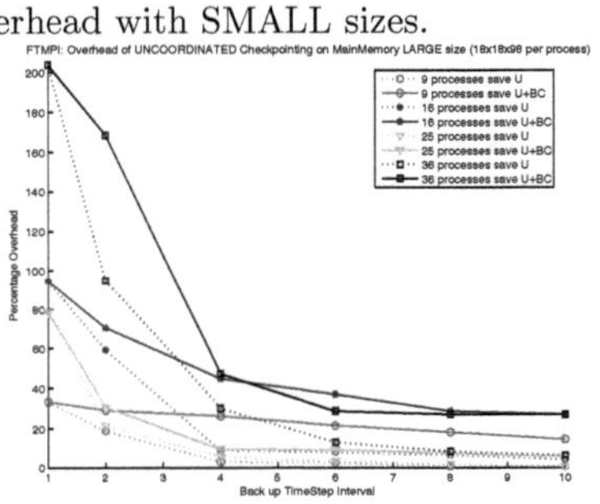

Figure 5. Uncoordinated Checkpointing Overhead with LARGE sizes.

term of computation. For example, the recovery time is staying below 0.3 seconds when running with 36 processes and large problem size for an execution time of 52.62 seconds. Furthermore, the recovery time does not really seem to depend on the number of running processes but rather on the number of failures. Indeed, the reconstruction algorithms performance scale pretty good when increasing the number of processes and as expected, it raises when the number of failures grows.

5. Conclusion

This paper focuses on two approaches on how to handle process failures for three dimensionnal parabolic problems. Based on distributed and uncoordinated checkpointing, numerical methods presented here can reconstruct a consistent state in the parallel application, despite of storing checkpoints of various processes at different time steps. The first method, the forward implicit scheme, requires for the reconstruction procedure the boundary variables of each time step to be stored along with the current solution. The second method, based on explicit space/time marching, only requires checkpointing the solution of each process every K time steps. Performance results comparing both methods with respect to the checkpointing overhead have been presented. We presented the results for recovery time of with the 3-D heat equation. Our ongoing work is focusing on the implementation of our recovery scheme for a 3D Reaction-Convection-Diffusion code simulating the Air Quality Model [11]. Furthermore, a distribution scheme for Grid environments is currently being developed, which chooses the checkpoint processes location in an optimal way, such that the lost solution can be reconstructed even if a whole machine is failing.

REFERENCES

1. MPI Forum : *MPI2 : Extensions to the Message-Passing Interface Standard.* Document for a Standard Message-Passing Interface, University of Tennessee, 1997.
2. Sriram Sankaran, Jeffrey M. Squyres, Brian Barrett, Andrew Lumsdaine, Jason Duell, Paul Hargrove, and Eric Roman, The LAM/MPI Checkpoint/Restart Framework: System-Initiated Checkpointing, in *International Journal of High Performance Computing Applications*, 2004.
3. G. Bosilca, A. Bouteiller, F. Cappello, S. Djilali, G. Fedak, C. Germain, T. Herault, P. Lemarinier, O. Lodygensky, F. Magniette, V. Neri, and A. Selikhov, MPICH-V: Toward a Scalable Fault Tolerant MPI for Volatile Nodes, in SC'2002 Conference CD, IEEE/ACM SIGARCH, Baltimore, MD, 2002.
4. M. Garbey, H. Ltaief, *Fault Tolerant Domain Decomposition for Parabolic Problems*, Domain Decomposition 16, New York University, January 2005. To appear.
5. M. Garbey and H. Ltaief *On a Fault Tolerant Algorithm for a Parallel CFD Application*, PARCFD 2005, UNiversity of Maryland, Washington DC, USA, May 2005.
6. Graham E. Fagg, Edgar Gabriel, Zizhong Chen, Thara Angskun, George Bosilca, Jelena Pjesivac-Grbovic, and Jack J. Dongarra, 'Process Fault-Tolerance: Semantics, Design and Applications for High Performance Computing', in 'International Journal of High Performance Computing Applications', Volume 19, No. 4, pp. 465-477, Sage Publications 2005.
7. Beck, Dongarra, Fagg, Geist, Gray, Kohl, Migliardi, K. Moore, T. Moore, Papadopoulous, Scott, and Sunderam, 'HARNESS: a next generation distributed virtual machine', Future Generation Computer Systems, 15, 1999.
8. W. Eckhaus and M. Garbey, Asymptotic analysis on large time scales for singular perturbation problems of hyperbolic type, SIAM J. Math. Anal. Vol 21, No 4, pp867-883 (1990).
9. D.A. Murio, The Mollification Method and the Numerical Solution of Ill-posed Problems, Wiley, New York (1993).
10. H. Ltaief, M. Garbey and E. Gabriel *Parallel Fault Tolerant Algorithms for Parabolic Problems*, Euro-Par 2006, Dresden, Germany, Aug-Sept 2006.
11. F. Dupros, M.Garbey and W.E. Fitzgibbon, A Filtering technique for System of Reaction Diffusion equations, Int. J. for Numerical Methods in Fluids, Vol 52, pp1-29, 2006.

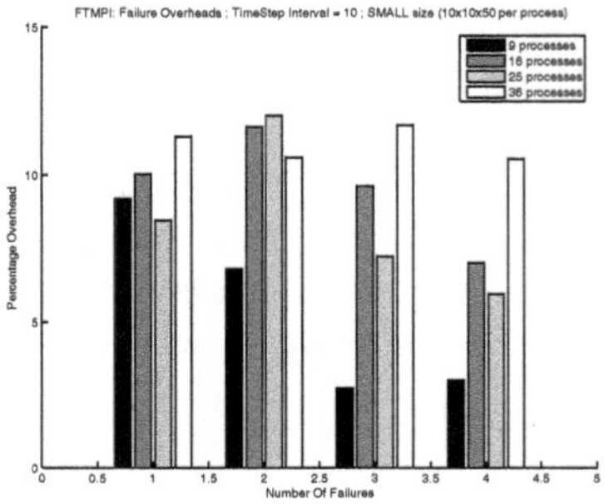

Figure 6. Failure Overheads with SMALL size.

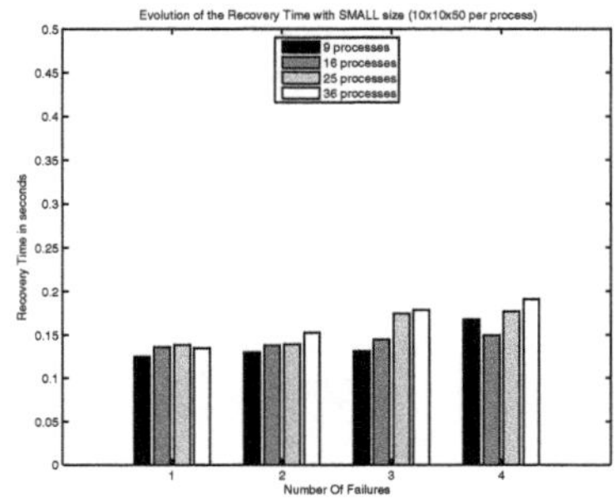

Figure 7. Recovery Time in Seconds with Small size.

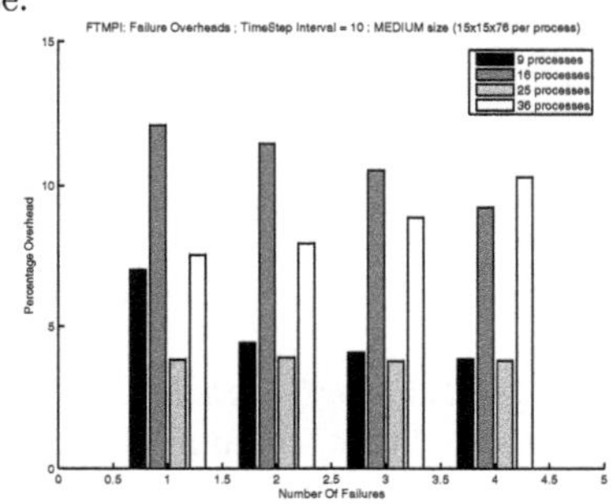

Figure 8. Failure Overheads with MEDIUM size.

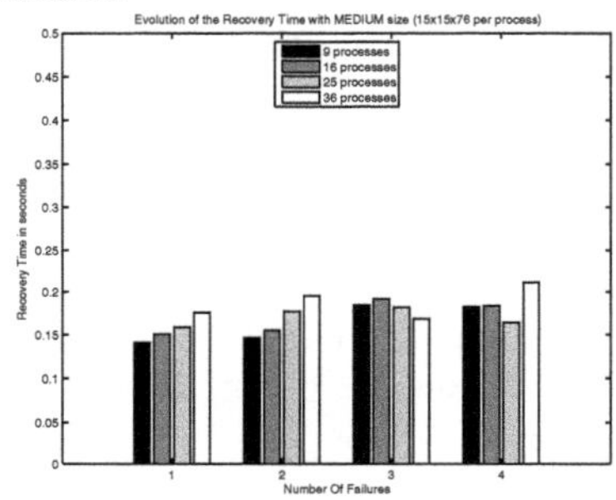

Figure 9. Recovery Time in Seconds with MEDIUM size.

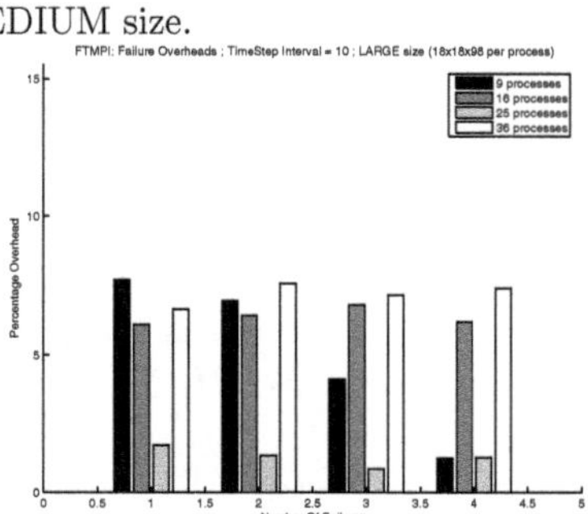

Figure 10. Failure Overheads with LARGE size.

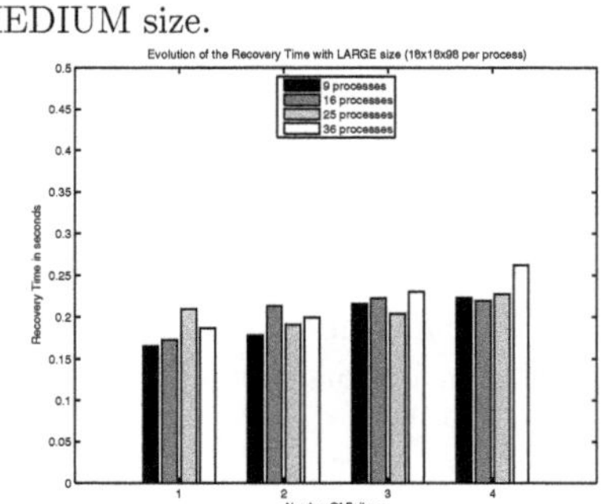

Figure 11. Recovery Time in Seconds with LARGE size.

Parallel Computational Fluid Dynamics – Parallel Computing and Its Applications
J.H. Kwon, A. Ecer, J. Periaux, N. Satofuka and P. Fox (Editors)
2007 Elsevier B.V.

Parallel Deferred Correction method for CFD Problems

D. Guibert [a] and D. Tromeur-Dervout[a b *]

[a]ICJ UMR5208 Université Lyon 1 - CNRS , Centre de Développement du Calcul Scientifique Parallèle Université Lyon 1

[b]IRISA/SAGE INRIA Rennes

1. Motivation of Time Decomposition in the Parallel CFD context

Time domain decomposition methods can be a complementary approach to improve parallelism on the forthcoming parallel architectures with thousand of processors. Indeed, for a given modelled problem, — e.g. through a partial differential equation with a prescribed discretization— the multiplication of the processors leads to a decrease of the computing load allocated to each processors, then the ratio of the computing time with the communication time decreases, leading to less efficiency. Nevertheless , for the evolution problems, the difficulty comes from the sequential nature of the computing. Previous time steps are required to compute the current time step. This last constraint occurs in all time integrators used in computational fluid dynamics of unsteady problems.

During these last years, some attempts have been proposed to develop parallelism in time solvers. Such algorithms as Parareal [6] or Pita [2] algorithms are multiple shooting type methods and are very sensitive to the Jabobian linearization for the correction stage on stiff non-linear problems as shown in [4].

Moreover, this correction step is an algorithm sequential and often much more time consuming to solve for the time integrator than the initial problem.

In this paper we investigate a new solution to obtain a parallelism in time with a two level algorithm that combines pipelined iterations in time and deferred correction iterations to increase the accuracy of the solution. In order to focus on the benefit of our algorithm we approach CFD problems as an ODE (eventually DAE) problem. From the numerical point of view the development of time domain decomposition methods for stiff ODE problems allows to focus on the time domain decomposition without perturbation coming from the space decomposition. In other side, classical advantages of the space decomposition on the granularity are not available. Indeed, from the engineering point of view in the design with ODE/DAE, the efficiency of the algorithm is more focus on the saving time in the day to day practice.

*This work was funded by the thema "mathématiques" of the Région Rhône-Alpes thru the project: "Développement de méthodologies mathématiques pour le calcul scientifique sur grille"

2. ODE approach for CFD problem

The traditional approach to solve CFD problems starts from the partial differential equation, which is discretized in space with a given accuracy and finally a time discretization with a first order scheme or a second order time scheme is performed to solve problems (usually implicitly due to Courant-Friedrich-Lax condition constraint).

Let consider the 2D Navier-Stokes equations in a Stream function - vorticity formulation:

$$\left\{ \begin{array}{l} \omega_t + \psi_x \omega_y - \psi_y \omega_x - \frac{1}{Re} \Delta \omega = 0 \\ \Delta \psi = -\omega \\ \psi = 0, \psi_n = v(t) \text{ on } \partial\Omega \end{array} \right. \tag{1}$$

then applying a backward (forward Euler or Runge-Kutta are also possible) time discretisation, we obtain that follows:

$$\left\{ \begin{array}{l} \omega^{n+1} + \Delta t(\psi_x^n \omega_y^{n+1} - \psi_y^n \omega_x^{n+1}) - \frac{\Delta t}{Re} \Delta \omega^{n+1} = \omega^n \\ \Delta \psi^{n+1} = -\omega^{n+1} \end{array} \right. \tag{2}$$

The discretization in space of the discrete in time equations leads to solve a linearized system with an efficient preconditionned solver such as Schwarz-Krylov method or multigrid:

$$\{ \; A(\psi^n)\omega^{n+1} = f(\psi^n), \;\; \Delta \psi^{n+1} = -\omega^{n+1} \tag{3}$$

Let us consider an another approach that is the method of lines or ODE to solve CFD problems. Firstly, the Navier-Stokes problem is written in a Cauchy equation form:

$$\frac{d}{dt} \begin{pmatrix} \omega \\ \psi \end{pmatrix} (t) = f(t, \omega, \psi), \begin{pmatrix} \omega \\ \psi \end{pmatrix} (0) = \begin{pmatrix} \omega_0 \\ \psi_0 \end{pmatrix} \tag{4}$$

then the discretization in space of the $f(t, \omega, \psi)$ leads to a system of ODE :

$$\frac{d}{dt} \begin{pmatrix} \omega_{i,j} \\ \psi_{i,j} \end{pmatrix} = f(t, \omega_{i[\pm 1],j[\pm 1]}, \psi_{i[\pm 1],j[\pm 1]}) \tag{5}$$

This last equation can be solved by an available ODE solver such as SUNDIALS (SUite of Nonlinear and DIfferential/ALgebraic equation Solvers (http://www.llnl.gov/casc/sundials/)) that contains the latest developments of classical ODE solvers as LSODA [8] or CVODE [1].

Figure 1 corresponds to the lid driven cavity results computed by the ODE method a Reynold number of 5000 and up to 10000. These results are obtained for the impulse lid driven with a velocity of 1. Let us notice that the convergence is sensitive to the singularities of the solution.

The advantages and drawbacks of the PDEs and ODE are summarized in Table 1

In conclusion the ODE approach put the computing pressure on the time step control while PDEs approach put the computing pressure on the preconditionned linear/nonlinear solver.

The high time consumming drawback for ODE explains why this approach is not commonly use in CFD. Nevertheless grid computing architecture could give a renewed interest to this ODE approach.

	PDEs	ODE
1	Fast to solve one time step	slow to solve one time step (control of time step)
2	Decoupling of variables	coupling of variables (closest to incompressibility constraint)
3	computing complexity is reduced with the decoupling	high computing complexity () control of error, whole system
4	B.C. hard to handle with impact on the linear solver (data structure/choice)	B.C. conditions easy to implement
5	CFL condition even with implicit solve (decoupling of variables)	CFL condition replaced by the control on time step
6	Building of good preconditionner is difficult	Adaptive time step allows to pass stiffness
7	Adaptive time step hard to handle	Adaptive time step is a basic feature in ODE

Table 1
Advantages and drawbacks of ODE and PDEs approaches for CFD problems

3. Spectral Deferred Correction Method

The spectral deferred correction method (SDC) [7] is a method to improve the accuracy of a time integrator iteratively in order to approximate solution of a functional equation with adding corrections to the defect.

Let us summarize the SDC method for the Cauchy equation:

$$y(t) \quad = \quad y_0 + \int_0^t f(\tau, y(\tau))d\tau \tag{6}$$

Let $y^0_m \simeq y(t_m)$ an approximation of the searched solution at time t_m. Then consider the polynomial q_0 that interpolate the function $f(t,y)$ at P points $\tilde{t}_i$ i.e:

$$\exists! q_0 \in R_P[x], q_0(\tilde{t}_i) = f(y^0(\tilde{t}_i)) \tag{7}$$

with $\{\tilde{t}_0 < \tilde{t}_1 < ... < \tilde{t}_P\}$ are points around t_m. Then if we define the error between the approximate solution y^0_m and the approximate solution using q_0:

$$E(y^0, t_m) = y_0 + \int_0^{t_m} q^0(\tau)d\tau - y^0_m \tag{8}$$

Then we can write the error between the exact solution and the approximation y^0_m:

$$y(t_m) - y^0_m = \int_0^{t_m} \left(f(\tau, y(\tau)) - q^0(\tau)\right) d\tau + E(y^0, t_m) \tag{9}$$

The defect $\delta(t_m) = y(t_m) - y^0_m$ satisfies:

$$\delta(t_{m+1}) = \delta(t_m) + \int_{t_m}^{t_{m+1}} \left(f(\tau, y^0(\tau) + \delta(\tau)) - q^0(\tau)\right) d\tau + E(y^0, t_{m+1}) - E(y^0, t_m) \tag{10}$$

For the spectral deferred correction method an approximation δ_m^0 of $\delta(t_m)$ is computed considering for example a Forward Euler scheme :

$$\delta_{m+1}^0 = \delta_m^0 + \Delta t(f(t_{m+1}, y_{m+1}^0 + \delta_{m+1}^0) - f(t_{m+1}, y_{m+1}^0)) - y_{m+1}^0 + y_m^0 + \int_{t_m}^{t_{m+1}} q^0(\tau)d\tau$$

Once the defect computed , the approximation y_m^0 is updated

$$y_m^1 = y_m^0 + \delta_m^0. \quad (11)$$

Then a new q_0 is computed with using the value y_m^1 for a new iterate. This method is a sequential iterative process as follows: compute an approximation y^0, then iterate in parallel until convergence compute δ and update the solution y by adding the correction term δ.

We propose in the next section a parallel implementation to combine deferred correction method with time domain decomposition.

4. Parallel SDC: a time domain decomposition pipelined approach

The main idea to introduce parallelism in the SDC is to consider the time domain decomposition and to affect a set of time slices to each processors. Then we can pipeline the computation, each processor dealing a different iterate of the SDC.

Algorithm 4.1 (Parallel Version). *1 spread the time slices between processors with an overlap $P/2$ slices and compute an approximation y^0 with a very low accuracy because this is a sequential stage. Each processor keeps only the time slices that it has to take in charge, and can start the pipelined as soon it has finished.*

2 Iterate until convergence

- *2.1 prepare the non blocking receiving of the last $P/2$ delta value of the time interval take in account by the processor.*
- *2.2 If the left neighbor processor exist wait until receiving the $P/2$ δ value from it in order to update the y_m^i in the overlap of size $P/2$.*
- *2.3 Compute q_0 from y_m^i in order to compute δ_m^i When the δ values between $P/2+1$ and P of time slices are computed send to left neighbor processor in order it will be able to start the next iterate (corresponding to the stage 2.1) .*
- *2.4 Send the $P/2$ δ values corresponding to the overlap position of the right neighbor processor.*
- *2.5 Update the solution $y_m^{i+1} = y_m^i + \delta_m^i$*

Remark 4.1. *The stage 1 initialization should not be parallelized in order to keep the time integrator history and avoid jump between the initial value inside the domain and the value in the overlap on the neighbor processor.*

Remark 4.2. *The communication stages [2.1] and [2.2] are needed to compute the q_0 polynomial that uses the behavior of the function $f(t, y)$ before and after the point t_m.*

#proc	1	2	4	8
[1] (s)	17.12	18.68	17.36	16.41
[2.2] (s)	1.35e-5	72.39	113	129
[2.3] $f(t,y)$ eval (s)	1.66e-2	7.25e-3	3.13e-3	1.126e-3
[2.3] q_0 computing (s)	5.99e-3	5.5e-3	5.90e-3	5.94e-3
euler	0.288	0.25	0.19	0.16
Total(s)	2062	1113	742	423
Efficiency	100%	92.6%	69.4%	60.9%
Speed-up	1	1.85	2.77	4.87
Speed up Corrected	1	1.98	3.27	7.01

Table 2
Time, Speed-up and Efficiency of the Time domain decomposition pipelined SDC

Table 2 gives the times to run the time domain decomposition pipelined SDC method on an ALTIX350 with IA64 1.5Ghz/6Mo processors and with numalink communication network. The lid driven parameters are a Reynolds number of 10 and a regular mesh of 20×20 points, the final time to be reach is T=5. The number of time slices has been set equal to 805. The number of SDC iteration is 25. The stage [1] correspond to the sequential first evaluation of the solution on the time slices (with an rtol of 10^{-1}). The time [2.2] is the time of the pipelined to reach the last processor, results show that the ratio of this time [2.2] on the computation is increasing up to 1/4 for 8 processors case. The Speed-up corrected line gives the parallel capability of the parallel implementation when the pipeline is full.

It exhibits quite reasonable speed-up up to 8 processors for the present implementation.

Remark 4.3. *Some improvement of the presented implementation can be performed. firstly we implement the forward Euler to compute the δ_m^{i+1} that needs a Newton solving at each iteration with the Jacobian computed by a numerical differentiation. We also can use the sundials solver with blocking at first order the BDF. Secondly, all the value $f(\tilde{t}_m, y_m^i)$ are computed before and could be progressively computed in order to enhance the chance of parallelism. Thirdly, a better load balancing could be acheived with a cyclic distribution of the time slices among the processors.*

Figure 2 gives the convergence of the defect on the time interval for different iteration numbers (1,5,9,13,17). Symbols represent the runs on 1 processor (x), 2 processors (o) , 8 processors (□). Firstly, we remark that the number of processors does not impact the convergence properties of the method. Secondly, the convergence behavior increases for the last times due to the steady state of the solution. Nevertheless, a good accuracy is obtained on the first times in few SDC iterates.

5. conclusions

In this work we have investigated the numerical solution of 2D Navier-stokes with ODE solvers in the perspective of the time domain decomposition which can be an issue for the grid computing.

we proposed to combine time domain decomposition and spectral deferred correction in order to have a two level algorithm that can be pipelined. The first results proved some parallel efficiency with a such approach.

Nevertheless, the arithmetical complexity of the ODE solver for the lid driven problem consider here, is too much costly. Some scheme such as the $C(p, q, j)$ [3] schemes can decouple the system variables in order to reduce the complexity.

REFERENCES

1. S. D. Cohen and A. C. Hindmarsh. *CVODE, a Stiff/Nonstiff ODE Solver in C.* Computers in Physics, 10(2):138-143, 1996.
2. C. Farhat and M. Chandesris. *Time-decomposed parallel time-integrators: theory and feasibility studies for fluid, structure, and fluid-structure applications.* Int. J. Numer. Meth. Engng., **58**(9):1397–1434, 2003.
3. M. Garbey and D. Tromeur-Dervout,*A parallel adaptive coupling algorithm for systems of differential equations*, J. Comput. Phys.,161(2):401–427,2000
4. D. Guibert and D. Tromeur-Dervout, *Parallel Adaptive Time Domain Decomposition for Stiff Systems of ODE/DAE*, Computer and Structure, 2006 to appear.
5. E. Hairer, S. P. Norsett and G. Wanner, *Solving Ordinary Differential Equations II. Stiff and Differential-Algebraic Problems*, Springer Series in Computational Mathematics, Springer, Berlin, 1991.
6. J.-L. Lions, Y. Maday, and G. Turinici. *Résolution d'EDP par un schéma en temps "pararéel"*, C.R.A.S. Sér. I Math., **332**(7):661–668, 2000.
7. Minion, Michael L., *Semi-implicit spectral deferred correction methods for ordinary differential equations*, Commun. Math. Sci.,1(3):471–500,2003.
8. L. R. Petzold, *A Description of DASSL: A Differential and Algebraic System Solver* SAND82-8637 (September 1982).

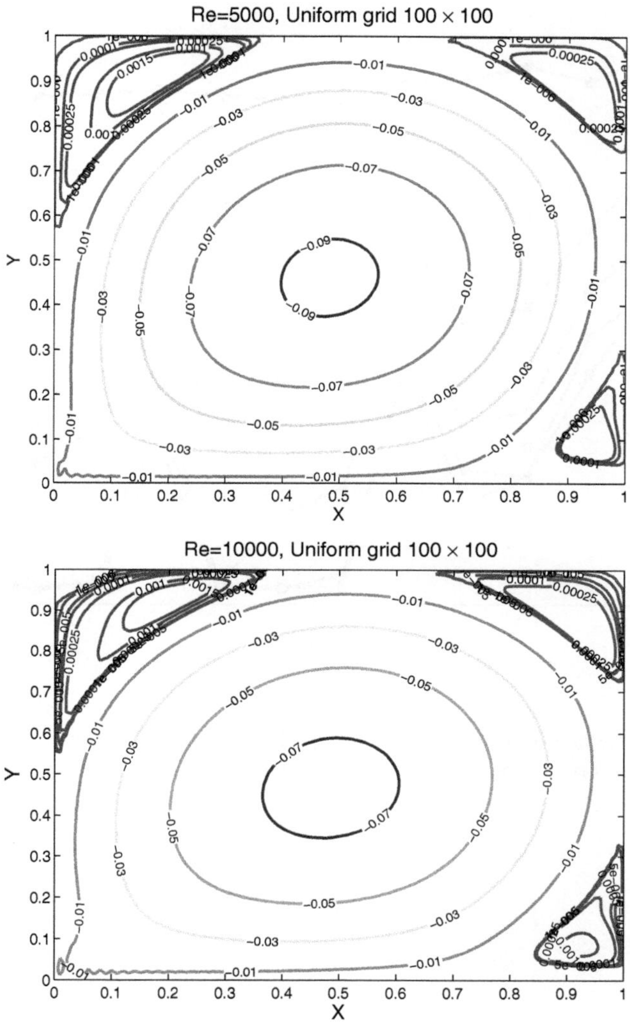

Figure 1. Impulse lid driven cavity results for a Reynold number of 5000 (top) and 10000 (bottom) with the ODE approach

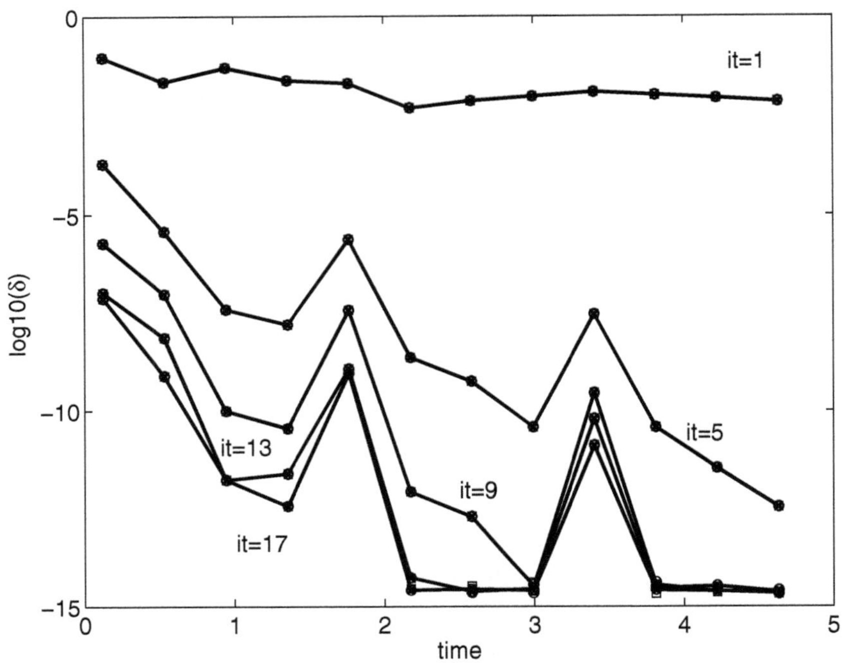

Figure 2. Convergence on the time interval of the defect δ for different iteration numbers (1,5,9,13,17). Symbols represent the runs on 1 processor (x), 2 processors (o) , 8 processors (□)

Parallel Computational Fluid Dynamics – Parallel Computing and Its Applications
J.H. Kwon, A. Ecer, J. Periaux, N. Satofuka and P. Fox (Editors)

Performance Evaluation and Prediction on a Clustered SMP System for Aerospace CFD Applications with Hybrid Paradigm

Yuichi Matsuo [a], Naoki Sueyasu [b], and Tomohide Inari [b]

[a]*Japan Aerospace Exploration Agency, 7-44-1 Jindaijihigashi, Chofu-shi, Tokyo 182-8522, Japan*

[b]*Fujitsu Limited, Makuhari Systems Laboratory, 1-9-3 Nakase, Mihama-ku, Chiba-shi, Chiba 261-8588, Japan*

Japan Aerospace Exploration Agency has introduced a terascale clusterd SMP system as a main compute engine of Numerical Simulator III for aerospace science and engineering research purposes. The system is using Fujitsu PRIMEPOWER HPC2500; it has computing capability of 9.3Tflop/s peak performance and 3.6TB of user memory, with 56 compute nodes of 32-way SMP. In this paper, we first present the performance evaluation results for aerospace CFD applications with hybrid programming paradigm used at JAXA on the clusterd SMP system. Next, we propose a performance prediction formula for hybrid codes based on a simple extension of Amhhal's law, and discuss about the predicted and measured performances for some typical hybrid CFD codes.

Keywords: Performance evaluation and prediction, Aerospace CFD, Hybrid paradigm, SMP

1. Introduction

Remarkable developments in microprocessor technology are pushing up computing power everyday. Parallel processing is becoming inevitable for large-scale numerical simulations such as used in Computational Fluid Dynamics (CFD) in aerospace. Meanwhile, shared-memory architecture is becoming widespread because of its programming ease.

From 1993 through the third quarter of 2002, the Japan Aerospace Exploration Agency (JAXA), formerly known as the National Aerospace Laboratory of Japan (NAL), operated a distributed-memory parallel supercomputer system called *Numerical*

Wind Tunnel (NWT) [1] that consists of 166 vector processors with 280Gflop/s of peak performance. In October 2002, it was replaced with a large clustered SMP system(i.e., a distributed parallel system consisting of SMP nodes) with approximately 1,800 scalar processors, peak performance of 9.3Tflop/s, and 3.6TB of user memory. The new system is called *Numerical Simulator III* (NS-III).

In this paper, we first show the results of performance evaluation for our current parallel aerospace CFD applications. Next, we discuss about the performance prediction for the parallel CFD applications on the clustered SMP system by proposing a simple prediction formarism based on Amdhal's law.

2. System Overview

The computing subsystem in NS-III is called *Central Numerical Simulation System* (CeNSS). We have 18 cabinets; each is Fujitsu PRIMEPOWER HPC2500 [2], where a cabinet is the physical unit of hardware. Each cabinet has 128 CPUs with 256GB of shared memory and can act as a 128-way symmetric-multi-processor (SMP) system in its maximum configuration. The CPU is the SPARC64 V scalar chip with 1.3GHz clock. Thus, the theoretical peak performance per CPU becomes 5.2Gflop/s and 665.6Gflop/s per cabinet. L2 cache is 2MB on chip. For computation, 14 cabinets are dedicated, giving a total peak computing performance of 9.3Tflop/s and 3.6TB of memory. A cabinet can be partitioned into either 2 or 4 nodes according to need, where a node is the logical unit from the operating system's point of view. In the CeNSS, each compute cabinet is partitioned into 4 nodes where each node is a 32-way SMP with 64GB shared memory, giving a total of 56 compute nodes. For the remaining cabinets, 3 are partitioned into 2 nodes each, for a total of six 64-way SMP nodes and used for service, login and I/O purposes. All nodes are connected to a crossbar interconnect network with high 4GB/s bi-directional bandwidth and low 5μsec latency through one data transfer unit per node.

The programming environment is crucial for any parallel system. In principle, we adopt the so-called hybrid programming paradigm; that is, we use the 'thread parallel' model within a node, with automatic parallelism or OpenMP, whereas between nodes, we use the 'process parallel' model with MPI or XPFortran, as shown in Fig. 1. Codes that use 'process parallel' only can also be run. Since we have already been accustomed to our own data parallel language NWT-Fortran (similar to HPF), we use its successor XPFortran. This programming style is very convenient for us, because if we think one NWT processor is mapped onto a node of CeNSS, the transition in programming from NWT to CeNSS is quite natural, particularly in parallel coding. For more details on the system description, see Ref.[3].

3. Performance evaluation for JAXA aerospace CFD applications on the CeNSS

We firstly measured the performance of several parallel CFD applications written with the hybrid programming paradigm actually being used at JAXA. As indicated in Table 1, six parallel CFD applications with different memory-access/data-transfer features are chosen. In Fig. 2, each code's characteristics are plotted schematically where the horizontal line corresponds to the memory access ratio and the vertical line to the data transfer ratio. Linpack and NPB (NAS Parallel Benchmark) are also plotted for reference. Here, all the values are from the same job configuration (process*thread). As shown in Fig. 2, the selected codes can be classfied into the three types, that is, Type 1; CPU intensive with moderate memory access and light communication but with large amounts of floating point operations, Type 2; data transfer intensive with moderate memory, and Type 3; with heavy memory access. For more details on the applications, see Ref. [4].

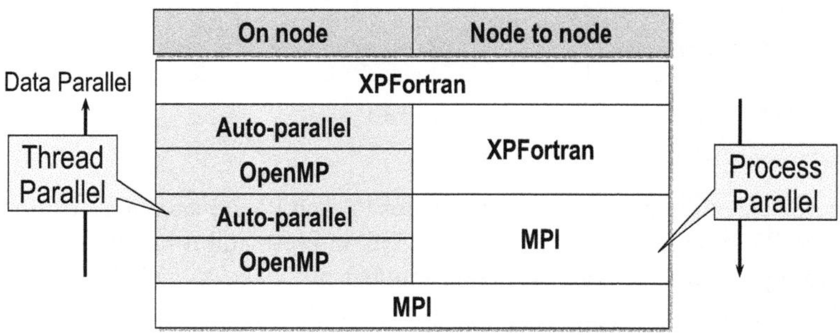

Figure 1: Programming paradigm in CeNSS.

Table 1: Specifications of JAXA parallel CFD applications.

Code: Name	Application	Simulation model	Numerical method with features	Parallel strategy	Language
P1:LES	Aircraft component	LES	FDM	OpenMP + MPI	F77
P2:HJET	Combustion	DNS	FDM with chemistry	OpenMP + MPI	F77
P3:CHANL	Turbulence	DNS	FDM with FFT	OpenMP + XPF	F77
P4:HELI	Helicopter	URANS	FDM with overlapped mesh	Auto-parallel + XPF	F77
P5:UPACS	Aeronautics	RANS	FVM with multi-block mesh	OpenMP + MPI	F90
P6:JTAS	Aeronautics	RANS	FVM with unstructred mesh	OpenMP + MPI	F77

LES: Large-Eddy Simulation, DNS: Direct Numerical Simulation, URANS: Unsteady RANS, RANS: Reynolds-Averaged Navier-Stokes, FDM: Finite Difference Method, FVM: Finite Volume Method

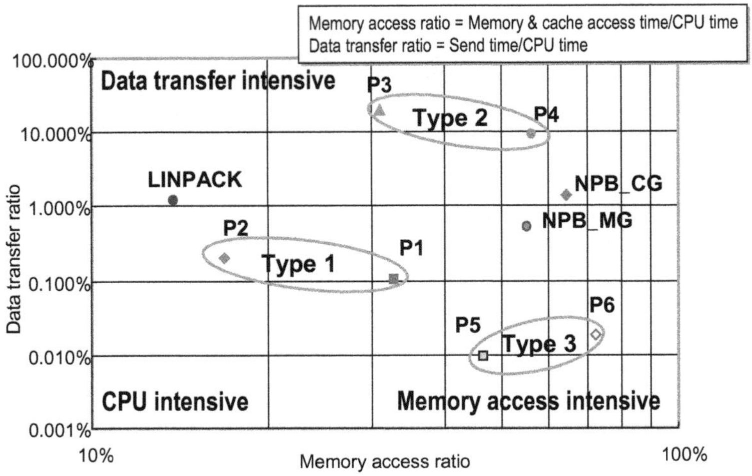

Figure 2: Features of JAXA hybrid CFD codes.

The measured speed-up performances for codes P3 and P5 are shown as examples in Fig. 3. At code P3, a large grid of 2048*448*1536 is used, and it can be seen that the process scalability tends to saturate for larger number of CPUs. At code P5, meanwhile, a 40*20*80*512 grid is used, and the thread scalability tends to saturate. To see the hybrid performances, the speed-up ratios are plotted in Fig. 4 with the number of CPUs constant where (*,*) means (process, thread). We found that in code P3 even with the same number of CPUs, the performance is better for the job with 4 threads by 56 processes than that with 1 thread by 224 processes. This is a typical example on the performance of the hybrid model while generally pure MPI is faster than hybrid [5] as in Fig. 4(b). For more details on the benchmark results, see Ref. [4].

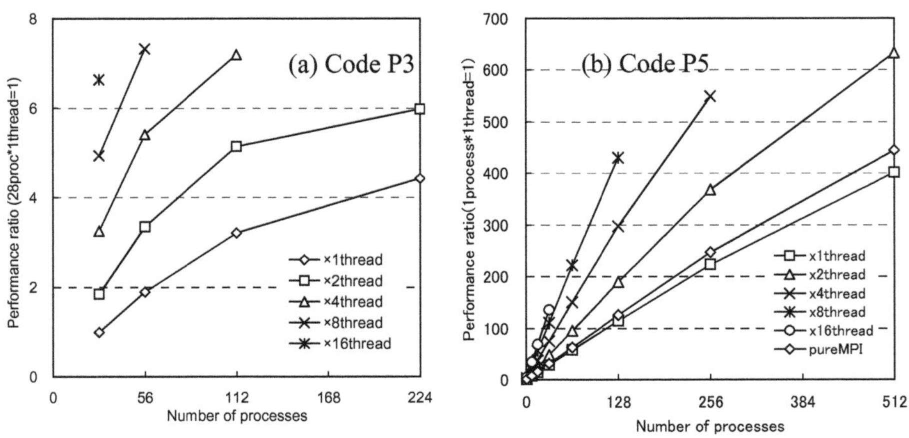

(a) Code P3 (b) Code P5

Figure 3: Speed-up performance for codes P3 and P5.

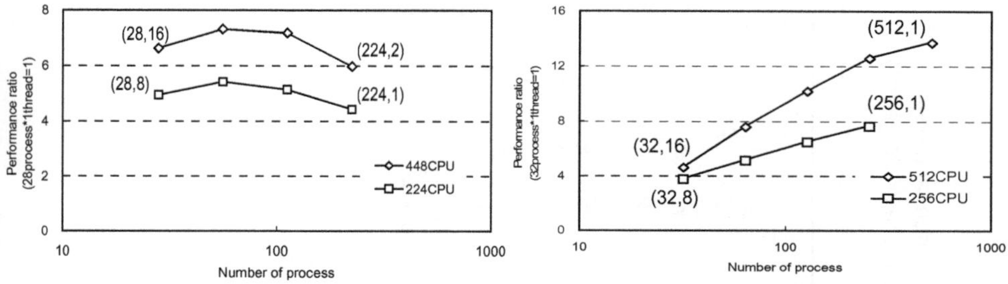

(a) Code P3 (b) Code P5

Figure 4: Speed-up performance for codes P3 and P5 shown with constant CPUs.

4. Performance prediction for the JAXA CFD applications with hybrid programming

It is well known that the speed-up performance of a parallel code is predicted by Amdhal's law, saying that the speed-up ratio is written by

$$S(n) = T_{serial}/T_{parallel}, \text{ with } T_{parallel} = T_{serial} \times \{(1-a) + a/n\} \quad (1)$$

where $T_{serial/parallel}$ is serial/parallel CPU time, a parallel ratio, and n number of CPUs. For the hybrid parallel programming paradigm, the Amdhal's law can be extended with a straightforward manner and the speed-up is written by

$$S(n_p,n_t) = T_{serial}/T_{hybrid}, \text{ with } T_{hybrid} = T_{serial} \times \{(1-a_p) + a_p/n_p\} \times \{(1-a_t) + a_t/n_t\} \quad (2)$$

where $T_{serial/hybrid}$ is serial/hybrid CPU time, a_p process parallel ratio, n_p number of processes, a_t thread parallel ratio, and n_t number of threads. Also we find that when the communication volume is not negligible, the speed up can be written to

$$S(n_p,n_t) = T_{serial}/T_{hybrid}, \text{ with } T_{hybrid} = T_{serial} \times [\{(1-a_p-c_t-c_n) + a_p/n_p\} \times \{(1-a_t) + a_t/n_t\} + (c_t + n_p \times c_n)] \quad (3)$$

where $T_{serial/hybrid}$ is serial/hybrid CPU time, a_p process parallel ratio, n_p number of processes, a_t thread parallel ratio, n_t number of threads, c_t ratio of communication independent to the number of processes , and c_n ratio of communication proportional to the number of processes.

In Fig. 5, the predicted and measured performances are compared for codes P3 and P5 with the number of CPUs constant. We found that for code P5 the performance can be predicted well with Eq. (2) while for code P3 the performance can be predicted well with Eq. (3).

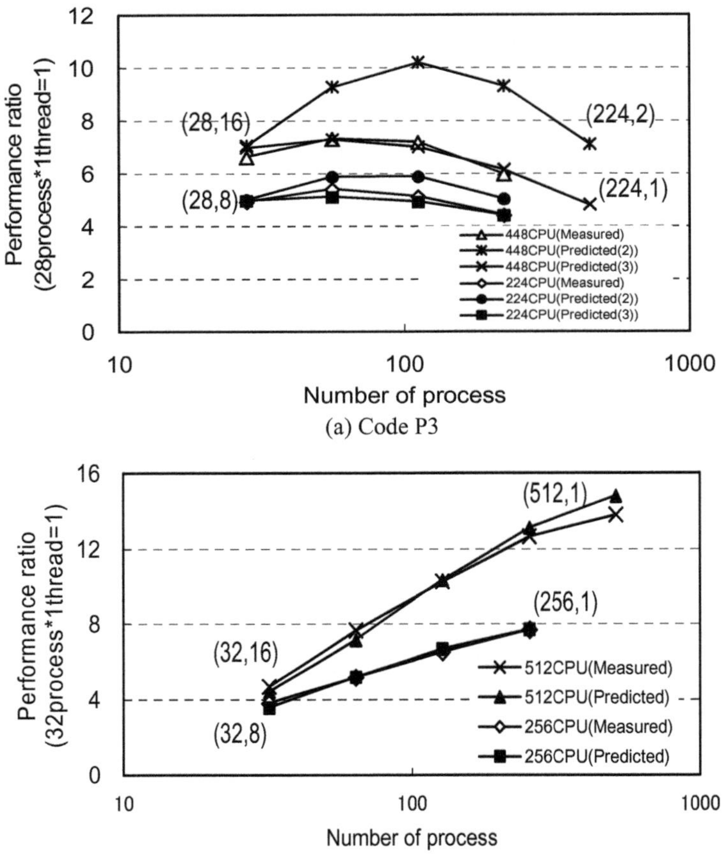

(a) Code P3

(b) Code P5

Figure 5: Speed-up performance for codes P3 and P5 shown with constant CPUs.

5. Summary

In this paper, we discussed about the performance evaluation and prediction results for JAXA hybrid CFD applications by selecting some typical codes. We found that the benefit of hybrid programming was visible for the application with large communication cost, e.g. code P3, and also found that by using the extended Amdhal's law the speed-up performances of the hybrid CFD codes can be predicted well even when the communication cost is high. This shows that the best process*thread combination of a code could be determined without a close benchmark if the memory-access and communication cost is known. However, the present formalism is strongly based on empiricism so we plan to confirm it theoretically as future work.

References

[1] Miyoshi, H., et al., "Development and Achievement of NAL Numerical Wind Tunnel (NWT) for CFD Computations," In *Proceedings of SC1994,* November 1994.

[2] Fujitsu Limited, http://www.fujitsu.com.

[3] Matsuo, Y., et al., "Numerical Simulator III – Building a Terascale Distributed Parallel Computing Environment for Aerospace Science and Engineering," *Parallel Computational Fluid Dynamics, New Frontiers and Multi-Disciplinary Applications*, Elsevier (2003), 187-194.

[4] Matsuo, Y., et al., "Early Experience with Aerospace CFD at JAXA on the Fujitsu PRIMEPOWER HPC2500," In *Proceedings of SC2004,* November 2004.

[5] Cappelo, F. and Etiemble, D.,"MPI versus MPI+OpenMP on IBM SP for the NAS Benchmarks," In *Proceedings of SC2000,* November 2000.

Parallel Computational Fluid Dynamics – Parallel Computing and Its Applications
J.H. Kwon, A. Ecer, J. Periaux, N. Satofuka and P. Fox (Editors)

Non-Intrusive Data Collection for Load Balancing of Parallel Applications

Stanley Y. Chien, Gun Makinabakan, Akin Ecer, and Hasan U. Akay
Purdue School of Engineering and Technology
Indiana University-Purdue University Indianapolis
Indianapolis, Indiana 46202 USA
E-mail: schien@iupui.edu

Key words: job scheduler, dynamic load balancing

Abstract: It is highly desirable to have a tool to support dynamic load balancing for parallel computation in a large network of heterogeneous computers. Many schedulers and mid-wares developed to date can only support dynamic load balancing in homogeneous computation environments. Although our previously developed scheduler and load balancer can support heterogeneous environment, they are only applicable to domain decomposition based parallel applications. The main problem of the slow progress in the development of general load balancing tool is due to the difficulty of obtaining the needed information of the parallel application for load balancing. This paper describes new application execution information gathering methods that use an MPI profiler or use an available system tool, PROC. PROC is supported by some versions of UNIX operating systems and Linux. These two new methods are different from our previously used time stamp method that inserts time measuring library functions into the application code. These new execution gathering approach can support dynamic load balancing with general parallel applications and support network of heterogeneous computers. This new approach also makes a lot convenient for user to take the advantage of dynamic load balancing.

1. INTRODUCTION

Job scheduling and dynamic load balancing for parallel jobs have been studied for many years. Various algorithms and tools were developed [1-5]. Most of these schedulers can provide good process scheduling for single jobs and loosely coordinated parallel jobs. For tools that do dynamic load balancing for tightly coordinated parallel jobs [1], they were developed for specific type of computation environments, e.g., machines use the same operating systems, or developed for systems with homogeneous computers dedicated to single users.

A unique property of tightly coordinated parallel job, in which the parallel processes must communicates periodically, is that the execution times of all processes of the parallel job are determined by the slowest process since the fast processes must wait for the slowest process periodically to get needed information to proceed. Therefore, the load balancing of tightly coordinated parallel job is to reduce the execution time of the slowest process by assigning it to a fast computer or reducing the load of the computer that runs the slowest processes. Many factors affect the execution time of the parallel processes of a parallel job. The application related information include the speeds of different computers used for the parallel job execution, the computation work of different processes of a parallel job, the size of messages among parallel processes, etc. The environment related information includes the different loads on different computers, the network communication speed among computers, operating modes of the computation nodes (single user/multiple user, single job/multiple job), etc. Environment related information usually can be obtained by system tools. The application related information may be provided by the user but usually takes a lot user's effort that reduces user's interest of using the load balancing tool. It is much desirable that the application related information to be obtained by the software tools automatically.

We have developed an algorithm for dynamic load balancing of tightly coordinated parallel CFD jobs [6-8]. The algorithm uses all aforementioned application related information. All the measured data except the network communication speed are obtained by inserting time stamp library functions in the parallel application code. Since the use of the time stamp library requires that the parallel CFD code be organized in specific formats, it was not generally accepted by the scientific computation community even though the algorithm did provide good load balancing results. In order to make our dynamic load balancing method applicable to general parallel application programs, we developed new execution information gathering methods that eliminate the use of time stamp library while obtaining the same specific information of the application execution. The paper is organized as follows. Section2 describes the use of an MPI profiler to obtain the application specific information. Section 3 explains the method of using the PORC system tool to obtain the application specific information. Section 4 shows the method of interfacing between application code and dynamic load balancer. Section 4 demonstrates the experimental results.

2. MEASURING TIMING INFORMATION USING A MPI PROFILING LIBRARY

Since a large portion of parallel applications use MPI communication library, moving the application execution information gathering from within the application code to the MPI library will require much less restrictions to the parallel application code development and hence make the load balancing more applicable to users. In this approach, we measure the application related information in an MPI profiling library. Every time an MPI function is called by the application program, the MPI profiler records the communication time and topology in each MPI call. Therefore, the information of execution time, communication topology, and size of messages can be obtained. This approach is convenient to the parallel code developer. What the user needs to do is to link the MPI profiling library to the application code and the MPI library. Since we are collecting the same information as we did using the previously developed timing library, our previously developed load balancer and scheduler can still be used with slight modification.

Since there is not an existing MPI profiling library that provides exact information we need, we modified the FPMPI library developed by Argonne National Laboratory [9]. While using many tools provided by original FPMPI, such as, FPMPI data structure, methods of mapping Fortran MPI calls, mapping process numbers in user structured groups to MPI_COMM_WORLD, etc., we made many modifications to the FPMPI.

The original FPMPI generates an output file that has the information of average CPU time of all the processes, average communication and synchronization time for each MPI call, and communication topology. However, our dynamic load balancing algorithm requires the information of elapsed execution time and the CPU time of each parallel process, the communication time for each process, and the average communication message size between each pair of parallel processes. Therefore, the methods for constructing the original output are removed. More timers are implemented and initialized at profiler's MPI_Init() method in order to calculate the needed communication and execution times,.

Original FPMPI writes the profiling information only once right before the program stops execution. Since our load balancing algorithm needs the profiling information while the parallel application is running; our profiling library constructs the output files periodically. Each time the application makes an MPI call, the elapsed time from last writing to the output file is checked. If the elapsed time is greater than 5 minutes, new profiling measurements are written to the output file and the information accumulation restarts from zero; otherwise the profiling information is accumulated. Since different processes make MPI calls at different times, it cannot be guaranteed that the profiler associated with all parallel processes generates new output at the same 5 minute period. Our solution is that all profilers associated with parallel processes scales the measurement to the same time unit. This scaling method provides an illusion that all process writes the timing files simultaneously.

Writing the timing files periodically creates the deadlock problem in original FPMPI. The original FPMPI sends all the timing measurements to one of the parallel processes using MPI_Gather() and writes all information into one file at the end of the program execution. However, when we tried to gather this information in the middle of execution, deadlocks occurred. They occurred when a process A was waiting a message from process B, and B was in the writing timing routine. Since process A needed process B's message to continue executing and process B waited for process A in MPI_Gather() to write the measurements to the file, deadlocks were inevitable and halted the execution. In order to avoid asking the modification of the application code, we solved this problem by not using the master process to collect information from other processes but let the profiler of each parallel process writes to its own files. We also modified the dynamic load balancing tool to make it read all these files.

The method used by the original FPMPI for measuring synchronization time is inconvenient for dynamic load balancing. FPMPI uses MPI_Probe() in a while loop to wait the asynchronous information from message sender. The total time of the busy waiting is considered by FPMPI as the synchronization time. Since the while loop takes the CPU time, the waiting CPU time is counted as part of the applications computation time. The concept of busy waiting is does not allow the other user to use the CPU therefore it is undesirable in an multi-user environment. Therefore, our profiling library does not use MPI-Probe to measure synchronization time and eliminated the waiting loop. In this way the synchronization time is counted as part of total communication time. The modified MPI profiling library supports MPICH, MPICH2, and LAM.

3. MEASURE TIMING INFORMATION USING PROC

Considering that many parallel applications do not use MPI communication library, we developed another approach that uses a system tool, PROC, to collect the application related information for dynamic load balancing. PROC measures the CPU usage, memory usage, and the port usage during the execution of any computer process and stores the measured information in a predefined format in a specific directory. We developed an information gathering program that searches all needed information in PROC's output file directory located in each computer, derive needed information for dynamic load balancing, and put the search results in the same format as the output of our old time stamp library. This information gathering program is executed periodically by the load balancer. The PROC based approach has wider applicability than that of the MPI profiling based approach since it is not restricted by using MPI in the parallel applications. There are some restrictions for this approach. PROC is currently supported in Sun Solaris and Linux operating systems but not supported by many other operating systems. Besides, the information gathering program is more difficult to maintain since different versions of PROC on different computers may have different PROC file structure. This means that the information gathering program may need to be modified for each version of PROC.

4. HOW THE LOAD BALANCER COMMUNICATE WITH THE APPLICATION PROGRAM

Since the load balancer runs in parallel with the application code and provides load balancing suggestions while the parallel application is running, the parallel application program needs a way to know if there is a suggestion from the load balancer. Once the parallel application stops execution for load balancing, it is the duty of the load balancer to restart the parallel application. Therefore, the dynamic load balancer also needs to communicate with the parallel application code in order to know if the parallel application stopped due to program completion or due to the load balancing.

Assuming that all parallel programs need the check-pointing ability to take the advantage of dynamic load balancing, we developed a pair of communication library functions, check_balance() and stop_application() to facilitate the communication between the parallel application and the load balancer. These two functions are provided as part of the profiling library package when MPI profiler is used or as a separate library functions when other methods are used. The parallel program developer needs to insert these two library functions into the parallel application program.

The library function check_balance() is inserted to the application code right before where the check pointing takes place. The task of this library function is simply to read a load-balance flag in a text file generated by the dynamic load balancer. If the load-balance flag is one, there is a benefit to do load balancing so that the application code does check-pointing and stop its execution; otherwise, the application code continues its execution.

The dynamic load balancer restarts the parallel application with a new load distribution only after the application is stopped for load balancing. Since the load balancer should not restart the application code after the application completes the execution, the load balancer needs to know if the application code stopped due to normal ending of the computation or for load-redistribution. Therefore, we created another communication flag between the load balancer and the application code through a text file. Before the program starts, the load balancer creates this file and set the flag to indicate that program needs to be executed. The load balancer restart the parallel application every time the parallel application stops and finds this flag not changed. A library function, stop_application(), is inserted right before final stop of the application program. This function set the flag in the text file to indicate the application execution is stopped due to completion.

5. EXPERIMENTAL RESULTS

In order to show that the newly developed non-intrusive timing measurement methods provide the same results as our old timing library, we run a test parallel program and get the program execution information using all three methods at the same time. The test application was a simple parallel program developed in C language and using MPICH for communication. The test program uses dummy loops to simulate computation, and sends dummy messages to simulate communication. The testing program runs eight equal sized parallel processes on four Linux workstations (not evenly distributed). To make the experiment result easier to understand, the speed of all four computers were measured. The measurement is based on the execution CPU time of a benchmark program. Table 1 shows the execution CPU time (in seconds) of the benchmark program on each computer. It is clear that all workstations had approximately the same CPU speeds.

Table 1. Execution CPU time of the bench mark program on each computer.

MACHINE NUMBER	1	2	3	4
Execution time (sec)	11.65	11.61	11.61	11.61

For fair comparison of the three timing measurement methods, our old timing stamp library, new profiling library, and new PROC reader mechanism, all methods were used in the same program at the same time. The timing library routines were inserted into the application code and measurements were taken in every internal iteration of the application program. The MPI profiling library and PROC took measurements periodically (every 5 minutes) to capture information from outside of the application code. Since the measurements using timing library and other two methods were in different time interval, their results could not be compared directly. However, due to the fact that load balancing was not based on the actual measured execution times but based on the relative ratio of the actual measured execution times for all processes, we convert all the measurement into the same scale by dividing the execution CPU time and elapsed time of all parallel process by the execution CPU time of process seven. Table 2 shows the scaled execution CPU time of all processes measured by all three measurement methods as process seven runs one second. The fact that the CPU execution time off all processes are close to each other in any measurement method demonstrates that all blocks are the very similar size.

Table 3 shows the percentage differences of CPU time measurements using time stamp library as reference based on Table 2. The results had proved that our new ways of acquiring timing information are accurate. For the profiling library the biggest difference from that of the time stamp library was 0.32 %, while for PROC the difference was 0.47%. The difference was under 1% for both methods. These errors are acceptable and our new ways of acquiring timing measurements can be considered successful.

To demonstrate that the elapsed time measurement of the two new methods also agrees with our previous time stamp library measurement, we calculated the ratio of the elapsed time of every processes and CPU execution time of process 7 obtained by all three application information gathering methods (as shown in Table 4). It can be seen that the ratios for each parallel process using the three information gathering methods are very close (less than 1% difference).

Table 2. Scaled CPU time of all processes measured by all three methods

PROCESS NUMBER	0	1	2	3	4	5	6	7
MACHINE NUMBER	1	1	1	2	3	3	4	4
SCALED TIME MEASUREMENT USING MPI PROFILING LIBRARY	1.036	1.044	1.045	1.018	1.023	1.02	1.011	1
SCALED TIME MEASUREMENT USING PROC	1.033	1.041	1.041	1.018	1.020	1.023	1.010	1
SCALED TIME MEASUREMENT USING TIME STAMP LIBRARY	1.033	1.041	1.046	1.020	1.021	1.022	1.009	1

Table 3. Percentage differences of CPU timing measurements

PROCESS NUMBER	0	1	2	3	4	5	6	7
MACHINE NUMBER	1	1	1	2	3	3	4	4
% DIFFERENCE BETWEEN MEASUREMENT OF MPI PROFILING AND TIME STAMP LIBRARY	0.324	0.268	0.125	0.210	0.259	0.273	0.174	0
% DIFFERENCE BETWEEN MEASUREMENT OF PROC AND TIME STAMP LIBRARY	0.017	0.023	0.477	0.224	0.083	0.093	0.066	0

Table 4. Ratio of elapsed time of all processes measured by all three methods and the CPU time of process 7.

PROCESS NUMBER	0	1	2	3	4	5	6	7
MACHINE NUMBER	1	1	1	2	3	3	4	4
ELAPSED TIME / CPU TIME OF PROCESS 7 USING PROFILING LIBRARY	3.155	3.155	3.155	3.155	3.155	3.155	3.155	3.155
ELAPSED TIME / CPU TIME OF PROCESS 7 USING PROC	3.140	3.147	3.142	3.146	3.137	3.139	3.146	3.153
ELAPSED TIME / CPU TIME OF PROCESS 7 USING TIME STAMP LIBRARY	3.145	3.145	3.145	3.145	3.145	3.145	3.145	3.145

6. CONCLUSION

We have greatly improved the applicability of our scheduler and dynamic load balancer (SDLB) by changing the application execution information gathering method. The improvement enables the load balancing of general parallel applications that use MPI communication library or the parallel applications that runs on the computers that has PROC system tool. Therefore our SDLB can support dynamic load balancing and fault tolerance of general parallel applications on net work of heterogeneous computers running in multiple user mode.

REFERENCES

1. Condor – High Throughput Computing, http://www.cs.wisc.edu/condor/
2. An introduction to PORTABLE BATCH SYSTEM, http://hpc.sissa.it/pbs/pbs.html
3. Load Sharing Facility, http://wwwinfo.cern.ch/pdp/bis/services/lsf/
4. "MOAB Cluster Suite™," Cluster Resources Inc., 2005. http://www.clusterresources.com/products/moabclustersuite.shtml
5. The Globus project, http://www.globus.org/toolkit/
6. Y.P. Chien, A. Ecer, H.U. Akay, S. Secer, R. Blech, , "Communication Cost Estimation for Parallel CFD Using Variable Time-Stepping Algorithms," *Computer Methods in Applied Mechanics and Engineering,* Vol. 190, (2000), pp. 1379-1389.
7. S. Chien, J. Zhou, A. Ecer, and H.U. Akay, "Autonomic System for Dynamic Load Balancing of Parallel CFD," *Proceedings of Parallel Computational Fluid Dynamics 2002*, Nara, Japan (May 21-24, 2002), Elsevier Science B.V., Amsterdam, The Netherlands, pp. 185-192, 2003.
8. S. Chien, Y. Wang, A. Ecer, and H.U. Akay, "Grid Scheduler with Dynamic Load Balancing for Parallel CFD," *Proceedings of 2003 Parallel Computational Fluid Dynamics,* May 2003, Moscow, Russia, pp. 259-266.
9. Fast Profiling library for MPI The Message Passing Interface, http://www-unix.mcs.anl.gov/fpmpi/WWW/

Parallel Computational Fluid Dynamics – Parallel Computing and Its Applications
J.H. Kwon, A. Ecer, J. Periaux, N. Satofuka and P. Fox (Editors)

Reuse Procedure for Open-Source Software

Jong Bae Kim,[a] Sung Yul Rhew[b]

[a] *First affiliation, Dep. Computer Science of Soongsil University, Seoul 156-743, Korea*

[b] *Second affiliation, Dep. Computer Science of Soongsil University, Seoul 156-743, Korea*

Keywords: Open Source Software (OSS); Reuse; Development Procedures

1. Introduction

Recently, open source softwares (OSSs) are widely used in various domains through open, copy, revise, and release. The companies are trying to apply to software development approach by utilizing open source software. The OSSs are new alternatives to solve the limits of the previous software developments such as quality of software, time and cost of development. Accordingly, various aspect analyses of OSSs should be performed. However, the researches about the detailed procedures and methods to utilize OSSs in practical industry are immature [1][2][3][4][5][6].
Most of OSS-Based Applications (OBA) are similar to COTS (Commercial-Off-The-Shelf) based on development of COTS-Based Applications (CBA). That is, suitable selections of OSS or COTS, modification, integration are similar. However, detailed procedures between OBA and CBA development are different. When we develop based on OSS, we use the directly modification but CBA is not use the directly modification since COTS doesn't provide a source code. When we develop an application based on OSS, we may not apply to the traditional process models. Therefore, newly defined procedures should be required to reuse the OSSs.
In this paper, we propose the comprehensive procedures based on Meng Huang's process, component based development about OSSs. We define 4 steps and 11 activities. To derive an effectiveness and improvement for each activity, we apply proposed procedures to real projects.

2. Related Work

2.1. T.R. Madanmohan & Rahul De's Work

This work [7] proposed the selection model of open source component. This model is composed of three basic steps as collection, incubation and critical revision. This work restricted in selection of open source component. Also, this work is able to apply to overall selection of OSSs. However, detailed procedures and activities to utilize this work are required. And, this work needs to extend overall OSS development not redistricted component development.

2.2. Meng Huang' work

This paper [8] offered a development process of building open source based application that is analogous to the process of developing component based application. The proposed process emphasized the early assessment to improve the architecture stability and project manageability by assessing available OSS. However, Meng Huang' work has limitations to utilize overall procedure in application based on modification and integration of OSSs. Also, this paper described the artifacts, but relationships between artifacts are not specified.

3. Reuse Procedures of OSSs

In this section, we propose the reuse procedures of OSSs that based on Ming's research and component based development. Reuse procedures are composed of 4 steps and 11 activities. Also, these procedures define input/output artifacts for each step and activity. As shown in figure 1, 4 steps and 11 activities are defined.

Figure 1. Reuse Procedures of OSSs

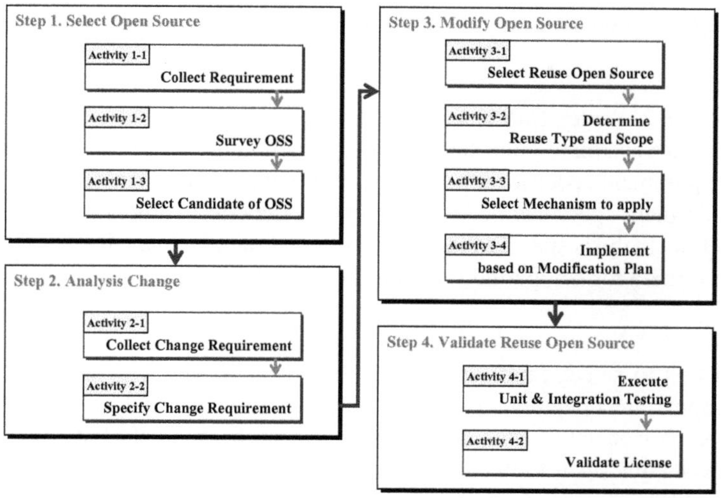

3.1. Step 1. Select Open Source

Presently, open sources are provided by many communities and sites. Therefore, step to select suitable candidates from various open sources are preceded. As shown the figure 2, this step defines three activities for identifying the correct requirements for target software and for surveying OSS, and for selecting candidate OSS.

Figure 2. Activities of Step 1

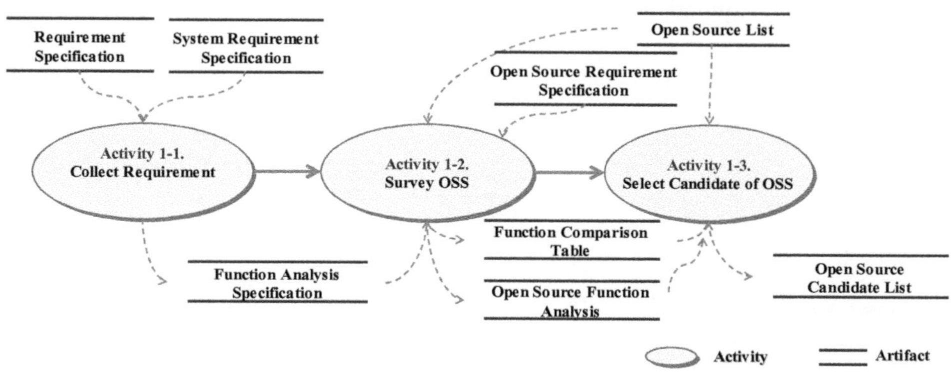

3.1.1. Activity 1-1. Collect Requirement

This activity is to analyze the functional and nonfunctional requirements of target software. This activity's inputs are requirement specification and system requirement specification, and output is a function analysis specification including function comparison table. Function analysis specification is used to survey OSS for reuse as a baseline.

3.1.2. Activity 1-2. Survey OSS

This activity surveys main open source site and community, examine reusable software based on requirement specification. This activity's inputs are open source list and open source requirement specification, and outputs are function comparison table and open source function analysis specification. Open source function analysis specification is not detailed performed for comparing functions between OSS and target software.

3.1.3. Activity 1-3. Select Candidate of OSS

This activity is to select candidate of OSS based on function comparison table. This activity's inputs are open source list, open source function analysis specification and function comparison table, and output is open source candidate list. In this activity, criteria for selecting open source candidate are established. These criteria have different selection standard by discretion of project leader. Generally, since it is difficult to find that open source softwares are satisfied with all required functions, we may determine the criteria as 70% and over by discretion of project leader. Therefore, these criteria are different from characteristics of project and discretion of project leader.

3.2. Step 2. Analysis Change

It is difficult to find fully consistency OSS satisfying requirement specification since characteristics and objectives of target software are various. Therefore, candidate OSS may use not without modification and use through modification of partially functions. To modify and reuse of candidate OSS, identifying and analyzing of change requirements are required. As shown in figure 3, these activities are defined.

Figure 3. Activities of Step 2

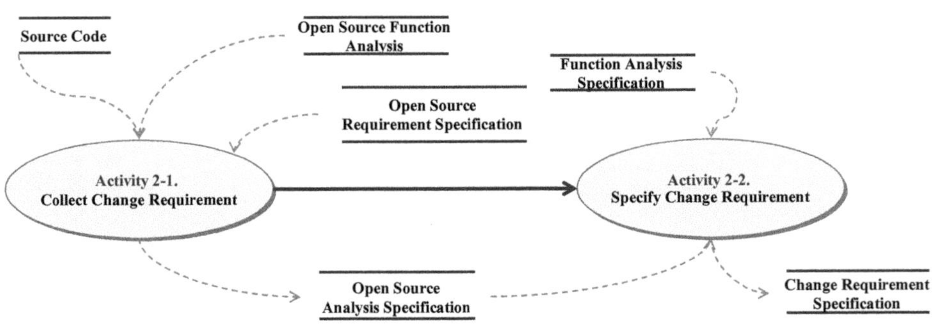

3.2.1. Activity 2-1. Collect Change Requirement

OSSs may provide the partially specification and no specification. Therefore, we may analyze the change requirement using only source codes. This work is a very not easy. In this activity, detailed information of candidate OSSs should be collected. To do this, we can reference evaluations and opinions of the OSS experienced user through communities as forums and news are provided by open source sites or homepages are operated by project teams or companies of relevant open source. This activity's inputs are open source function analysis specification, open source requirement specification and source code, output is open source analysis specification. Open source analysis specification includes the detailed information such as input value, output value, parameter type of functions.

3.2.2. Activity 2-2. Specify Change Requirement

This activity is to specify through analyzing the change requirement. Chang requirement is a specification of changeable part for reusing each candidate open source. This activity's inputs are open source analysis specification and function analysis specification, and output is a change requirement specification.

3.3. Step 3. Modify Open Source

After evaluating the candidate open source list based on various elements, selection of reusable OSS and determination of type, scope and mechanism of selected OSS are performed. Activities for modification of open source are defined in figure 4.

Figure 4. Activities of Step 3

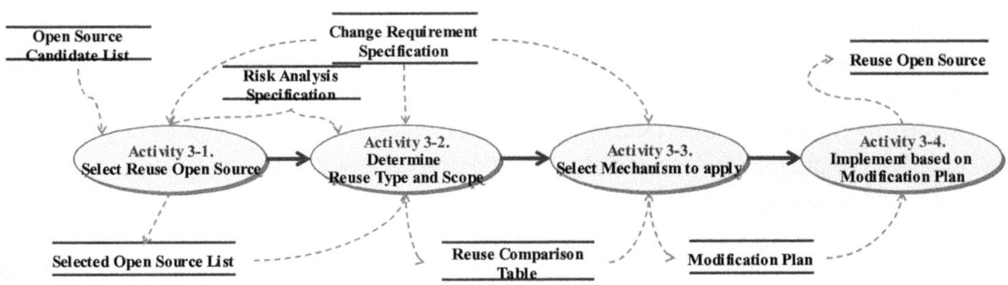

3.3.1. Activity 3-1. Select Reuse Open Source

This activity's inputs are open source candidate list, change requirement specification and risk analysis specification, and output is a selected open source list. To select an open source using reuse through change, risk and cost by changing or modifying are considerate. That is, we may select the open source having a minimum cost and maximum effectiveness after risk analysis when open source are change.

3.3.2. Activity 3-2. Determine Reuse Type and Scope

Since open source are reused by various types, reuse types should be determined. Selected reuse open source partially reused as one function or class. Therefore, scope of reusable open source is determined. This activity's inputs are risk analysis specification and change requirement specification, and output is a reuse comparison table.

3.3.3. Activity 3-3. Select Mechanism to apply

After determined reuse type and scope, mechanism for changing is also determined. This activity's inputs are change requirement specification and reuse comparison table, output is a modification plan. Reuse type of open source is a source code reuse, framework reuse, and component reuse [9]. According to this type, detailed activities are divided. However, we don't describe the detailed activities following reuse type.

- **Source code reuse**. Source code reuse is the simplest case of reuse possible, because it doesn't use any particular mechanism. However, source code reuse use when part of program is a useful and but not individually packaged. In this case, code reuse needs to suitable change from copied a part of code. However, one advantage acquired from component reuse such as maintenance of reused coded is missed. Therefore, only a little source code has to change, reference code instead of code copy, use inheritance mechanism for modification.
- **Component reuse**. Since COTS component provides only a binary code and specification, change of code are not performed. However, since inner code of open source is known, direct change of source code may perform. Hence, because this modification makes low an effectiveness of component, modification of component is restricted performed. Therefore, customization or connector without directly modification is used. Customization [11] uses to change the attribute and workflow of open source component. Attribute of open source component is changed by getter or setter function of interface. When workflow is changed by interface, class name of workflow transmit to interface of open source component. Connector [11][12] uses to solve the mismatch such as association or dependency between existing component and open source component The open source component requests a method invocation on the existing component through the connector. The request is delivered to the connector through port, and the connector performs a designated role such as data transformation or interface mediation, and transmits a modified request to the target component through port.
- **Framework reuse**. Framework is to allow its users to reuse an existing structure providing different functionalities or coding advices across different programs. It can be considered a kind of software reuse, similar to components reuse, but bringing different functionalities together in a single framework.

3.3.4. Activity 3-4. Implement based on Modification Plan

This activity's input is a modification plan and output is a reuse open source. In this activity, real modification of open source performs with modification plan.

3.4. Step 4. Validate Reuse Open Source

This activity is to execute testing to validate not only unit testing caused by changing open source but also integration testing to validate whether or not has a problems when integrate between modified reuse open sources and existed sources. This activity defines executing unit and integration testing and validating licenses.

3.4.1. Activity 4-1. Execute Unit & Integration Testing

This activity validates elements such as testing of modified source code, integration testing between existing code and modified code, and integration testing between reuse open sources.

3.4.2. Activity 4-2. Validate License

To utilize the open source, we have to need a compliance of license. After development and test are finished, the developed final artifact as source code is need to validating. In spite of development plan with license issue, it is possible to using in development step without validating license.

4. Case Study

In this section, we applied the proposed procedures to project as supporting tool for small scale software development methodology based on OSS. First, we identified functions from the requirement for surveying reusable open source. The main functionalities of supporting tool domain are following as the artifact management, schedule management, resource management, etc. The total number of functions described in this domain was 30. Also, identified non-functional were 10. Next, we specified the function analysis table to survey suitable OSS for reuse. We surveyed the well-known open source site as http://sourceforge.net using keywords as functions through function analysis table. Surveyed list is categories including ***Project Management***, ***Groupware***, and ***Community*** [13][14]. Open source sites provided the following information as briefly function descriptions, Project Details, Development Status, License, Registered, Activity Percentile, Operating System, Programming Language, and User Interface. As shown in table 1 is a function comparison table.

Table 1. Function Comparison Table

No.	Requirements	MetisProject	dotProject	…	Weights
1	Project Life Cycle	0.5	0.8	…	60
2	Note	0.4	0.7	…	15
3	BookMark	0.5	0.6	…	15
4	BugTracking	0.5	0.6	…	15
5	help requests	0.6	0.6	…	15
6	Doc Management	0.3	0.8	…	40
…	…	…	…	…	…
Sub Total		104	150	…	200

Through colleting and specifying the detailed information of candidate open source, we specified the change requirement specification of each candidate open source. The table 2 is an example of PM01's *change requirement specification.*

Table 2. Change Requirement Specification of PM01

Function		Agreement/Disagreement	Change Requirement
PM01	F01.Artifact Management	Disagreement of Parameter	Parameter Coherence
	F02.Schedule Management	Disagreement of Operation	Operation Modification
	F03.Resource Management	Disagreement of Data	Data Coherence
	……	None	None

To examine numerous source codes for eliciting information from OSS only having source code without specification has a limitation. Therefore, it is required to using re-engineering tool, which elicit from source code to specification information. Also, we selected 3 reusable open sources among open source candidate list based on risk and cost caused by change. We determined that PM01 as project management function integrated with GW02 as module management, schedule management, web mail function. To do this, we selected reuse type and scope through comparison of reuse possibility about selected open source. The table 3 is an example for comparison table of PM01.

Table 3. Comparison Table of PM01

Function		Reuse Type	Reuse Possibility
PM01	F01.Artifact Management	Source Code	O
	F02.Schedule Management	Source Code	X
	F03.Resource Management	Framework	O
	...	Component	O

Selected open source was not satisfied to requirement of target software. Therefore, we performed design and implementation through analyzing architecture and source code. Also, we defined interfaces of user and schedule management, and integrated new function and revised source code.
We experienced the continuously change and extend during modification of open source. Therefore, we need to determine change of modified project and complement of preliminary procedures when design and develop without repetition of the existing activities.

5. Conclusion

OSS based software development is a differ from general software reuse in some point selecting open source, collecting change requirement, determining reuse type and scope, and validating license.
Our paper proposed the 4 steps and 11 activities for software development procedures to utilize OSSs. By applying the proposed procedures, we reduced the development time to market.
However, we may study the metric for correctly and objective to select the reuse open source and need the detailed definition of procedure and mechanism according to reuse type. In the future work, we will perform the study on additional researches for more detailed and more practical reuse process development, and applying to reuse process for OSSs.

References

1. Research, FLOSS(Free/Libre and Open Source Software: Survey and Study), FINAL REPORT, 2002.
2. MICHAEL J. KARELS, "Commercializing Open Source Software", ACM Queue, vol.1, no.5, 2003.
3. Feller, J. and Fitzgerald, B., "A Framework Analysis of the Open Source Software Development Paradigm", Proceedings of the twenty first international conference on Information systems table of contents, pp.58-69, 2000.
4. James W. Paulson, Giancarlo Succi, Armin Eberlein, "An Empirical Study of Open-Source and Closed-Source Software Products", IEEE TRANSACTIONS ON SOFTWARE ENGINEERING, Vol.30, No.4, 2004.
5. Michel Ruffin and Christof Ebert, "Using Open Source Software in Product Development: A Primer", IEEE Software, vol.21, no.1, pp.82-86, 2004.

6. Diomidis Spinellis and Clemens Szyperski, “How is Open Source affecting software development?”, IEEE Software, Vol. 21-1, pp. 28–33, 2004.
7. T.R. Madanmohan and Rahul De’, "Open Source Reuse in Commercial Firms", IEEE Software, Vol. 21, No. 6. pp. 62-69, 2004.
8. Meng Huang, Liguang Yang, and Ye Yang, "A Development Process for Building OSS-Based Applications", SPW 2005, LNCS 3840, pp.122–135, 2005.
9. ODETTE Project, Software Reuse in Free software: State-of-the-art, 2005.
10. Keepence, B., and Mannion, M., “Using patterns to model variability in product families,” IEEE Software, Vol. 16, Issue. 4, pp. 102-108, July-Aug., 1999.
11. Clements, P., et al., Documenting Software Architectures: Views and Beyonds, Addison-Wesley, pp. 112-113, 2003.
12. D’Souza, D. F., and Wills, A. C., Objects, Components, and Frameworks with UML, Addison-Wesley, pp. 477-478, 1999.
13. Sourcefoge.net, http://www.sourceforge.net
14. Freshmeat.net, http://www.freshmeat.net

Parallel Computational Fluid Dynamics – Parallel Computing and Its Applications
J.H. Kwon, A. Ecer, J. Periaux, N. Satofuka and P. Fox (Editors)

Design of CFD Problem Solving Environment based on Cactus Framework

Soon-Heum Ko[a], Kum Won Cho[b], Chongam Kim[a], Jeong-su Na[b]

[a] *Department of Mechanical and Aerospace Engineering,*

Seoul National University,

San 56-1 Shillim-Dong, Kwanak-Gu, Seoul, 151-742 Republic of Korea

E-mail: {floydfan,chongam}@snu.ac.kr

[b] *Department of Supercomputing Applications,*

Korea Institute of Science and Technology Information(KISTI),

52 Eoeun-Dong, Yuseong-Gu, Daejeon, 305-333 Republic of Korea

E-mail: {ckw,ninteger}@kisti.re.kr

Key Words: Cactus; PSE(Problem Solving Environment); Grid Computing

1. INTRODUCTION

Accuracy and efficiency are two key issues to CFD(Computational Fluid Dynamics) researches. Inaccurate solution will bring about the wrong resultant, while inefficiency will result in the failure of manufacturing within demanded time. Researchers usually increase mesh size, conduct turbulent analysis, increase the order of spatial accuracy and tune turbulent parameters for specific applications in order to get an accurate solution. But, all these techniques except parameter tuning are inevitably the enemy to efficiency. On the other hand, faster and inaccurate solution requires increasing safety factor to satisfy the safety standard and it increases manufacturing cost.

Apparently, parallel and Grid computing can be the answer to this conflict. More computing power guarantees faster solution. Besides, a group of moderate processors is far cheaper than single processor with the same capacity in total. The only difficulty is that progress of computer technology enforces application scientists to have additional knowledge on computing techniques. For example, researchers have to include MPI library[1] on their application program for parallelization. And they need to be well aware of many additional softwares like Globus[2] to conduct Grid computing.

Nowadays, many computer scientists are trying to reduce the above difficulty by supporting easy computing environment. This kind of specialized computational convenience is called PSE(Problem Solving Environment) and Nimrod/G[3], Triana[4], and Cactus[5] are the examples. Of these PSEs, Cactus was firstly developed for supporting large-scale astrophysics simulations[6,7], but it is applicable to other applications like CFD, quantum relativity, chemical reaction and EHD(Electro-Hydro-Dynamics)[8,9].

Therefore, present research focuses on supporting CFD research environment on the basis of Cactus framework. Some of computational modules have been improved and some are newly added for CFD simulation through the collaboration among computer scientists and application researchers. And, compressible / incompressible flow solvers are modularized and implemented into the current Cactus framework. In addition, a new driver supporting unstructured mesh system is developed and validated within the range of current Cactus structure.

2. CACTUS FRAMEWORK

Cactus is an open source PSE designed for scientists and engineers. Its modular structure easily enables parallel computation across different architectures and collaborative code development among different groups. As can be inferred from its name, Cactus is composed of a central core (flesh) and application modules (thorns) which are connected to flesh through an extensible interface. Through the continuous researches of Cactus framework, various computational toolkits including parallel I/O, data distribution, or checkpointing are standardized. So, application researchers only need to implement their application solver in the form of a thorn (Figure 1).

As is shown in Figure 2, Cactus operates by connecting the flesh, thorns, drivers and other linkable softwares. Among the modules, flesh arranges application thorns, drivers and linkable softwares and interacts with all the others. Thorns are the modules that are applicable to specific researches. Driver supports the Grid service, parallel processing including automatic domain partitioning or communication across inter-processor boundary. Finally, users can link Globus, PETSc and other softwares to the existing Cactus framework.

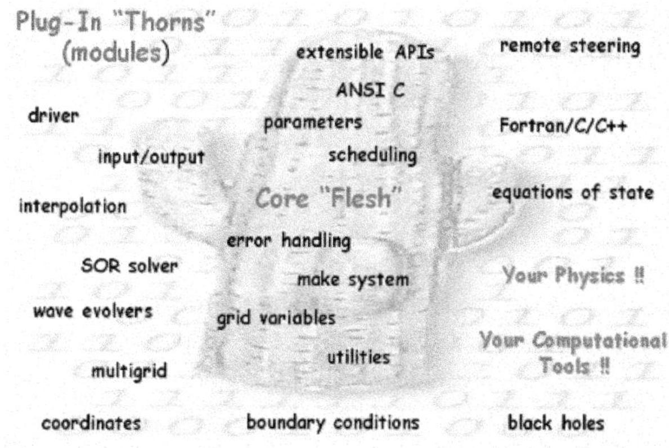

Figure 1. Cactus Flesh & Thorns

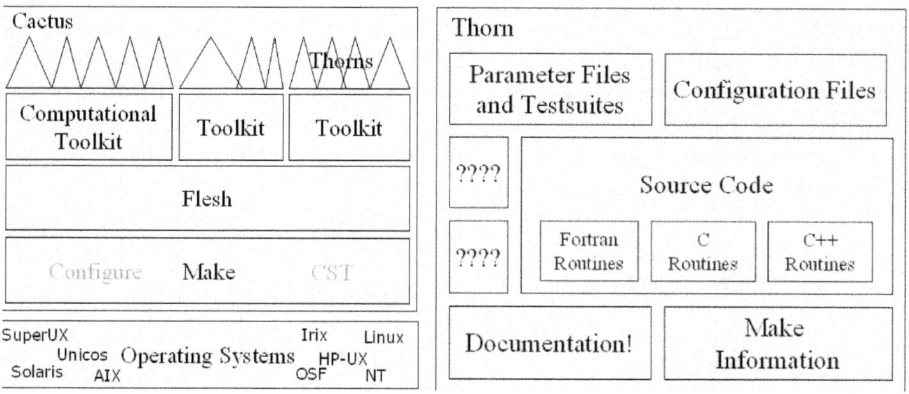

Figure 2. Structure of Cactus(L) / Structure of Thorn(R)

Cactus has the following advantages. Firstly, users do not have to learn a new programming language and convert their application solver to a specific language since Cactus supports multiple languages including Fortran and C. Secondly, Cactus runs on a wide range of architectures and operating systems. And, Cactus automatically performs parallel programming, so users do not need to have additional knowledge of parallel processing. Additionally, as Cactus supports object-oriented environment, researchers can work with collaborators on the same code and avoid having your programs fragmented. Finally, users can make use of the latest software technologies, like the fastest way of transmitting simulated data and the optimal visualization technique, by adopting associated modules from supported Cactus thorns.

On the other hand, Cactus framework is not sufficient as a problem solving environment for CFD researches currently. First of all, Cactus is not an easy framework for application scientists. Though it gives monitoring/steering as well as visualization on the web, all of the research processes including programming and scheduling of the modules are conducted on the text interface, not the graphic interface. Secondly, because of its general-purpose, Cactus cannot give the optimized research environment for specific application. As Cactus framework gives the base functions for the applications, users need to convert supported functions to the optimized form of their own researches.

3. NUMERICAL RESULTS

3.1. Compressible Flow Analysis Using Cactus Framework

Compressible CFD toolkit on Cactus framework has been developed since 2003 and the details of previous works are presented at Refs. 8 and 10. Comparing with previous work, 2-equation k-ω SST model[11] is adopted to existing flow solver by collaboration with Cactus main developers at 'Center for Computation and Technology' and 'Albert Einstein Institute'.

The flowfield around a wing is analyzed using Cactus-based CFD analyzer. As the first example, RAE-2822 airfoil is analyzed to show the possibility of CFD simulations on body-fitted coordinates. The present configuration is basically 2-dimensional, but the mesh is extended along the z-(spanwise) direction, making a 3-D mesh with symmetric boundary condition along the z-direction. Mach number is 0.729, angle of attack is 2.79 degrees and Reynolds number of 1.6972×10^5. As is shown from the results in Figure 3, the present CFD toolkit can accurately analyze the flowfield around the body.

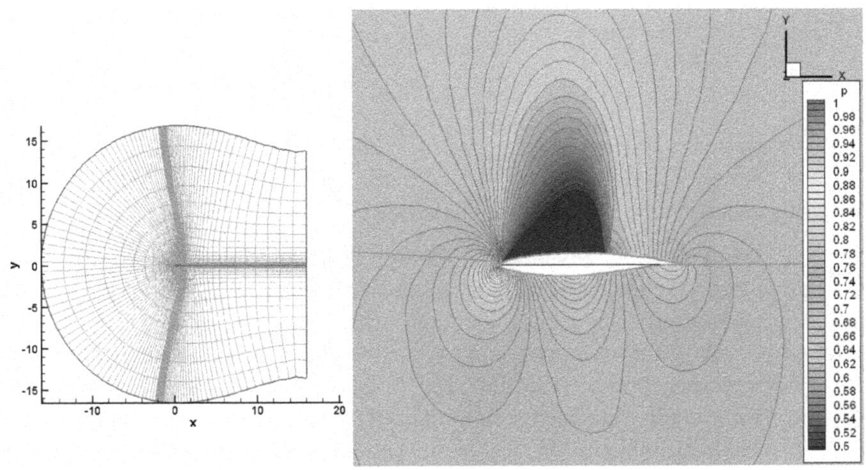

Figure 3. Mesh System(L) and Pressure Contour(R) of RAE2822 Airfoil

3.2. Development of Incompressible CFD Toolkit

A major objective of incompressible toolkit development is to develop a CFD educational system based on Cactus framework. Thus, an incompressible solver with FVM(Finite Volume Method) as well as additional boundary condition modules are implemented on the Cactus system. Currently, developed CFD toolkit can analyze the problems with rectangular mesh system such as flat plate flow, channel flow and cavity-driven flow. However, the current toolkit can be easily expanded for problems with complex geometries by adopting coordinate transformation routine which is already developed from compressible CFD researches.

Giverning equations are the three-dimensional incompressible Navier-Stokes equations. Current study adopted the method of artificial compressibility method to guarantee mass conservation in incompressible flow solver. For spatial discretization, Roe's flux scheme[12] is adopted and Euler explicit time integration is applied. For pressure correction by artificial compressibility, the artificial speed of sound is set to be 3.

The solution procedure of the developed CFD module is shown in Figure 4. At first, mesh reading is conducted. Flow initialization and coordinate transformation routines are called at initial step. After time scale is determined at pre-evolution level, the iterative process of flux calculation, time integration, pseudo-compressibility analysis along with velocity and pressure boundary conditions are conducted at evolution step. Finally, resultant data are visualized and analyzed. Those Routines are built in the 'CFDIncomp' thorn. And some modules commonly used for both compressible and incompressible simulations are adopted from compressible analysis toolkit.

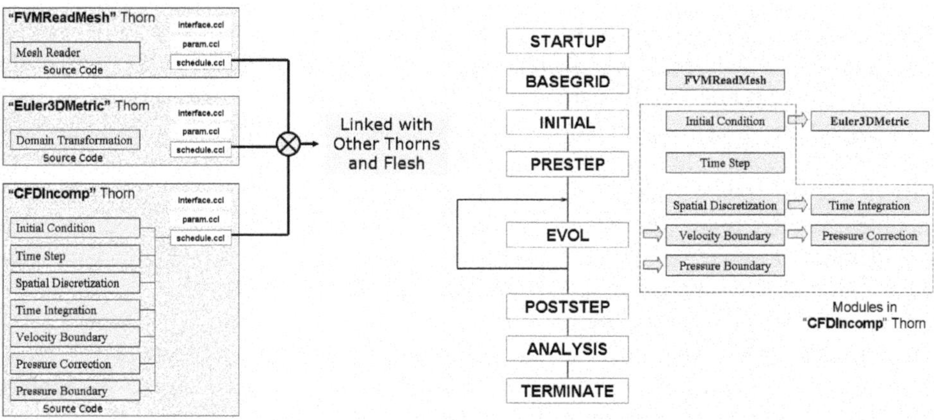

Figure 4. Structure of Incompressible CFD Toolkit and Scheduling Bins

As the applications, channel flow and cavity flow are simulated. Reynolds number is 9.97 and all the other properties are the same as water at 20℃. Velocity at the free stream is set to be 1.0. From the result of channel flow analysis, the velocity profile at the exit is the same as analytical solution, $U(y) = 6.0 \times U_0 \times y(1.0 - y)$. And cavity flow analysis shows flow rotation by viscosity very well. Core of rotating flow is positioned in (0.52, 0.765). Since the Reynolds number is sufficiently low, it does not shows secondary circulation at the corner.

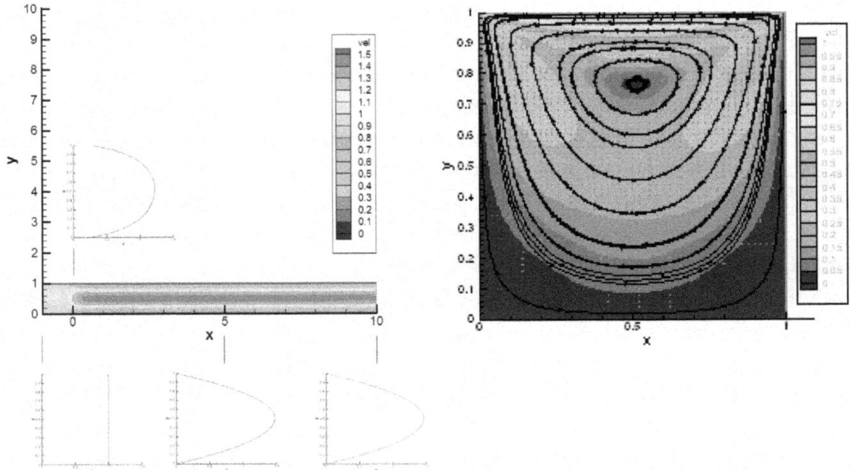

Figure 5. Flow Analysis in the Channel(L) and Cavity(R)

3.3. Cactus Simulations on Unstructured System

Up to now, two-dimensional unstructured driver which supports both FEM(Finite Element Method) and FVM techniques has been developed. For an unstructured mesh support, unstructured driver defines the data types of vertex, edge and face and declares the relations among those domains. For the implementation of unstructured driver, current Cactus flesh has been modified and, unstructured mesh reader and visualization module are newly developed. A solver for two-dimensional wave equation on unstructured domain is attached on current system for validation. Currently, unstructured driver supports serial processing only.

For the analysis of wave equation on unstructured system, a thorn ‘FVMWaveToy’ is developed as well as flesh and driver modules. In the ‘FVMWaveToy’ thorn, unstructured data types are called and the relations among those data types are adopted from an unstructured driver, named ‘UMDriver’. When program starts, unstructured driver calls unstructured mesh reader and gets vertex position and the relation between faces and vertices from mesh file. Then, driver automatically arranges the edges and judges the relation among those domains. From the additional supports of edge length and face area from unstructured driver, application thorn can finally start an analysis. The application thorn is composed of initial condition, application, boundary condition and visualization functions. Currently, boundary condition module can impose Neumann and Derichlet boundary conditions and, output data from visualization function is formatted for GNUPlot software. In near future, those routines are to be separated from the application thorn.

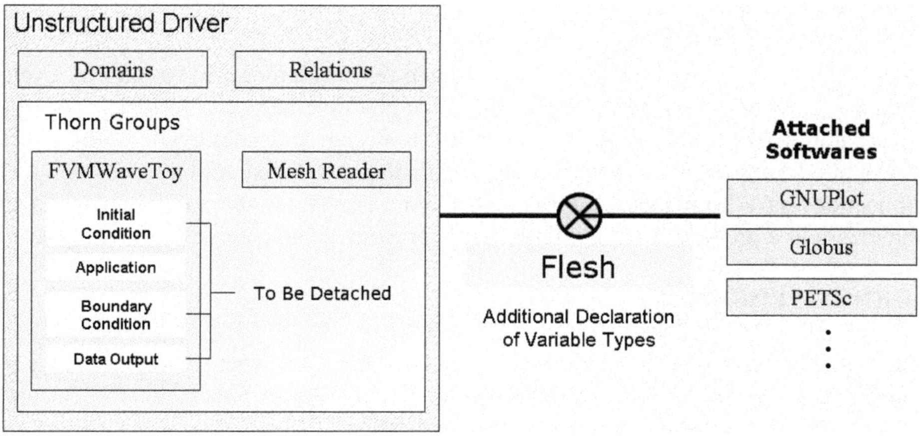

Figure 6. Improvement of Cactus Framework for Unstructured Approach

As a validation problem, the propagation of a planar wave is simulated by using both structured and unstructured Cactus drivers. For comparison, unstructured mesh has the same computational domain as structured mesh. By splitting each rectangular cell in structured mesh into two triangles, unstructured mesh could be generated. As initial condition, two periods of sine-function wave is located in the diagonal direction of computational domain. The amplitude of the wave is 1.0 and length is 0.1. Simulations of wave propagation on structured and unstructured mesh system are shown in Figures 7 and 8. As shown in the results, two analyses show the same wave propagation and reflection pattern. In both cases, all the boundaries are set to be wall. In unstructured analysis, more numerical dissipation is observed and it is due to the inaccuracy of interpolated data at each vertex. Anyway, both the results show a good agreement with each other and are physically acceptable.

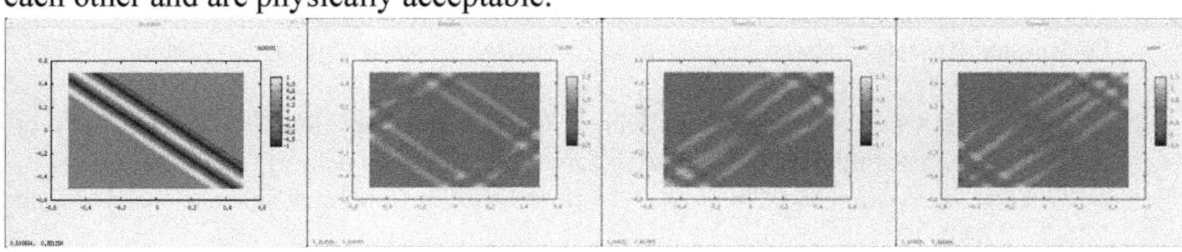

Figure 7. Structured Simulation at t=0.0, 0.2, 0.4 and 0.6

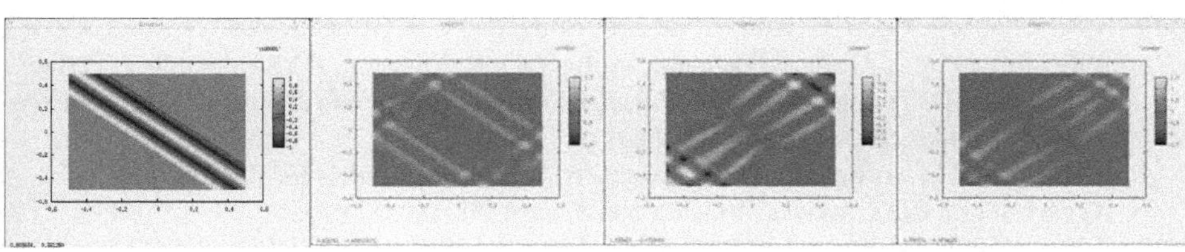

Figure 8. Unstructured Simulation at t=0.0, 0.2, 0.4 and 0.6

4. CONCLUSION

International and multi-disciplinary collaboration for developing CFD toolkit on Cactus framework has been conducted. Up to now, compressible and incompressible modules are separately implemented on structured mesh system and additional researches for unstructured support are accomplished. Using the developed toolkits, various CFD applications have been successfully simulated and they show the possibility of Cactus framework as a problem solving environment for general and complex CFD analyses.

ACKNOWLEDGEMENT

The authors would like to appreciate the support from the Korea National e-Science project. And the first and third authors also would like to appreciate the financial support from BK(Brain Korea)-21 research program.

References

1. http://www-unix.mcs.anl.gov/mpi/
2. http://www.globus.org/
3. D. Abramson, K. Power, L. Kolter, "High performance parametric modelling with Nimrod/G: A killer application for the global Grid," *Proceedings of the International Parallel and Distributed Processing Symposium*, pp. 520–528, 2000
4. http://www.triana.co.uk/
5. http://www.cactuscode.org/
6. G. Allen, D. Angulo, I. Foster, G. Lanfermann, C. Liu, T. Radke, E. Seidel, J. Shalf, "The Cactus Worm: Experiments with Dynamic Resource Discovery and Allocation in a Grid Environment," *International Journal of High Performance Computing Applications*, Vol.15, No.4, pp. 345-358, 2001
7. R. Bondarescu, G. Allen, G. Daues, I. Kelley, M. Russell, E. Seidel, J. Shalf, M. Tobias, "The Astrophysics Simulation Collaboratory Portal: A Framework for Effective Distributed Research", *Future Generation Computer Systems*, Vol.21, No.2, pp. 259-270, 2005
8. S.H. Ko, K.W. Cho, Y.D. Song, Y.G. Kim, J. Na, C. Kim, "Development of Cactus Driver for CFD Analyses in the Grid Computing Environment", *Advances in Grid Computing - EGC 2005: European Grid Conference*, Lecture Notes in Computer Science, Vol.3470, pp. 771-777, 2005
9. K. Camarda, Y. He, K. Bishop, "A Parallel Chemical Reaction Simulation Using Cactus," *Linux Clusters: The HPC Revolution*, 2001
10. K.W. Cho, S.H. Ko, Y.G. Kim, J. Na, Y.D. Song, and C. Kim, "CFD Analyses on Cactus PSE," Edited by A. Deane, G. Brenner, A. Ecer, D. R. Emerson, J. McDonough, J. Periaux, N. Satofuka, D. Tromeur-Dervout : *Parallel Computational Fluid Dynamics 2005: Theory and Applications*, 2006 (To be published)
11. F. R. Menter, "Two-Equation Eddy-Viscosity Turbulence Models for Engineering Applications," *AIAA Journal*, Vol. 32, No. 8, pp. 1598-1605, 1994
12. J. Tannehill, D. Anderson and R. Pletcher, *Computational Fluid Mechanics and Heat Transfer*, Second Edition, pp.388~398, 1997

Parallel Computational Fluid Dynamics – Parallel Computing and Its Applications
J.H. Kwon, A. Ecer, J. Periaux, N. Satofuka and P. Fox (Editors)

Prediction of secondary flow structure in turbulent Couette-Poiseuille flows inside a square duct

Wei Lo[a] and Chao-An Lin[a*]

[a]Department of Power Mechanical Engineering, National Tsing Hua University, Hsinchu 300, TAIWAN

Turbulent Couette-Poiseuille flows inside a square duct at bulk Reynolds number 9700 are investigated using the Large Eddy Simulation technique. Suppression of turbulence intensities and a tendency towards rod-like axi-symmetric turbulence state at the wall-bisector near the moving wall are identified. The turbulence generated secondary flow is modified by the presence of the top moving wall, where the symmetric vortex pattern vanishes. The angle between the two top vortices is found to correlate with the ratio of moving wall velocity to duct bulk velocity.

1. Introduction

The turbulent Poiseuille or Couette-Poiseuille flows inside a square or rectangular cross-sectional duct are of considerable engineering interest because their relevance to the compact heat exchangers and gas turbine cooling systems. The most studied flow is the turbulent Poiseuille type inside a square duct and is characterized by the existence of secondary flow of Prandtl's [1] second kind which is not observed in circular ducts nor in laminar rectangular ducts. The secondary flow is a mean circulatory motion perpendicular to the streamwise direction driven by the turbulence. Although weak in magnitude (only a few percent of the streamwise bulk velocity), secondary flow is very significant on the momentum and hear transfer.

Fukushima and Kasagi [2] studied turbulent Poiseuille flow through square and diamond ducts and found different vortex pairs between the obtuse-angle and acute-angle corners in the diamond duct. A pair of larger counter-rotating vortices presented near the corners of acute-angle with their centers further away from the corner. Heating at the wall in a square duct was found to have evident effect on the secondary flow structure as well ([3] [4] [5]). Salinas-Vázquez and Métais [5] simulated a turbulent Poiseuille flow in a square duct with one high temperature wall and observed that the size and intensity of secondary flow increase in tandem with the heating. At the corner with heated and non-heated walls, there exhibited a non-equalized vortex pair where vortex near the heated wall is much larger than the other. The combined effects of the bounding wall geometry and heating on the secondary flow in a non-circular duct is studied by Salinas-Vázquez et al. [6]. By computing the heated turbulent flow through a square duct with one ridged wall, new

*Email address: calin@pme.nthu.edu.tw

secondary flows were revealed near the ridges. Rotation around an axis perpendicular to the streamwise flow direction was also found to greatly modify the secondary flow structure in a LES calculation conducted by Pallares and Davidson [7]. The above investigations have implied that, with careful manipulation, the secondary flow is very much promising on enhancement of particle transport or single phase heat transfer in different industrial devices. They also demonstrated that the turbulence generated secondary flow in non-circular ducts could be modified by bounding wall geometry, heating and system rotation.

However, little is known about the effect of the moving wall on the secondary flow structure. Lo and Lin [8] investigated the Couette-Poiseuille flow with emphasis on the mechanism of the secondary flow vorticity transport within the square duct. However, the relationship between the secondary flow structure and the moving wall is not addressed. In the present study, focus is directed to the influences of the moving wall on the secondary flow pattern and hence turbulence structure.

2. Governing Equations and Modeling

The governing equations are grid-filtered, incompressible continuity and Navier-Stokes equations. In the present study, the Smagorinsky model[9] has been used for the sub-grid stress(SGS).

$$\tau_{ij}^{s} = -(C_s\Delta)^2\frac{1}{\sqrt{2}}\sqrt{(S_{kl}S_{kl})}S_{ij} + \frac{2}{3}\rho k_{sgs}\delta_{ij} \tag{1}$$

where $C_s = 0.1, S_{ij} = \frac{\partial \overline{u}_i}{\partial x_j} + \frac{\partial \overline{u}_j}{\partial x_i}$, and $\Delta = (\Delta x \Delta y \Delta z)^{1/3}$ is the length scale. It can be seen that in the present study the mesh size is used as the filtering operator. A Van Driest damping function accounts for the effect of the wall on sub-grid scales is adopted here and takes the form as, $l_m = \kappa y[1 - exp(-\frac{y^+}{25})]$, where y is the distance to the wall and the length scale is redefined as, $\Delta = Min[l_m, (\Delta x \Delta y \Delta z)^{1/3}]$. Although other models which employed dynamic procedures on determining the Smagorinsky constant (C_s) might be more general and rigorous, the Smagorinsky model is computationally cheaper among other eddy viscosity type LES models. Investigations carried out by Breuer and Rodi [10] on the turbulent Poiseuille flow through a straight and bent square duct have indicated that, the difference between the predicted turbulence statistics using dynamic models and Smagorinsky model was only negligible.

3. Numerical and parallel Algorithms

A semi-implicit, fractional step method proposed by Choi and Moin [11] and the finite volume method are employed to solve the filtered incompressible Navier-Stokes equations. Spatial derivatives are approximated using second-order central difference schemes. The non-linear terms are advanced with the Adams-Bashfoth scheme in time, whereas the Crank-Nicholson scheme is adopted for the diffusion terms. The discretized algebraic equations from momentum equations are solved by the preconditioned Conjugate Gradient solver . In each time step a Poisson equation is solved to obtain a divergence free velocity field. Because the grid spacing is uniform in the streamwise direction, together with the adoption of the periodic boundary conditions, Fourier transform can be used to reduce the 3-D Poisson equation to uncoupled 2-D algebraic equations. The algebraic equations are solved by the direct solver using LU decomposition. In all the cases considered here the grid size employed is (128x128x96) in the spanwise, normal, and streamwise direction, respectively.

In the present parallel implementation, the single program multiple data (SPMD) environment is adopted. The domain decomposition is done on the last dimension of the three dimensional computation domain due to the explicit numerical treatment on that direction. The simulation is conducted on the HP Integrity rx2600 server (192 Nodes) with about 80 percent efficiency when 48 CPUs are employed in the computation shown in Figure 2. Linear speed-up is not reached in present parallel implementation mainly due to the global data movement required by the Fast Fourier Transform in the homogenous direction.

4. Results

Schematic picture of the flows simulated is shown in Figure 1, where D is the duct width. We consider fully developed, incompressible turbulent Couette-Poiseuille flows inside a square duct where the basic flow parameters are summarized in Table 1. Reynolds number based on the bulk velocity (Re_{bulk}) is kept around 9700 for all cases simulated and the importance of Couette effect in this combined flow field can be indicated by the ratio $r = (W_w/W_{Bulk})$ and $-\frac{D}{\rho W_w^2}\frac{\partial P}{\partial z}$. Due to the lack of benchmark data of the flow filed calculated here, the simulation procedures were first validated by simulating a turbulent Poiseuille flow at a comparable bulk Reynolds number (case P). The obtained results (see Lo [12]) exhibit reasonable agreement with DNS data from Huser and Biringen [13].

4.1. Mean secondary flow structure

Streamlines of mean secondary flow are shown in Figure 3. The vortex structure is clearly visible, where solid and dashed lines represents counter-clockwise and clockwise rotation, respectively. The two clockwise rotating vortices are observed to gradually merge in tandem with moving wall velocity. Focus is directed to the top vortex pair which consists of a small and a larger vortex. Examination of the vortex cores by the streamline contours indicates that the distance between the vortex cores of vortex pair is approximately constant. Therefore, the angle formed by the horizonal x axis and the line joining the two vortex cores might become a good representation of the relative vortex positions. This angle is calculated and plotted against the parameter r defined by W_w/W_{bulk} which can be interpreted as the non-dimensional moving wall velocity. It is interesting to note that a linear relation exits between the angle and the parameter r, as shown in Figure 4. The least square fit of the simulation data reveals a relation of $\Theta = -13.56 + 38.26r$.

4.2. Turbulence intensities

Figures 5(a)-(c) show the cross-plane distributions of the resolved turbulence kinetic energy. The distribution shows locally maximum near the stationary wall. Near the moving wall, on the other hand, turbulence level is greatly reduced by the insufficient mean shear rate which is caused by the moving wall. The distribution of the turbulence kinetic energy near the top and bottom corners are highly related to the presence of mean secondary flow which has been revealed in Figure 3. Near the bottom corner high turbulence kinetic energy is been convected into the corner. Near the top corner, on the other hand, high turbulence kinetic energy generated near the side wall has been convected into the core region of the duct horizontally. The predicted normal Reynolds stresses and primary shear stress profiles normalized by the plane average friction velocity at bottom wall along wall-bisector ($x/D = 0.5$) are given in Figure 5(d). Here, the DNS data of plane Couette-Poiseuille (r=1.16, Kuroda et al. [14]) flows and plane channel flows (Moser et al. [15]) are also included for comparisons. The r value of the plane Couette-Poiseuille flow is close to 1.17 of case CP3. Near the stationary wall, it is not surprising to observe that predicted turbulence intensities are similar of all cases considered, indicating that the influence of moving wall is not significant. On the contrary, near the top moving wall, gradual reductions of turbulence are predicted in tandem with the increase of the moving wall velocity. The profiles in the plane Couette-Poiseuille flow of Kuroda et al. [14] also show this differential levels and agree with present results for case CP2 and CP3, where the operating levels of r are similar. Near the stationary wall ($y/D < 0.05$), $< v'w' >$ appears to be unaffected by the moving wall. The gradients of the $< v'w' >$ of case CPs in the core region are less than case P and the position of $< v'w' >= 0$ are shifted toward the moving wall. The same behaviors were found in the experimental study of Thurlow and Klewicki [16] and they concluded that the position of zero Reynolds shear stress would shift to the lower stress wall if the wall stresses in a channel are not symmetric. The shifting of the zero Reynolds stress location implies that there are stronger vertical interactions of fluctuations across the duct center.

4.3. Anisotropy invariant map

The above discussion has revealed that the turbulence inside the turbulent Couette-Poiseuille duct flows is very anisotropic. The anisotropy invariant map (AIM) is introduced here in order to provide more specific description of the turbulence structures. The invariants of the Reynolds stress tensors are defined as $II = -(1/2)b_{ij}b_{ji}$, $III = (1/3)b_{ij}b_{jk}b_{ki}$ and $b_{ij} =< u_i'u_j' > / < u_k'u_k' > -1/3\delta_{ij}$. A cross-plot of $-II$ versus III forms the anisotropy invariant map (AIM). All realizable Reynolds stress invariants must lie within the Lumley triangle [17]. This region is bounded by three lines, namely two component state, $-II = 3(III{+}1/27)$, and two axi-symmetric states, $III = \pm(-II/3)^{3/2}$. For the axi-symmetric states, Lee and Reynolds [18] described the positive and negative III as disk-like and rod-like turbulence, respectively. The intersections of the bounding lines represent the isotropic, one-component and two-component axi-symmetric states of turbulence. An demonstration of the Lumley triangle [17] is given in Figure 6.

The AIM at two horizonal locations across the duct height is presented in Figure 7. Near the stationary wall ($y/D \leq 0.5$), turbulence behaviors of different Couette-Poiseuille flows resemble those of the Poiseuille flow. In particular, along x/D=0.5, the turbulence structure is similar to the plane channel flow, where turbulence approaches two-component state near the stationary wall due to the highly suppressed wall-normal velocity fluctuation (Moser et al. [15]). It moves toward the one-component state till $y^+ \sim 8$ (Salinas-Vázquez and Métais [5]) and then follows the positive III axi-symmetric branch (disk-like turbulence, Lee and Reynolds [18]) to the isotropic state at the duct center. However, near the moving wall due to the reduction of the streamwise velocity fluctuation at the moving wall, turbulence structure gradually moves towards a rod-like axi-symmetric turbulence (negative III) as r increases. It should be noted that for case CP2 and CP3, turbulence close to the top wall at x/D=0.5 has reached the two-component axi-symmetric state. The invariant variations of the case CP3 are similar to the plane Couette-Poiseuille flow of Kuroda et al. [14], though the latter case is at a lower Reynolds number and therefore the isotropic state is not reached at the center of the channel. Comparison between AIM at x/D=0.5 from case CP1 to CP3 reveals that the anisotropy near the moving wall reduces with the wall velocity increased. At x/D=0.2 the turbulence is influenced by the side wall and deviates from the plane channel flow behavior near the stationary wall, where at y/D=0.5 the isotropic state is not reached. It is also noted that the difference between turbulence anisotropy near the stationary and moving wall is most evident at wall-bisector.

5. Conclusion

The turbulent Couette-Poiseuille flows inside a square duct are investigated by present simulation procedures. Mean secondary flow is observed to be modified by the presence of moving wall where the symmetric vortex pattern vanishes. Secondary flow near the top corner shows a gradual change of vortex size and position as the moving wall velocity increased. The vortex pair consists of a dominate (clock-wise) and relatively smaller (counter-clockwise) vortex. For the three r considered, the relative positions of vortex cores have a linear dependence with r, which offers an easy estimation of the secondary flow structure. In addition, distance between the vortex cores of each vortex pair is also observed to remain approximately constant.

The turbulence level near the moving wall is reduced compared to other bounding walls which is caused by the insufficient mean shear rates. The distribution of the turbulence kinetic also demonstrated that the capability of mean secondary flow on transporting energy on the cross-plane of the square duct. The reductions of turbulence intensities along the wall-bisector are found to be similar with Kuroda et al. [14] when the operating value of r is comparable.

The anisotropy invariant analysis indicates that along the wall bisector at the top half of the duct, turbulence structure gradually moves towards a rod-like axi-symmetric state (negative III) as the moving wall velocity increased. It is observed that the turbulence anisotropy level is reduced along the wall-bisector near the moving wall. On the other hand, the anisotropy states near the bottom stationary wall resemble those presented in turbulent channel flow or boundary layers. The difference between turbulence anisotropy near the stationary and moving wall is most evident at wall-bisector.

6. Acknowledgments

This research work is supported by the National Science Council of Taiwan under grant 95-2221-E-007 -227 and the computational facilities are provided by the National Center for High-Performance Computing of Taiwan which the authors gratefully acknowledge.

REFERENCES

1. L. Prandtl, Uber die ausgebildete turbulenz, Verh. 2nd Intl. Kong. Fur Tech. Mech., Zurich,[English translation NACA Tech. Memo. 435, 62]. (1926)
2. N. Fukushima, N. Kasagi, The effect of walls on turbulent flow and temperature fields, in: Proceedings of Turbulent Heat Transfer III, Anchorage, USA (2001).
3. M. Hirota, H. Fujita, H. Yokosawa et al, Turbulent heat transfer in a square duct, Int. J. Heat and Fluid flow **18**, (1997) 170-180.
4. M. Piller, E. Nobile, Direct numerical simulation of turbulent heat transfer in a square duct, Int. J. Numerical methods for heat and fluid flow, **12** (2002) 658-686.
5. M. Salinas-V*á*zquez and O. M*é*tais, Large eddy simulation of the turbulent flow through a heated square duct, J. Fluid Mech. **453** (2002) 201-238.
6. M. Salinas-V*á*zquez, W. Vicente Rodriguez, R. Issa, Effects of ridged walls on the heat transfer in a heated square duct, Int. J. Heat and Mass Transfer **48**, (2005) 2050-2063.
7. J. Pallares and L. Davidson, Large eddy simulations of turbulent heat transfer in stationary and rotating square ducts, Physics of Fluids **14**, (2002) 2804-2816.
8. W. Lo and C. A. Lin, Large Eddy Simulation of turbulent Couette-Poiseuille flows in a square duct, in: Proceedings of Parallel Computational Fluid Dynamics, Washington USA (2005).
9. J. Smagorinsky, General circulation experiments with the primitive equations. I. The basic experiment., Mon. Weather Rev. **91**, (1963) 499-164.
10. M. Breuer, W. Rodi, Large eddy simulation of turbulent flow through a straight square duct and a 180 degree bend, Direct and Large-Eddy Simulation I, eds. P. R. Voke et al., Kluwer Academic Publishers, (1994) 273-285.
11. H. Choi and P. A. Moin, Effects of the Computational Time Step on Numerical Solutions of Turbulent Flow, J. Comput. Phys, **113** (1994) 1-4.
12. W. Lo, Large Eddy Simulation of turbulent Couette-Poiseuille flows inside a square duct, PhD thesis, National Tsing Hua University, TAIWAN (2006).
13. A. Huser and S. Biringen, Direct numerical simulation of turbulent flow in a square duct, J. Fluid Mech. **257** (1993) 65-95.
14. A. Kuroda, N. Kasagi, and M. Hirata, Direct numerical simulation of turbulent plane CouetteVPoiseuille flows: Effect of mean shear rate on the near wall turbulence structures, Proceeding of the 9th Symposium on Turbulent Shear Flow, Kyoto, Japan pp.8.4.1-8.4.6 (1993).
15. R. D. Moser, J. Kim, and N. N. Mansour, Direct numerical simulation of turbulent channel flow up to $Re_\tau = 590$, J. Fluid Mech. **177** (1999) 943-945.
16. E. M. Thurlow and J. C. Klewicki, Experimental study of turbulent Poiseuille-Couette flow, Physics of Fluids, **4** (2000) 865-875.
17. J. L. Lumley, Computational modeling of turbulent flows, Advances in Applied Mechanics, **18** (1978) 123-176.
18. M. J. Lee and W. C. Reynolds, Numerical experiments on the structure of homogeneous turbulence," Report TF-24, Thermoscience Division, Stanford University (1985).

Table 1
The flow conditions for simulated cases; Re_τ is defined by mean friction velocity averaged over four solid walls (t=top,b=bottom wall); W_w denotes the velocity of the moving wall and W_{Bulk} is the bulk velocity; $Re_c = \frac{W_w D}{\nu}$; $r = \frac{W_w}{W_{Bulk}}$.

	$Re_{\tau t}$	$Re_{\tau b}$	Re_τ	Re_{Bulk}	Re_c	r	$-\frac{D}{\rho W_w^2}\frac{\partial P}{\partial z}$
Case P	600	600	600	9708	0	0	∞
Case CP1	441	605	571	9716	4568	0.47	0.0621
Case CP2	305	591	538	9760	9136	0.94	0.0138
Case CP3	284	587	533	9770	11420	1.17	0.0083
Kuroda et al. (1993)	35	308	172	5178	6000	1.16	0.0026

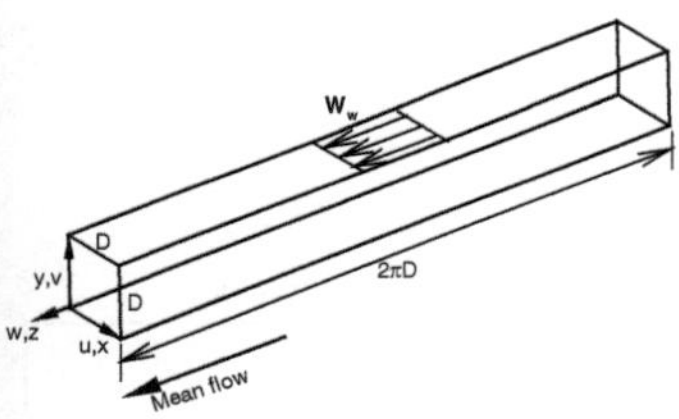

Figure 1. Schematic plot of the square duct with a moving wall.

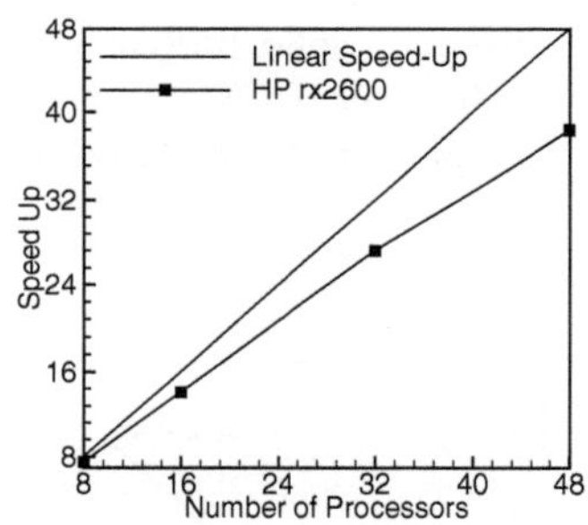

Figure 2. Parallel performance of present simulation.

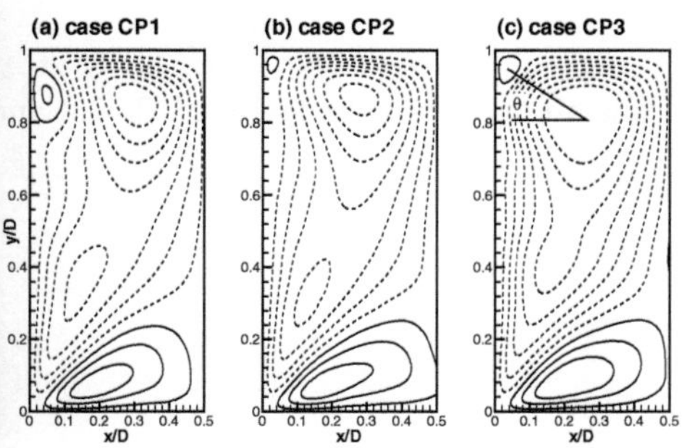

Figure 3. Streamlines of mean secondary flow; solid lines for counter-clockwise rotation, dashed lines for clockwise rotation.

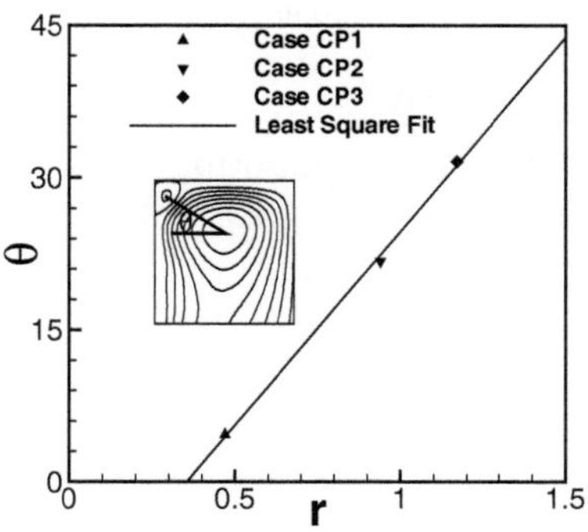

Figure 4. Angles between vortex cores near the top corners in Figure 3.

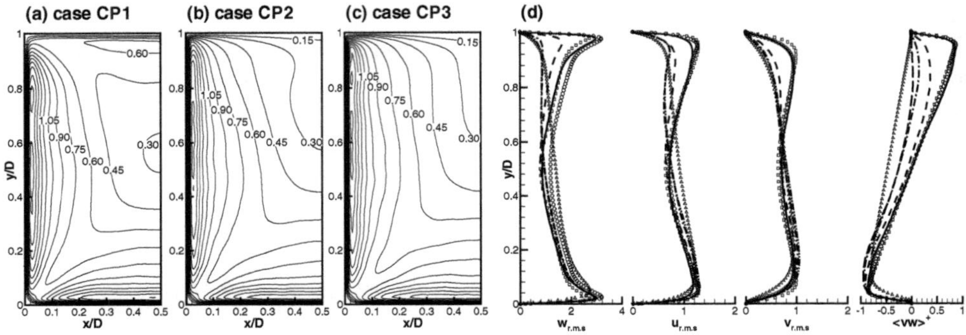

Figure 5. (a)-(c) Turbulence kinetic energy. (d) Reynolds stress components along the wall-bisector. Solid line:case P, Dashed line: case CP1, Dash-dot line:case CP2, Dash-dot-dot line:case CP3, △ symbol: Kuroda et al. [1993], ○ symbol: Huser and Biringen [1993]. *square* symbol: Moser et al. [1999].

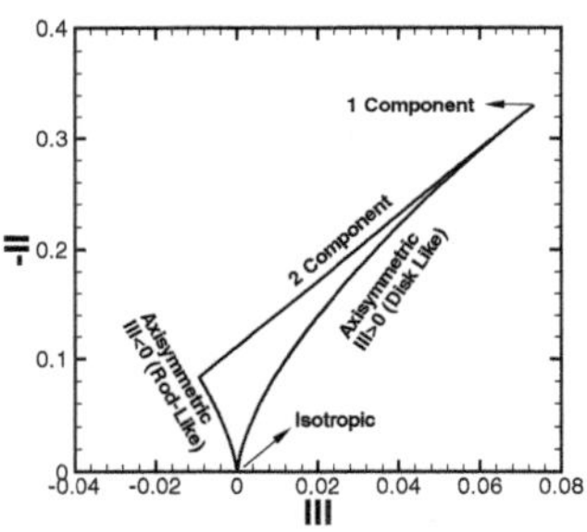

Figure 6. The anisotropy invariant map.

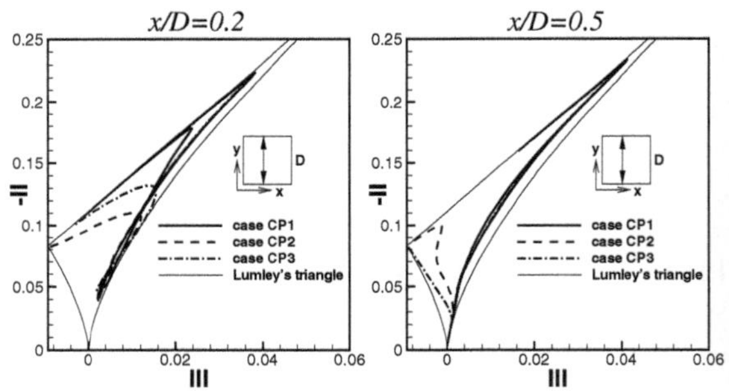

Figure 7. Anisotropy invariant map at x/D=0.2 and x/D=0.5.

Parallel Computational Fluid Dynamics – Parallel Computing and Its Applications
J.H. Kwon, A. Ecer, J. Periaux, N. Satofuka and P. Fox (Editors)

The Prediction of the Dynamic Derivatives for the Separated Payload Fairing Halves of a Launch Vehicle in Free Falling

Younghoon Kim,[a] Honam Ok,[a] Insun Kim[a]

[a]*Korea Aerospace Research Institute, 45 Eoeun-Dong, Youseong-Gu, Daejeon, 305-333, Korea*

Keywords: Dynamic derivatives; Payload fairing; Aerodynamic damping; Launch vehicle; Free falling

1. Introduction

After the separation of the PLF(Payload Fairing) halves, each PLF half is in free falling apart. If both PLF halves are completely destroyed by fire due to the aerodynamic heating , it dosen't necessary to analysis the accurate dispersed range of the impact point for the PLF halves. However, unfortunately, the PLF halves are not burned out due to the lack of the aerodynamic heating. The easiest approach to predict the dispresed range of the impact point is neglecting the effect of the aerodynamic force and assuming that the PLF half is a particle. If the predicted impact point with that assumption is near the dangerous region, the aerodynamic characteristics of the PLF halves are additionally needed to make the precise prediction. The aerodynamic characteristics of the PLF halves consist of the static aerodynamic characteristics as well as the dynamic characteristics. The static aerodynamic characteristics of the PLF healves are already investigated.[3] A number of approach methods are applied to obtain the dynamic damping parameters. The methods to predict the dynamic derivatives consist of the wind tunnel test, using the empirical data base and the CFD analysis. The wind tunnel test for the PLF halves in free falling is so much time and cost consuming. The prediction approach using the empirical data base is the simplest but there are no data base for the PLF halves. In this paper, predicted are the dynamic damping parameters for the configuration of the three dimensional PLF halves using the most complicated approach, CFD analysis even though it spends a lot of time since the configuration of the PLF halves is so unique and the range of the angle-of-attack and the side slip angle are highly wide.

2. Numerical Approach

The CFD approch to obtain the dynamic derivatives of the PLF halves is similar to the approach for the wind tunnel test using the forced harmonic motion. Fig. 1 shows the simple schematic diagram of the forced harmonic motion. The forced harmonic motion is the sum of the pitching motion and the plunging motion. For the pitch-axis oscillation, the pitch damping sum is obtained only by the forced harmonic motion. If the dynamic damping for the pure plunging motion is obtained, the forced plunging motion should be applied. And if the dynamic damping for the pure pitching motion is predicted, the forced harmonic motion and the forced plunging motion should be applied simuataneously.

Obtained are the dynamic derivatives with a CFD code implementing the forced harmonic oscillation of a grid system.[1] The unsteady Euler equation as a governing equation is applied to predict more economically the aerodynamic characteristics neglecting the viscous effects even though using Navier-Stokes equation is more precise approach. And the simple equation to simulate the forced harmonic oscillation is as below.

$$\theta = \theta_0 + \theta_1 \sin(2M_\infty kt)$$

where θ_0 is the initial angular displacement, θ_1 is the amplitude of the oscillation, M_∞ is the free stream Mach number, k is the reduced frequency and t is the nondimensional time.

It is assumed that the rotation center for the oscillation motion is the center of gravity. θ means the angular displacement containing the effect of the roll rate only for the oscillation with respect to the x-axis. θ, however, includes the effect of both the pitch rate and the variation of the angle-of-attack for the oscillation with respect to the y-axis and θ also contains the effect of both the yaw rate and the variation of the side slip angle for the oscillation with respect to the z-axis. It is also assumed that the dynamic derivatives are constant during the oscillatory motion. In this paper, θ_1 is 0.25 deg since the amplitude of oscillation is small in order that the dynamic derivatives are constant during the oscillatory motion and k is arbitrary value 0.1 because the dynamic derivatives are not highly dependent on the variation of the damping frequency. θ_0 is 0 deg since the harmonic motion starts at the initial position. The dynamic damping coefficients are obtained integrating the moment coefficients for a period for the forced harmonic oscillation as below. [2]

$$\textit{Dynamic Derivative } (C_{m_{\dot{\theta}}}) = (2M_\infty / \pi\theta_0) \int_0^T C_m \cos(2M_\infty kt)dt$$

The configurations of Near and Far PLF are quite different from each other. Far PLF has more complex geometry consisting of spherical nose, cone and cylindrical part. Near PLF, however, doesn't have a spherical nose.(Fig. 2) And it is assumed that the thickness of the PLF halves is uniformly 0.2 m. In this research, the dynamic derivatives of PLF halves are presented for Mach number ranging from 0.6 to 2.0, the angle-of-attack ranging from -180 deg. to 180 deg. and the side slip angle ranging from -90 deg. to 90 deg. The reference length and area are the diameter 2.0 m and the area of base π

m^2, respectively. The aerodynamic moments are calculated about the c.g point of the PLF halves. Fig. 2 shows the configurations of the PLF halves and the reference axis system.

3. Code Validation

In the CFD code, the forced harmonic oscillation of the PLF halves is simulated. And the dynamic derivatives are predicted separately using the aerodynamic moment coefficients. Validated below for the simple basic finner is, therefore, the numerical method to predict the dynamic damping paramters.[2]

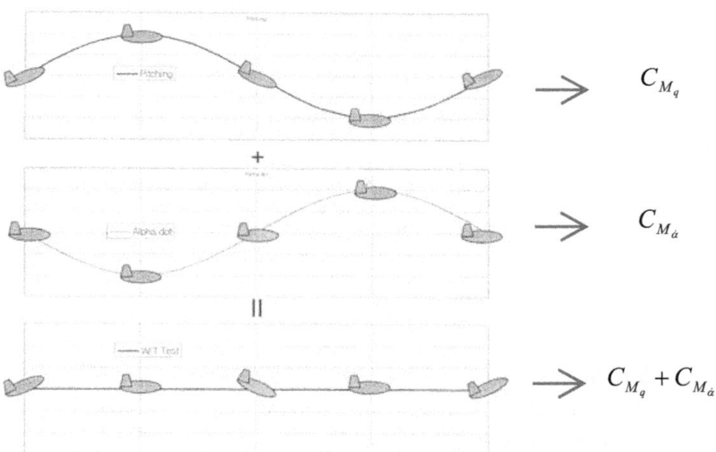

Fig. 1 The schematics for the forced oscillation

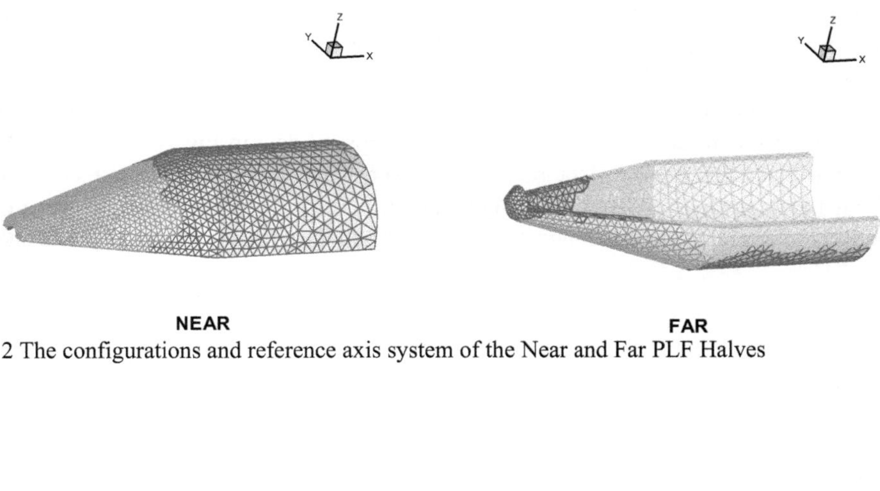

Fig. 2 The configurations and reference axis system of the Near and Far PLF Halves

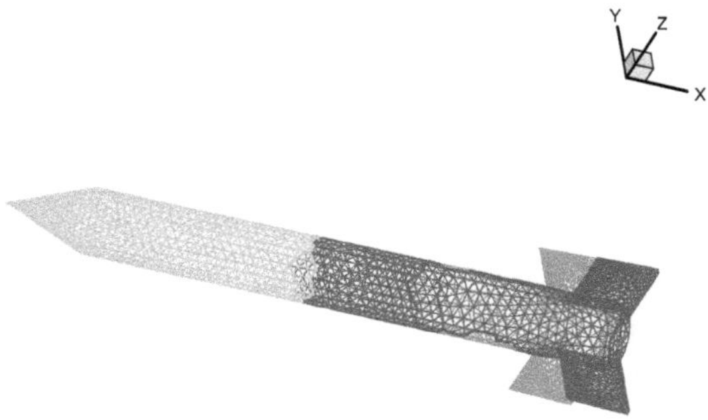

Fig. 3 The configuration of the basic finner

Table 1. The validation for the configuration of the basic finner

M=1.1, k=0.05	Pitching-damping moment coefficient ($C_{M_q} + C_{M_{\dot{\alpha}}}$)	Roll-damping moment c oefficient (C_{L_p})
Ref. CFD result [2]	-399	-20.5
Present	-378.360	-25.996

4. Numerical Results

The dynamic derivatives for Near half which has simpler configuration are firstly predicted. And then those of Far half are obtained. The parallel computaion for the economic computaion is used with 4 CPUs in KISTI(Korea Institute of Science and Technology Information) High Performance Computing system. It is certain that the flow characteristics of Far PLF are different from those of Near PLF due to the existence of the spherical nose. The six components of the aerodynamic static coefficients, however, are similar to each other.[3] Thus, the dynamic derivatives are also similar each other since the dynamic derivatives are obtained integrating the aerodynamic moment coefficients. Fig. 4 shows the comparison of the dynamic derivatives between Near PLF and Far PLF for the angle-of-attack of 0 deg. The dynamic damping coefficients in Fig. 4 with respect to the variation of both the side slip angle and Mach number imply that the dynamic damping parameters for Near PLF have very similar magnitude with those of Far PLF. Most of data in Fig. 4 show the PLF has dynamic stability. A few data points, however, have positive damping coefficients which mean the possibility of the dynamic instability. And moreover, there are the changes of the sign at some points. It is uncertain that this is physically reasonable.

Thus, more calculations will be made for these conditions with another approach to confirm their physical meaning and uncertainty.
Fig. 6 and Fig. 7 respresent the pitch and the yaw oscillaition for Near PLF half respectively. The variations of the dynamic damping parameter for the pitch and yaw oscillation are also complicated and irregular. The forced plunging motion should be applied to predict the effect of the pitch rate q only for the pitch oscillation(or the yaw rate r only for the yaw oscillation) since the forced harmonic motion includes the effect of both the pitch rate and the variation of the angle-of-attack for the pitch oscillation(or both the yaw rate and the variation of the side slip angle for the yaw oscillation).

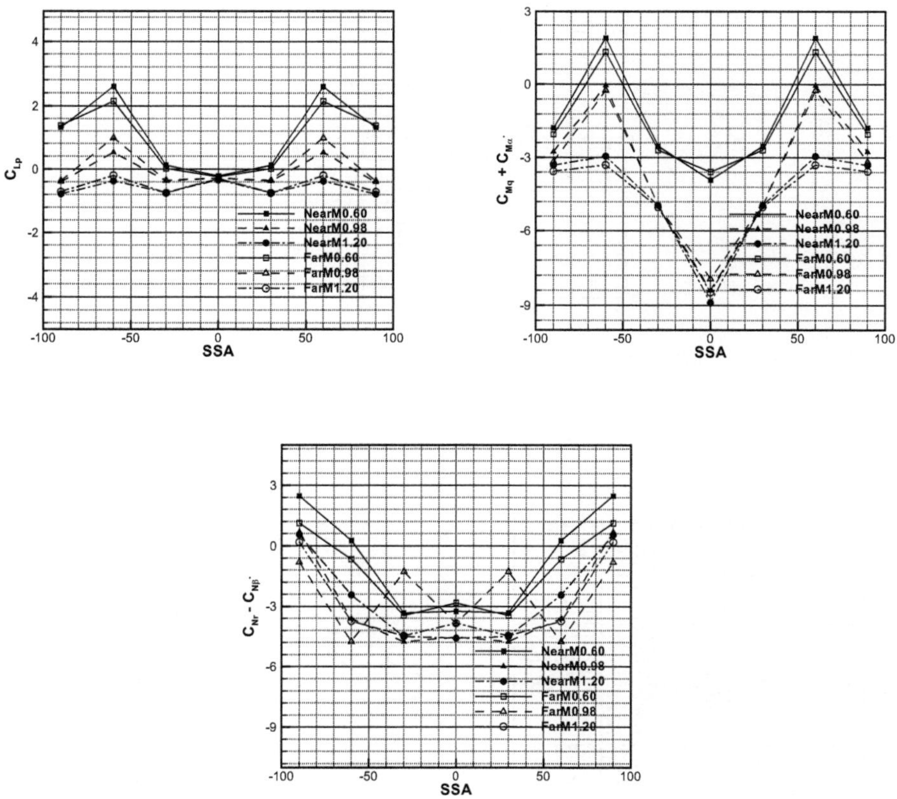

Fig. 4 The comparison of the dynamic derivatives betweed Near and Far PLF half

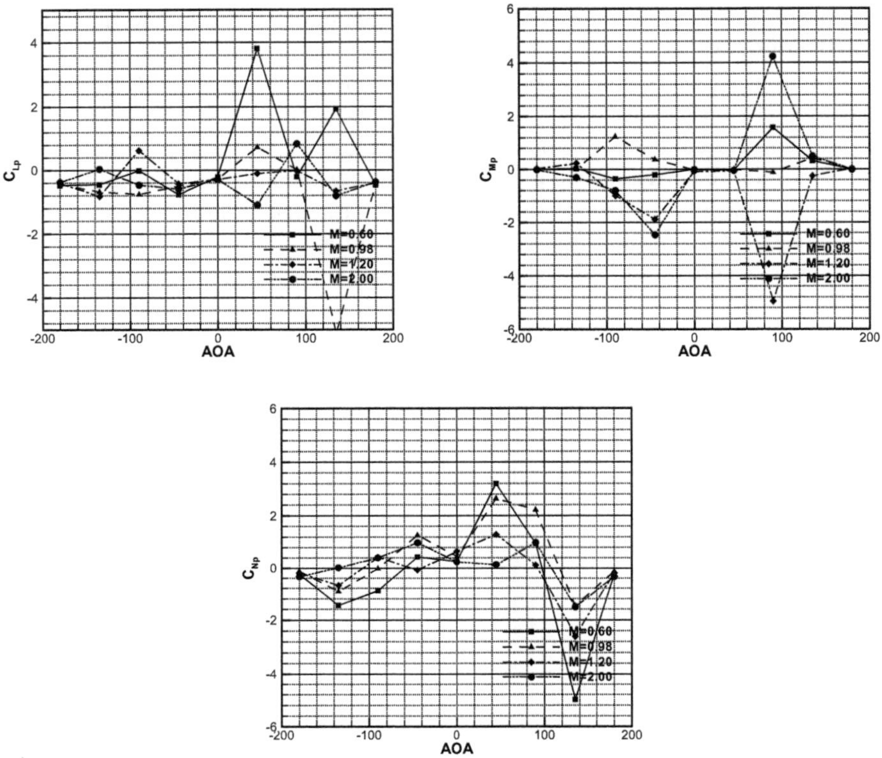

Fig. 5 The variation of the dynamic derivatives for the roll oscillation

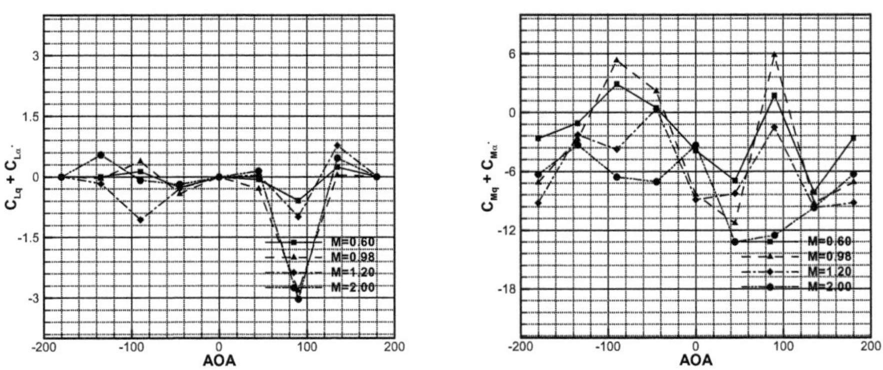

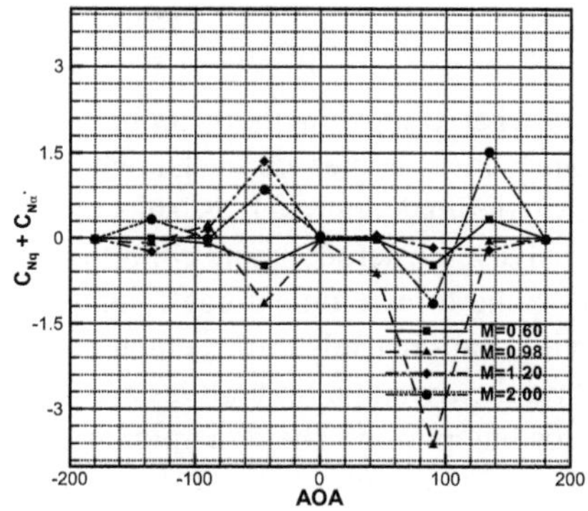

Fig. 6 The variation of the dynamic derivatives for the pitch oscillation

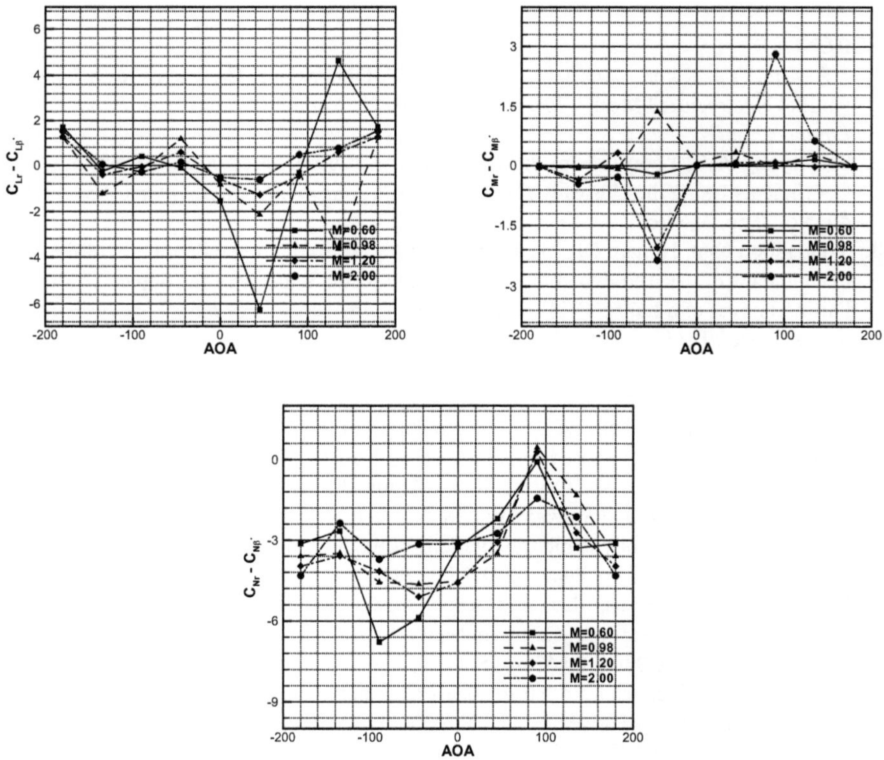

Fig.7 The variation of the dynamic derivatives for the yaw oscillaiton

5. Conclusion

A review has been made on what kind of method can be applied to predict the dynamic derivatives of the separated PLF halves of a launch vehicle in consideration of technology and budget. An optimal approach is selected considering the geometric characteristics of the PLF halves, the aerodynamic conditions and the required accuracy. The time history of aerodynamic force/moment coefficients are obtatined for the forced harmonic motions by solving the unsteady Euler equations derived with repect to the inertial reference frame. And the dynamic derivatives are deduced by integration of the aerodynamic coefficients for one period. The dynamic derivatives are presented for $0.6 \leq M \leq 2.0$, $-180° \leq \alpha \leq 180°$ and $-90° \leq \beta \leq 90°$. A few damping parameters are positive. And the change of the sign exists in some ranges of the aerodynamic angles. It is not sure that these physical phenomenon are reasonable. Their physical meaning and uncertainty will be confirmed for these conditions with another approach.

REFERENCES

1. J. S. Kim, O. J. Kwon, (2002). "Computaion of 3-Dimensional Unsteady Flows Using a Parallel Unstructured Mesh", *Proceedings of the 2nd National Congress on Fluid Engineering*.
2. Soo Hyung Park, Yoonsik Kim, Jang Hyuk Kwon, (2003). "Prediction of Damping Coefficients Using the Unsteady Euler Equation", *Journal of Spacecraft and Rocket,* Vol. 40, No. 3, pp. 356-362.
3. Younghoon Kim, Honam Ok, Insun Kim, (2006). "A Study on the Prediction of the Six Components of the Aerodynamic Forces and the Static Stability of the Separated Payload Fairing Halves of a Launch Vehicle Using CFD", *Proceedings of the 7th Symposium on the Space Launch Vehicle Technology*, pp. 63-67.
4. Oktay, E. and Akay, H. U., " CFD Prediction Of Dynamic Derivatives For Missiles," AIAA Paper 2002-0276,2002.
5. Weinacht, P. and Sturekek, W. B., " Computation of the Roll Characteristics of a Finned Projectile," *Journal of Spacecraft and Rockets,* Vol. 33, No. 6, 1996, pp. 796-775.
6. Pechier, M., Guillen, P., and Cayzac, R., " Magnus Effect over Finned Projectiles," *Journal of Spacecraft and Rockets,* Vol.38, No. 4, July-August 2001, pp.542-549.
7. Almosnino, D., " Aerodynamic Calculations of the Standard Dynamics Model in Pitch and Roll Oscillations," AIAA Paper 94-0287, 1994.
8. Murman, S. M., Aftosmis, M. J., and Berger, M. J., " Numerical Simulation of Rolling Airframes Using a Multilevel Cartesian Method, "*Journal of Spacecraft and Rockets,* Vol. 41, No. 3, May-June 2004, pp.426-435.
9. Park, M. A. and Gren, L. L., " Steady-State Computation of Constant Rotational Rate Dynamic Stability Derivatives," AIAA 2000-4321, 2000.
10. Limache, A. C. and Cliff, E. M., " Aerodynamic Sensitivity Theory of Rotary Stability Derivatives, " AIAA 1999-4313, 1999.
11. Silton, S. I., " Navier-Stokes Computations for a Spinning Projectile from Subsonic to Supersonic Speeds," AIAA Paper 2003-3936

Variability of Mesoscale Eddies in the Pacific Ocean Simulated by an Eddy Resolving OGCM

B.Y. Yim[a], Y. Noh[a], S. H. You[b], J. H. Yoon[c], and B. Qiu[d]

[a]*Department of Atmospheric Science, Yonsei University, 134 Sinchon-dong, Seodaemun-gu, Seoul 120-749, Korea*

[b]*Marine Meteorology & Earthquake Research Lab., Meteorological Research Institute, 460-18 Sindaebang-2dong, Dongjak-gu, Seoul 120-749, Korea*

[c]*Research Institute for Applied Mechancis, Kyushu University, 6-1 Kasuga, Fukuoka, 816-8580, Japan*

[d]*Department of Oceanography, University of Hawaii, 1000 Pope Road Honolulu, HI 96822, U.S.A.*

Key Words: OGCM; mesoscale variability; eddy kinetic energy; Pacific Ocean

The mesoscale eddy field in the North Pacific Ocean, simulated by a high resolution eddy-resolving OGCM (1/12° horizontal resolution), was analyzed, and compared with satellite altimetry data of TOPEX/Poseidon. A good agreement was found between the simulated and observed data in which high levels of eddy kinetic energy (EKE) appear near the Kurosho, North Equatorial Current (NEC), and Subtropical Countercurrent (STCC) in the western part of the subropical gyre. The spatially averaged EKE and its seasonal variation were also investigated in these three regions. Finally, we attempted to relate the magnitude of EKE with the parameter of baroclinic instability.

1. INTRODUCTION

Almost everywhere in the open ocean the kinetic energy of the flow field is dominated by mesoscale variability. The potential impact of these mosescale eddies on the mean properties of the ocean with the associated large-scale transports of heat, energy, mass, and biogeochemical tracers has been the subject of interest to physical oceanographers for a long time.

Mesoscale eddies have the spatial scales equivalent to the Rossby radius of deformation, varying roughly from some 10 km in the subtropical region to a few 100 km in the tropics. It means that the grid size of the ocean general circulation model (OGCM) must be smaller than $1/10^{\circ}$ to resolve mesoscale eddies globally, which becomes possible only recently with the advance of supercomputing technology [1,2,3]. Earlier OGCM results with the insufficient resolution usually suffered from the underestimated eddy kinetic energy (EKE) level [4].

The mesoscale variation of the global ocean have been analyzed recently using the global altimetry data from the satellite, such as TOPEX/Poseidon and ERS-1/2, and the high resolution OGCM [2,5,6,7,8,9,10]. However, the detailed comparison between an eddy-resolving OGCM with the grid size smaller than $1/10^{\circ}$ and the satellite data has been limited to the North Atlantic Ocean so far.

In the present study, we analyzed the characteristics of mesoscale eddies in the Pacific Ocean, simulated by an high-resolution OGCM of $1/12^{\circ}$, taking advantage of the Earth Simulator (ES) that was developed in 2002 in Japan. We investigated the spatial and temporal variations of EKE in the Pacific Ocean, and compared the results with the altimetry data of the TOPEX/Poseidon mission (October 1992 – December 1997).

2. MODEL

The OGCM used in this study (RIAM Ocean Model; RIAMOM) is the primitive general ocean circulation model with a free surface. The model covers from 95°E to 70° W in longitude and from 50°S to 65°N in latitude. The horizontal grid interval is 1/12° in both latitudinal and longitudinal directions and vertical layers are 70 levels. The advection of momentum is treated by the generalized Arakawa scheme [11], which conserves potential enstrophy as well as kinetic energy. It also used the MSQUICK scheme for an improved advection of tracers [12] and the biharmonic diffusion for both momentum and tracers. The subgrid-scale vertical mixing process is improved by the Noh scheme [13,14,15]. The detailed explanation and the general performance of the model can be found in You [16].

The model was started from the state of rest with the climatological temperature and salinity distributions by Levitus [17], and was forced by the NCEP reanalysis data of wind stress and heat flux. The combined boundary condition using both the climatological flux and the restoring term was used for the heat flux, similarly to Noh et al. [18], and the restoring boundary condition was used for salinity. The model was integrated for 25 years using ES, which is long enough for the upper ocean to reach quasi-equilibrium. Three-dimensional prognostic variables were archived every model day in the last year, which were used for analysis.

3. PARALLEL COMPUTATION

The simulation was carried out in ES, which is a highly parallel vector supercomputer system of the distributed-memory type, and consists of 640 processor nodes (PNs). Each PN is a system with a shared memory, consisting of 8 vector-tyoe arithmetic processrors, a 16 GB main memory system, a remote access control unit, and an I/O processors.

The domian decomposition of the model for the MPI process was made two-dimensionally in the horizontal direction. The whole values calculated at each CPU were communicated in the buffer zone. The model used 120 nodes (960 CPU) of ES, and one year integration needed 10 hours of total CPU time.

4. RESULTS

Figure 1 compared the rms sea surface height anomaly (SSHA) in the Pacific Ocean, obtained from the OGCM and the T/P data [9]. The SSHA h' is related to the horizontal velocity of mesoscale eddies $\mathbf{u}_g'$ by geostrophic balance, i.e., $\mathbf{u}_g' = (g/f)\mathbf{k} \times \nabla h'$, and thus EKE $< u_g^2 >$. Comparison reveals a good agreement not only in the spatial pattern of EKE but also in the magnitude. High levels of EKE appear near the Kurosho, North Equatorial Current (NEC), and Subtropical Countercurrent (STCC) in the western part of the subropical gyre, and EKE is much weaker in the eastern part of the ocean.

We compared the annual mean EKE averaged over the regions of Kuroshio Extension (KE; 140°E-170°W, 32°-38°N), STCC (135°E-175°W, 19°-25°N), and NEC (135°E-175°W, 10°-16°N) (Fig. 2). The simulated EKE in the Kuroshio region is stronger and more expanded than the observation data, which may reflect the coarser resolution of the satellite data [9]. Note that similar results were also obtained in the Atlantic Ocean, in which the simulated EKE is much stronger in the Gulf Stream region [10].

Qiu [9] found that the EKE of the STCC has a well-defined annual cycle with a maximum in April/May and a minimum in December/January with its peak-to-peak amplitude exceeding 200 cm^2s^{-2} but no such distinct annual cycle is found in any other current zone. Figure 3 compares the seasonal variation of EKE anomalies in the STCC, NEC and Kuroshio Extension (KE) regions. A good agreement is observed in the STCC region, expcept in Januray/February, where the simulated EKE is higher. On the other hand, the agreement is not so good in the regions with no clear annual cycle such as KE and NEC. It is not clear what causes the discrepancy, but we can expect that the weaker signal of seasonal variation and stronger inherent instability of the ocean circulation make it very difficult to reproduce the meaningful signal in KE and NEC.

The generation of mesoscale eddies in the ocean is dominated by baroclinic instability, which can be estimated by the temperature structure of the ocean. Therefore, we compared the temperature distributions in March and September in the meridional crosssection in three regions (Fig. 4). Although the climatological data filtered out the fluctuations associated with instantaneous mesoscale eddies, a good agreement is found in the mean feature in all three regions in spite of the disagreement in the seasonal variation of EKE in KE and NEC.

Finally, we attempted to parameterize the magnitude of EKE in terms of the oceanic condition. According to theoretical analyses, the growth rate of baroclinic instability is proportional to $(f/N)dU/dz$ [19]. Figure 5 showed the scatter plot between EKE and $f'(\Delta U)/\Delta T^{1/2}$ obtained from the whole domain of the North Pacific Ocean, along with that between EKE and ΔU [9,20]. Both show positive correlation between two variables in spite of large scatter, but higher correlation can be found between EKE and $f'(\Delta U)/\Delta T^{1/2}$, suggesting the importance of stratification in parameterizing the magnitude of baroclinic eddies.

5. CONCLUSION

In the present paper, we presented the results from an eddy resolving Pacific Ocean model (RIAMOM) with the horizontal resolution 1/12°, which was obtained by taking advantage of the Earth Simulator in Japan, with the main focus on the mesoscale variability. A good agreement is found in general between the mesoscale eddy field in the North Pacific Ocean between the simulated and observed data. In particular, we analyzed the seasonal variation of EKE in the regions of the Kutoshio Extension, North Equatorial Current and Subtropical Countercurrent. We also investigated the relationship between annual cycle of EKE and baroclinic instability to parameterize the magnitude of EKE.

REFERENCES

1. Masumoto, Y., H. Sasaki, T.Kagimoto, N.Komori, A. Ishida, Y. Sasai, T. Miyama, T. Motoi, H. Mitsudera, K. Takahashi, H. Sakuma, and T. Yamagata (2004), A fifty-year eddy-resolving simulation of the world ocean – Preliminary outcomes of OFES (OGCM for the Earth Simulator), *J. Earth Simulator*, 1, 35-56.
2. Smith, R. D., M. E. Maltrud, F. O. Bryan, and M. W. Hecht (2000), Numerical simulation of the North Atlantic Ocean at 1/10°, *J. Phys. Oceanogr.*, 30, 1532–1561.
3. Oschlies, A. (2002), Improved Representation of Upper-Ocean Dynamics and Mixed Layer Depths in a Model of the North Atlantic on Switching from Eddy-Permitting to Eddy-Resolving Grid Resolution, *J. Phys. Oceanogr.*, 32, 2277-2298.
4. Stammer, D., R. Tokmakian, A. J. Semtner, and C. Wunch (1996), How well does a ¼ global circulation model simulate large-scale oceanic observations? *J. Geophys. Res.*, 101, 25 779–25 811.
5. Stammer, D. (1997), Global characteristics of ocean variability estimated from regional TOPEX/POSEIDON altimeter measurements, *J. Phys. Oceanogr.*, 27, 1743-1769.
6. Ducet, N., P.-Y. Le Traon, and G. Reverdin (2000), Global high resolution mapping of ocean circulation from TOPEX/Poseidon and ERS-1/2, *J. Geophys. Res.*, 105(C5), 19,477–19,498.
7. Ducet, N., and P.-Y. Le Traon (2001), A comparison of surface eddy kinetic energy and Reynolds stresses in the Gulf Stream and the Kuroshio Current systems from merged TOPEX/Poseidon and ERS-1/2 altimetric data, *J. Geophys. Res.*, 106(C8), 16,603– 16,622.
8. Maes, C., M. Benkiran, and P. De Mey (1999), Sea level comparison between TOPEX/POSEIDON altimetric data and a global ocean circulation model from an assimilation perspective, *J. Geophys. Res.*, 104, C7, 15,575-15,585.

9. Qiu, B (1999), Seasonal eddy field modulation of the North Pacific Subtropical Countercurrent: TOPEX/Poseidon observations and theory, *J. Phys. Oceanogr.*, 29, 2471-2486.
10. Brachet, S., P. Y. Le Traon, C. Le Provost (2004), Mesoscale variabillity from a high-resolution model and from altimeter data in the North Atlantic Ocean, *J. Geophys. Res.*, 109, C12025, doi:10.0129/2004JC002360.
11. Ishizaki, H., and T. Motoi (1999), Reevaluation of the Takano-Oonishi scheme for momentum advection of bottom relief in ocean models, *J. Atmos. And Ocean. Tech.*, 16, 1994-2010.
12. Webb, D. J., B. A. de Cuevas, and C. S. Richmond (1998), Improved advection scheme for ocean models, *J. Atmos. And Ocean. Tech.*, 15, 1171-1187.
13. Noh, Y., and H. J. Kim (1999), Simulations of temperature and turbulence structure of the oceanic boundary layer with the improved near-surface process, *J. Geophys. Res.*, 104, 15621-15634.
14. Noh, Y., C. J. Jang, T. Yamagata, P. C. Chu, and C. H. Kim (2002), Simulation of more realistic upper ocean process from an OGCM with a new ocean mixed layer model, *J. Phys. Oceanogr.*, 32, 1284-1507.
15. Noh, Y., Y. J. Kang, T. Matsuura, and S. Iizuka (2005), Effect of the Prandtl number in the parameterization of vertical mixing in an OGCM of the tropical Pacific, *Geophys. Res. Lett.*, 32, L23609.
16. You, S. H. (2005), A numerical study of the Kuroshio system southwest of Japan (Ph.D. thesis, Kyushu Univ., Japan).
17. Levitus, S. (1982), Climatological Atlas of the World Ocean. NOAAProf. Paper No. 13, U.S. Govt. Printing Office, Washington, D. C.,173 pp.
18. Noh, Y., H. S. Min, and S. Raasch (2004), Large eddy simulation of the ocean mixed layer: The effects of wave breaking and Langmuir circulation, *J. Phys. Oceanogr.*, 34, 720–735.
19. Gill, A. E. (1982), Atmosphere-Ocean Dynamics (Academic Press, U.S.A.), pp. 549- 593.
20. Stammer, D., C. Wunsch, R. Giering, Q. Zhang, J., Marotzke, J. Marshall, and C. Hill (1997), The global ocean circulation estimated from TOPEX/POSEIDON altimetry and a general circulation model. Report of the Center for Global Change Science, Dept. of Earth, Atmospheric, and Planetary Sciences, MIT, Cambridge, MA, 40 pp. [Available from Center for Global Change Science, Rm. 54-1312, Massachusetts Institute of Technology, Cambridge, MA 02139.].

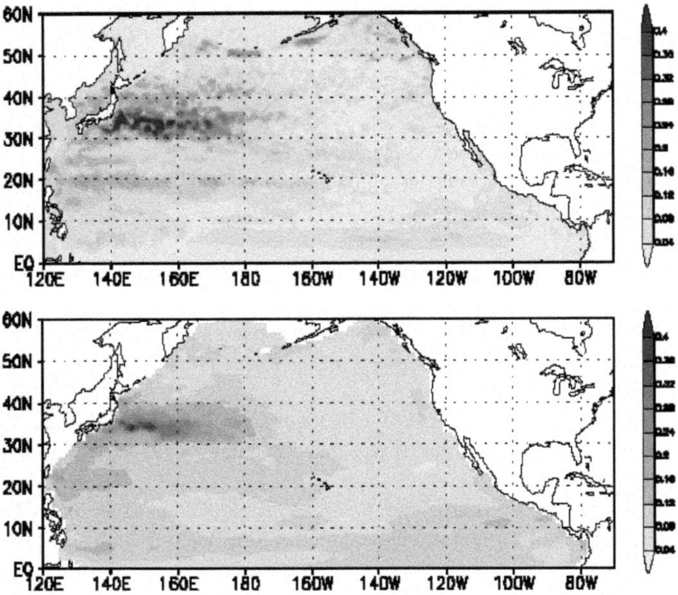

Fig. 1 Map of the rms sea surface height anomaly (SSH) variability in the North Pacific from RIAMOM (top) and T/P (bottom).

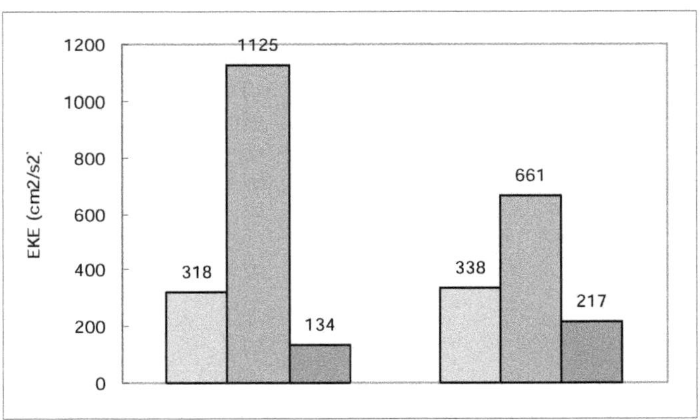

Fig. 2 Spatial average of the annual mean EKE from RIAMOM (left) and T/P (right): (a) STCC (light gray), (b) KE (gray), (c) NEC (dark gray).

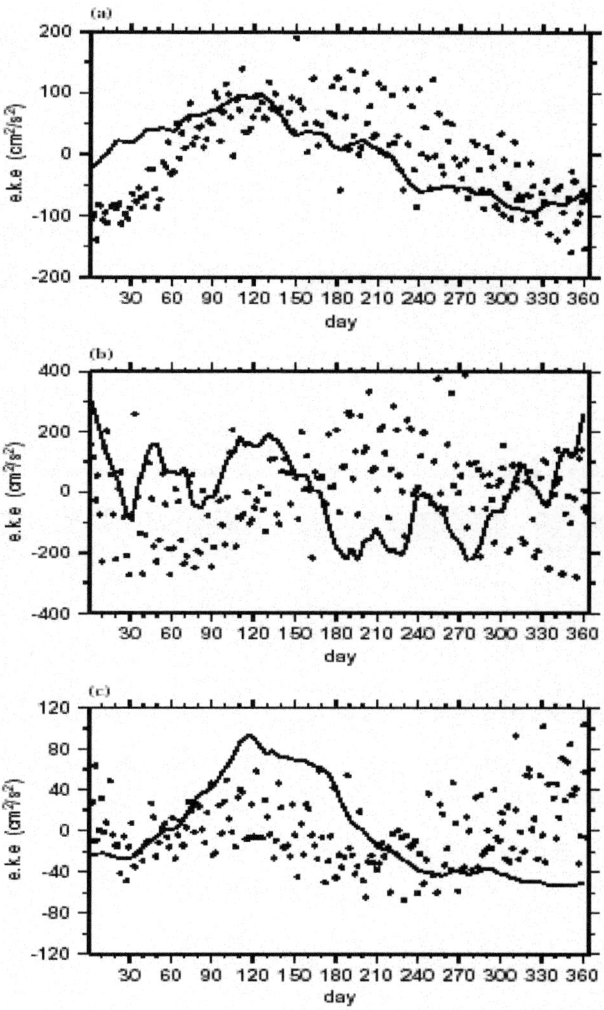

Fig. 3 Annual cycle of the EKE anomalies inferred from RIAMOM (solid line) and T/P (dots). The data from T/P are from October 1992 to December 1997: (a) STCC, (b) KE, (c) NEC.

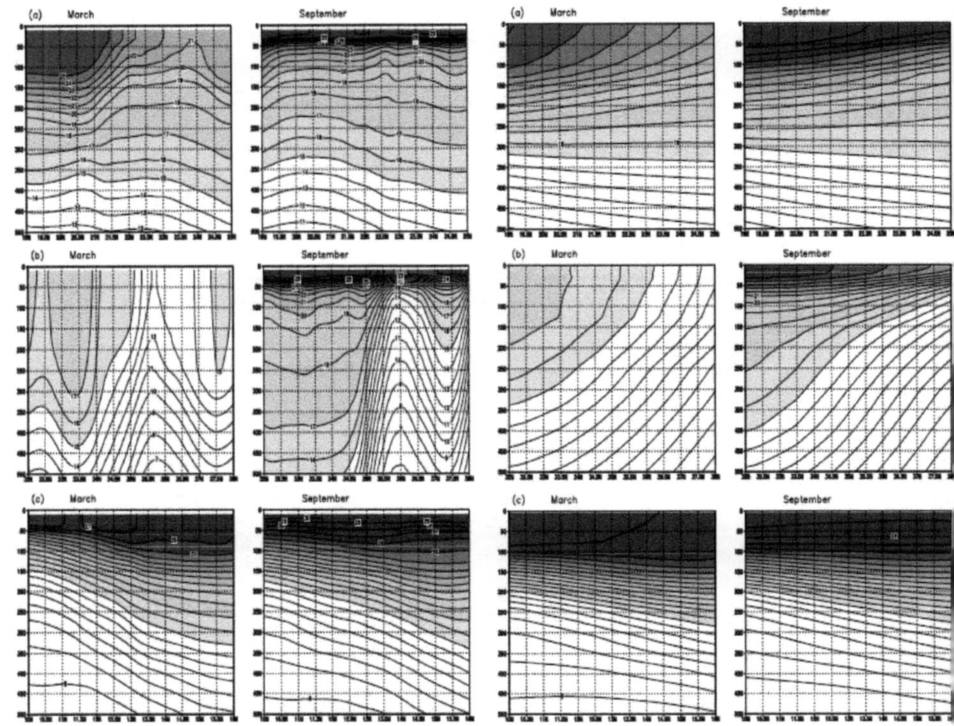

Fig. 4 Temperature distributions in March and September at the meridional (150°E) vertical crosssection from RIAMOM (left) and the climatological data (right): (a) STCC, (b) KE, (c) NE

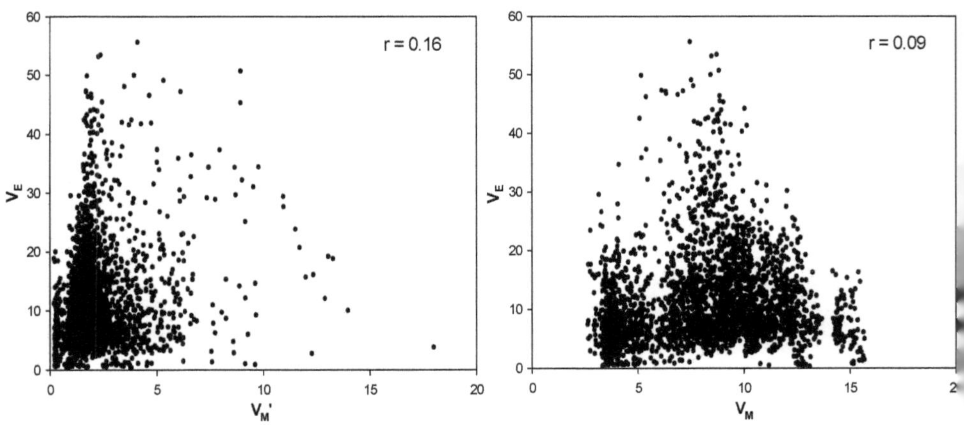

Fig. 5 Scatter plots between V_E ($= \sqrt{EKE}$) and V_M' $(= f'(\Delta U)/\Delta T^{1/2})$ (left) and between V_E an V_M $(= \Delta U)$ (right). f' is *sin (φ)* related Coriolis parameter. Here z = 200 m depth was used to evaluate ΔU and ΔT.

Parallel Computational Fluid Dynamics – Parallel Computing and Its Applications
J.H. Kwon, A. Ecer, J. Periaux, N. Satofuka and P. Fox (Editors)
2007 Elsevier B.V.

Sensitivity study with global and high resolution meteorological model

Paola Mercogliano[a], Keiko Takahashi[b], Pier Luigi Vitagliano[a], Pietro Catalano[a]

[a] *The Italian Aerospace Research Center (CIRA),Via Maiorise, 81043 Capua (CE), Italy*

[b] *Earth Simulator Centre (ESC), Japan Agency for Marine-Earth Science and Technology3173-25 Showa-machi, Kanazawa-ku, Yokohama Kanagawa 236-0001, Japan*

Keywords: weather forecast, sensitivity studies

1. Introduction and motivation

The aim of this paper is to present the numerical results obtained by global a non-hydrostatic computational model to predict severe hydrometeorological mesoscale weather phenomena on complex orographic area. In particular, the performed experiments regard the impact of heavy rain correlated to a fine topography. In the framework of an agreement between CIRA and the Earth Simulator Center (ESC), the new Global Cloud Resolving Model (GCRM)[1] has been tested on the European area The GCRM was developed by Holistic Climate Simulation group at ESC. More specifically, the selected forecast test cases were focusing on the North-Western Alpine area of Italy, which is characterised by a very complex topography [2].
The GCRM model is based on a holistic approach: the complex interdependences between macroscale and microscale processes are simulated directly. This can be achieved by exploiting the impressive computational power available at ESC.
In standard models the microscale phenomena have spatial and time scale smaller than the minimum allowed model grid resolution; thus, a parameterization is necessary in order to take into account the important effect that the interaction between phenomena at different length scales produces on weather. However, this parameterization introduces some arbitrariness into the model. On the other hand, by using the Earth

Simulator it is possible to adopt a resolution higher than that ever tested so far, thus decreasing the model arbitrariness.
Another advantage of using GCRM is that both global (synoptic) and local (mesoscale) phenomena can be simulated without introducing artificial boundary conditions (which cause some "side effects" in regional models); the latter being the approach adopted in the nested local model.
Computational meteorological models with higher resolution, as emphasized before, have a more correct representation of the terrain complex orography and also a more realistic horizontal distribution of the surface characteristics (as albedo and surface roughness). These characteristics are very interesting compared with those of LAMI (Italian Limited Area Model) [3] [4], the numerical model operatively used in Italy to forecast mesoscale-phenomena. The non-hydrostatic local model LAMI uses a computational domain covering Italy with a horizontal resolution of about 7 km . The initial and boundary conditions are provided by the European global models (ECMWF [5] and/or GME [6]), which have a resolution of about 40 km. In LAMI the influence of the wave drag on upper tropospheric flow is explicitly resolved, while in the global models the wave drag can only be simulated adopting a sub-grid orography scheme, as the small scale mountain orography is at subgrid scale level. LAMI model provides a good forecast of the general rain structure but an unsatisfactory representation of the precipitation distribution across the mountain ranges. An improvement of the rain structure was obtained [7] by adopting a not-operative version of the LAMI model with a higher resolution (2.8 km). Furthermore, the convection phenomena are explicitly represented in the LAMI version with higher resolution, and smaller and more realistic rainfall peak have been computed.
This background was useful to analyze and give suggestions in order to improve the GCRM. The performances of the model have been assessed in some test cases [8].
The focus of the analysis has been placed on the evolution of the Quantitative Precipitation Forecast (QPF), one of the most complex and important meteorological variables. An accurate estimation of spatial and temporal rainfall is also important to forecast floods [9] . The simulations performed aimed to investigate the QPF sensitivity with respect to some physical and numerical parameters of the computational model.

2. The test cases

The GCRM performances were studied by considering one event of intense rain occured on November 2002 from 23th to 26th in Piemonte, a region in the North-Western part of Italy. Piemonte is a predominantly alpine region, of about 25000 km^2, situated on the Padania plain and bounded on three sides by mountain-chains covering 73% of its territory (figure 1).
One problem to well forecast the meteorological calamitous in this area is due to its complex topography in which steep mountains and valleys are very close to each other (figure 2). In the event investigated, the precipitation exceeded 50 mm/24 hours over a vast Alpine area, with peaks above 100 mm over Northern Piemonte and also 150 mm in the South-Eastern Piemonte (figure 3).

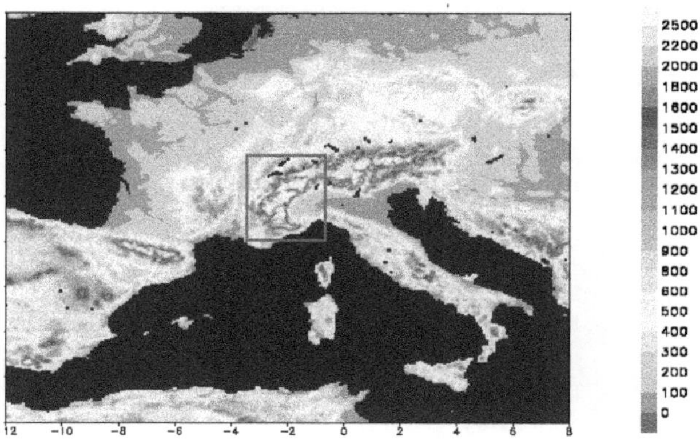

figure 1: studied domain: inside the red curve the heavy rainfall area for the test cases.

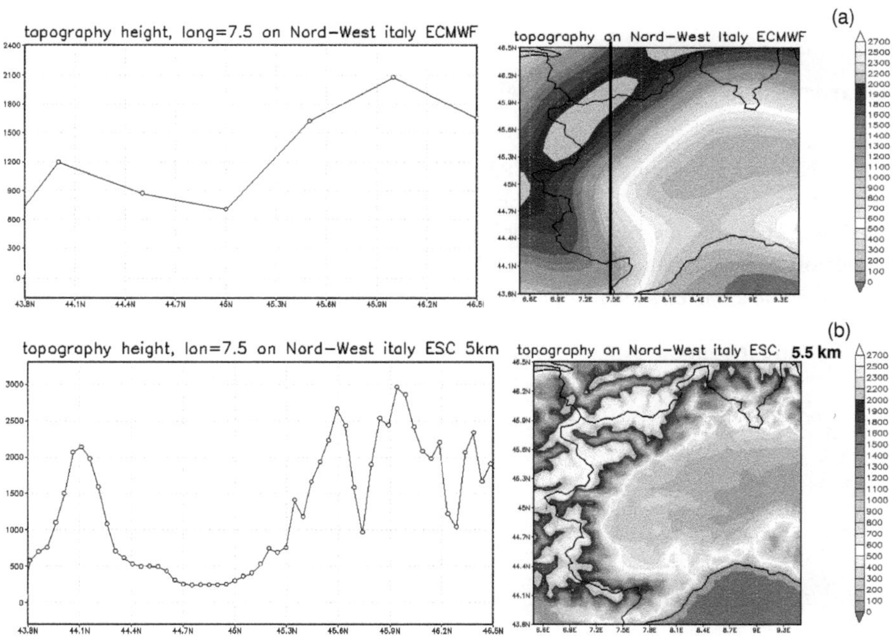

figure 2: particular of Piedmont topography: (a) ECMWF global model; about 40 km of horizontal resolution;(b)ESC global model; 5.5 km of horizontal resolution. In the first model there are not information on the actual peak and valleys

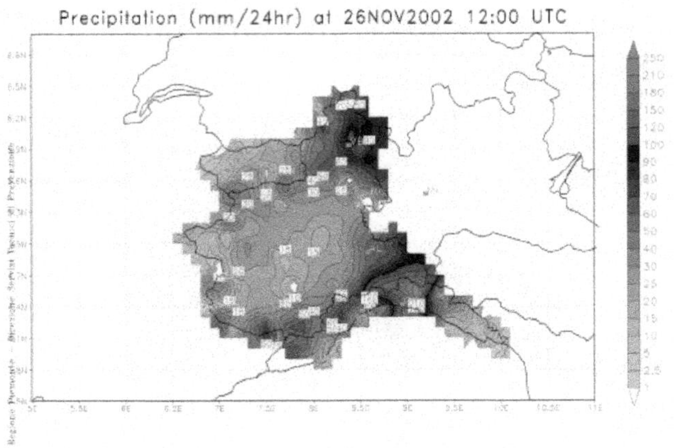

figure 3: Total amount of rain (mm) registered by ground stations in 24 hours: 25/11 12UTC-26/11 12UTC

3. Characteristic of numerical model and performed runs

A very special topic of the GCRM is the Yin-Yang grid system, characterized by two partially overlapped volume meshes which cover the Earth surface. One grid component is defined as part of the low-latitude region between 45N and 45S, extended for 270 degrees in longitude of the usual latitude-longitude grid system; the other grid component is defined in the same way, but in a rotated spherical coordinate system (ffigure 4)[10].

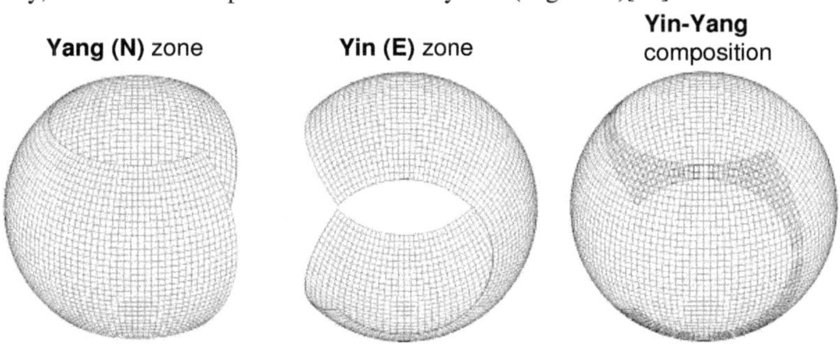

figure 4: Yin-Yang grid system

The upper surface of both meshes is located 30000 m above sea level. The flow equations for the atmosphere are non-hydrostatic and fully compressible, written in flux form [11]. The prognostic variables are: momentum, and perturbation of density and pressure, calculated with respect to the hydrostatic reference state. Smagorinsky-Lilly parameterization for sub-grid scale mixing, and Reisner scheme for cloud microphysics parameterization are used. A simple radiation scheme is adopted. The ground temperature and ground moisture are computed by using a bucket model as a simplified land model. No cumulus parameterization scheme is used, with the hypothesis that the largest part of the precipitation processes is explicitly resolved with a 5.5 km grid

resolution*. Silhouette topography is interpolated to 5.5 km from the 1 km USGS Global Land One-kilometer Base Elevation (GLOBE) terrain database. In the GCRM, as in LAMI, the influence of the wave drag on upper tropospheric flow is explicitly resolved. The GCRM programming language is Fortran90. On Earth Simulator (ES) 640 NEC SX-6 nodes (5120 vector processors) are available. The runs were performed using 192 nodes for a configuration with 32 vertical layers; the elapsed time for each run was about 6 hours. For each run the forecast time was 36 hours.
A particular hybrid model of parallel implementation is used in this software code ; the parallelism among nodes is achieved by MPI or HPF, while in each node microtasking or OpenMP is used. The theoretical peak performance using 192 nodes is about 12 TFlops, while the resulting sustained peak performance is about 6 TFlops; this means that the overall system efficiency is 48.3 %

4. The spin up

Since observation data are not incorporated into this version of GCRM, some period is required by the model to balance the information for the mass and wind fields, coming from the initial data interpolated by analysis* (data from Japan Meteorological Agency, JMA). This feature gives rise to spurious high-frequency oscillations of high amplitude during the initial hours of integration. This behaviour is called "spin up". For GCRM the spin up is also amplified because the precipitation fields (graupel, rain and snow) are set to zero at initial time. A more accurate interpolation and a reduction of the "spin up" phenomenon should be obtained by using a small ratio between the horizontal resolution of the analysis (data to initialize the model run) and of the model one. In our test cases the horizontal resolution of the analysis is 100 km while the GCRM resolution is 5,5 km (in some test cases also 11 km was used). Test performed with similar meteorological models have shown that a good resolution ratio is between 1:3-1:6 [12] [13]. The spin up "window" occurs for 6 to 12 hours (in the forecast time) after initialization; in order to avoid spin up problems the compared results are obtained by cumulating QPF data on 24 hours, beginning from +12 to +36 forecasts hours, then the first 12 hours of forecast are not reliable.

Without a sufficiently long spin-up period the output data may contain, as verified in our runs, strange transient values of QPF. In particular one typical spin up problem, consisting in an erroneous structure of rainfalls (with "spot" of rain maxima), has been identified in the runs performed.

---

* As Dare underlined [14] different research group noted that approximately 3 km, 4km, 5 km, may be the upper limit for highly-accurate grid scale resolution of convection.

* * the analysis is the best possible esteem of the atmosphere state; two pieces of information converge in it. The first piece derives from the observations (discontinuous and lacunoso field) and the second piece is obtained with a short term forecast of the model, called first guess, and it is a continuous field, physically coherent. The differentes tecniques for the merging of the two pieces are named "data assimilation".

4. Analysis of model performances

4.1. Visual verification forecast

The data verification is qualitative and has been obtained by the so-called called "visual" or "eyeball" technique. . This simply consists in the comparison between a map of forecast and a map of observed rainfall. This technique can also be applied to other meteorological variables as geopotential, relative humidity, wind and so on; but in these last cases it is necessary to compare the forecast maps with the analysis because no observation maps are available. As it will be demonstrated, also if this is a subjective verification, an enormous amount of useful information can be gained by examination of maps.

4.2. Results of the reference version

In the figure 5 has the forecast map of QPF obtained by the reference configuration, hereinafter CTRL is presented.

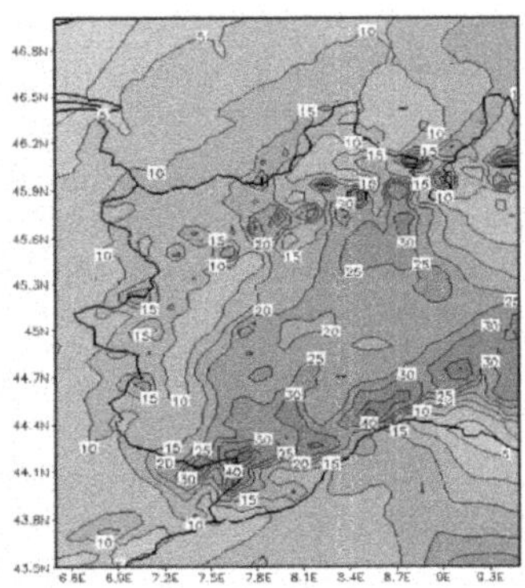

figure 5: forecast map of QPF cumulated on 24 hours (from 12UTC of 25th November to 12UTC of 26th November) for the CTRL run

Comparing the forecast (figure 5) with the observed map (figure 3) is clear that there is a general underestimation over mountain areas, and a weak overestimation on the central lowlands. The three QPF components, rain, graupel* and snow are shown in figure 6. The analysis of these maps is very useful to understand the main cause of the

* The graupel can be any of the following types of solid-ice precipitation: hail (large chunks of ice such as from a strong or severe thunderstorm), sleet (small pellets of raindrops that have frozen in mid-air, in winter or a thunderstorm), snow pellets-when freezing fog forms a 2.5 mm balls of rime ice around a centre of sleet or a snowflake. The graupel does not include other frozen precipitation such as snow or ice crystal.

rain underestimation . The forecasted values of snow and graupel are not realistic as it is possible to understand looking at Milan skew diagram*(figure7)

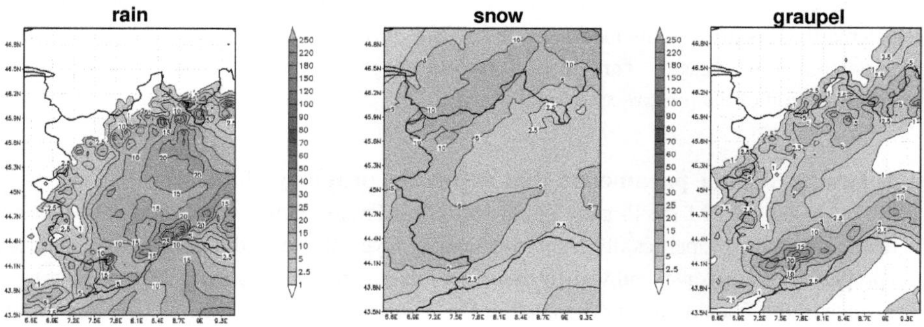

figure 6:forecast maps of the three component of QPF cumulated on 24 hours ((from 12UTC of 25th November to 12UTC of 26th November) for the CTRL run.

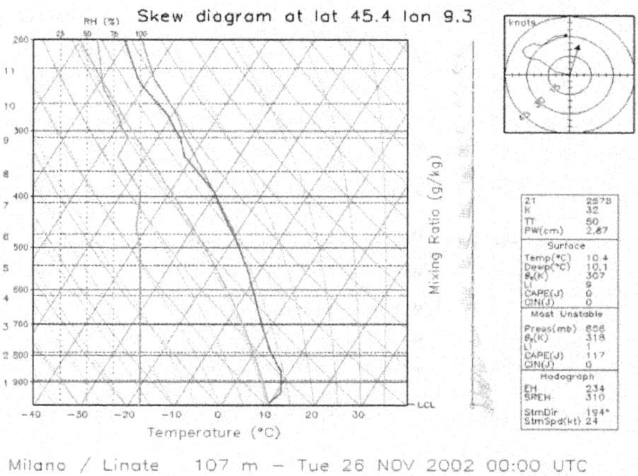

figure 7: skew diagram at 00 UTC of 26th November

* The skew diagram is derived from data collected by the radiosondes; it is a graph showing the vertical distribution of the temperature, dewpoint, wind direction, and wind speed of the radiosonde as it rises through the atmosphere near the site where the balloon was launched. Since pressure decreases with height logarithmically, the vertical axis of the graph shows higher pressure at the bottom of the chart decreasing to lower pressure at the top. The horizontal axis shows temperature in Celsius increasing from left to right but the temperature lines are "skewed" from lower left to upper right. The red and blue lines represent the temperature and dewpoint trace of the radiosonde. The temperature is always plotted to the right of the dewpoint because air temperature is always greater than the dewpoint temperature. Wind barbs representing wind speed and direction are always plotted on the right side margin. To the right of skew-T diagrams there is a list a parameters and indices. In particular, on the first line ZT is the pressure level at which the environmental sounding is exactly zero degrees Celsius.

The height of the freezing level was constantly increasing for the long duration of the southerly flow. The mean value over North-Western Italy was around 1900 meters on Saturday 23rd, and increased to 2600 meters on Tuesday the 26th. A further increase up to 2900 meters was registered the day after. This means, as observed by ground stations, that in this event only rain precipitation was present. This overestimation of graupel and snowfall can aexplain the underestimation of the forecasted amount of rain .

4.3. Tuning of some parameters that define the property of the air

In the GCRM the Prandtl and the turbulent Prandtl numbers are used. The first one is defined among the properties of planetary boundary layer; the second one is used to characterize the properties of horizontal and vertical turbulent fluxes (using Smagorinsky-Lilly scheme).

The value of the Prandtl number (hereinafter Pr) is defined by the fluid chemical composition and by its state (temperature and pressure), and is about 0.7 in the atmosphere. The turbulent Prandtl number (hereinafter Pr_turb) is defined by coefficients that are not air physical properties, but functions of flow and numerical grid characteristics. The turbulent Prandtl number depends on the horizontal numerical resolution and changes with the turbulence. In the reference version of the GCRM the Prandtl and the turbulent Prandtl numbers are set to the same value. Some runs to evaluate the model sensitivity to these two parameters has been performed [18]. The sensitivity to the thermal diffusivity (kappa) has also been checked. A Significant sensitivity was found. In particular the rain precipitation over the Mediterranean sea and over the southern coast (where the maximum was located) seems to be quite influenced by these parameters (figure 8).

	CTRL	Run1	Run2	Run3
Pr	0,0072	0,7	0,7	0,7
Pr_turb	7,2E+2	0,7	0,7	7,2E+2
kappa	1,00E-10	2,10E-5	1,00E-10	2,10E-5

table 1: some of the performed runs

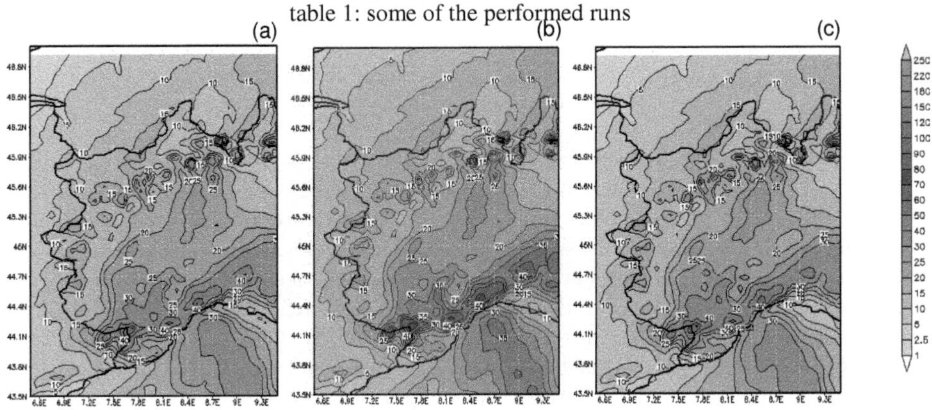

figure 8: QPF maps for some run performed: (a) run1, (b) run2, (c) run3.

5. Future developments

5.1. The microphysical schemes

The test cases performed by GCRM have shown a systematic overestimation of graupel and snow fall. This overestimation is one of the main cause of the strong underestimation of forecasted amount of rain .

The results obtained underline the necessity to improve the Reisner parameterization in GCRM. This sluod allow to obtain a better forecast of QPF. The Reisner explicit bulk mixed-phase model for microphysical parameterization consists (version used in GCRM) in five prognostic variables: water vapour mixing ratio (qv), cloud water mixing ratio (qc), rain water mixing ratio (qr), cloud ice mixing ratio (qi), snow mixing ratio (qs), graupel mixing ratio (qg), number concentration of cloud ice (Ni), number concentration of snow (Ns), number concentration of graupel (Ng) (figura 9).It is known from the literature that this scheme [15] has some overprediction of snow and graupel amount. Other teams developing meteorological models (as WRF model [16]), that previously used the Reisner parameterization, decided to upgrade the Reisner model following Thompson [17]. The principal improvements the replacement of primary ice nucleation and of auto-conversion formula, a new function for graupel distribution, and other corrections about formation of snow, rain and graupel.

5.2. The vertical grid system

Currently 32 vertical levels, defined by using the Gal-Chen terrain-following vertical-coordinate [19], are used in the reference version of GCRM, The top level is fixed at 30000 m (12 hPa in standard atmosphere). For meteorological models in which the upper boundary is supposed to be a rigid lid, the best choice, in order to avoid the spurious reflections of gravity waves, is 0 hPa; nevertheless this requirement is computationally quite expensive. A good compromise for the height-top in the global model is around 10 hPa for standard atmosphere (as ECMWF with 63 model levels [20]). Furthermore, to improve the representation of the near-bottom flow, it is necessary to increase the amount of vertical levels in the lower part of the troposphere (the bottom of the physical domain). This is useful especially in steep areas, as those involved in our test cases. This feature also requires an high computational cost. Looking at the literature, for meteorological model with a horizontal resolution of about 5-10 km, a good choice for the lowest level height is 40-80 m.

Taking into account all these considerations better results could be obtainded increasing the number of vertical level.

Acknowlegment

The authors wish to thank all the meteorological staff of the Environment Protection Regional Agency of Piedmont (ARPA) for having provided the data and maps and for help and support.

Reference

1. Takahashi K., Goto K., Fuchigami H., Yamada M., Peng X., Komine K., Ohdaira M., Sugimara T., Nagano K., 2006: Ultra high performance computation with coupled non-hydrostatic atmosphere-ocean-land GCM. Newsletter Earth Simulator Center.
2. Rabuffetti, D., Milelli, M., Graziadei, M., Mercogliano, P., Oberto, E.: Evaluation of the hydro-meteorological chain in Piemonte region, north western Italy- analysis of the Hydroptimet test case, Advances in geosciences, 2, 321-325, 2005.
3. COSMO Newsletter 5, 2005 www.cosmo-model.org.
4. G.Doms and U.Schattler, 2003: A description of the Nonhydrostatic Regional Model LM, Part 1- Dynamics and numerics. www.cosmo-model.org
5. http://www.ecmwf.int/research/ifsdocs/CY28r1/index.html
6. http://www.dwd.de/en/FundE/Analyse/Modellierung/Modellierung.htm
7. Elementi, M., Marsigli, C. Paccagnella, T.2005: High resolution forecast of heavy precipitation with Lokall Modell: analysis of the two test case studies in the Alpine area.
8. Lakhtakia, M. N., Yu, Z., Yarnal, B., White, R. A., Miller, D. A., 1999: Sensitivity of simulated surface runoff to mesoscale meteorological model resolution in a linked-model experiment. Climate Research, Oldendorf/Luhe, Germany, 12(1), 15-27.
9. Jasper, K., J. Gurtz and H. Lang, 2002. Advanced flood forecasting in Alpine watersheds by coupling meteorological observations and forecasts with a distributed hydrological model. Journal of Hydrology 267 (1-2): 38-50.
10. Kageyama, A.and Sato, T. 2004 " The Yin Yang Grid": An overset grid in spherical geometry. Geochem.Geophys.Geosyst., 5, Q09005, doi:10.1029/2004GC00734
11. Satomura, T. and Akiba, S. 2003. Development of high precision non-hydrostatic atmospheric model (1): Governing equations, Annual of Disas. Prev. Res. Inst., Kyoto Univ., 46B, 331-33.
12. http://atmet.com/html/docs/rams/rams_techman.pdf (RAMS technical description)
13. http://www.mmm.ucar.edu/mm5/documents/ , PSU/NCAR Mesoscale Modeling System Tutorial Class Notes and Users Guide (MM5 Modeling system version 3), MM5
14. Dare, R., 2003 : Modelling and parameterization of convection on the mesoscale. 15th Annual BMRC (Bureau of Meteorology Research Centre) Modelling Workshop.
15. Reisner, J., R.M. Rasmussen, R.T. Bruintjes, 1998: Explicit forecasting of supercooled liquid water in winter storms using the MM5 mesoscale model. Quart. J.Roy. Meteor. Soc., 124, 1071-1107
16. A description of the Advanced Research WRF version 2, NCAR technical note June 2005. http://www.wrf-model.org/wrfadmin/docs/arw_v2.pdf
17. Thompson, Gregory, R.M. Rasmussen, K. Manning, 2004: Explicit forecasts of winter precipitation using an improved bulk microphysics scheme. Part 1: description and sensitivity analysis. Mon. Wea. Rev., 132, 519-542
18. An introduction to boundary layer meteorology, Stull R.B., Atmospheric Sciences Library
19. Gal-Chen, T. and R.C.J. Sommerville, 1975: On the use of the coordinate transformation for the solution of the Navier-Stokes equations. J. Comput. Phys,.17, 209-228
20. http://www.ecmwf.int/products/data/technical/model_levels

Parallel Computational Fluid Dynamics – Parallel Computing and Its Applications
J.H. Kwon, A. Ecer, J. Periaux, N. Satofuka and P. Fox (Editors)

Computational Performance Evaluation of a Limited Area Meteorological Model by using the Earth Simulator

G. Ceci[a], R. Mella[a], P. Schiano[a], K. Takahashi[b], H. Fuchigami[b]

[a] Informatics and Computing Department, Italian Aerospace Research Center (CIRA), via Maiorise, I-81043 Capua (CE) - Italy
e-mail: g.ceci@cira.it - web page: www.cira.it

[b] Multiscale Climate Simulation Project Research, Earth Simulator Center/Japan Agency for Marine-Earth Science and Tech., 3173-25 Showa-machi, Kanazawa-ku, Yokohama Kanagawa 236-0001, Japan
e-mail: takahasi@jamstec.go.jp - web page: http://www.es.jamstec.go.jp

Key words: parallel performance, numerical discretization, Navier-Stokes equations

INTRODUCTION

The work described in this paper comes from an agreement signed between ESC/JAMSTEC (Earth Simulator Center/Japan Agency for Marine-Earth Science and Technology) [1] and CIRA (Italian Aerospace Research Center) [2] to study hydro-geological (landslides, floods, etc.) and local meteorological phenomena.
It is well known that meteorological models are computationally demanding and they require both accurate and efficient numerical models on high performance parallel computing. Such needs are strictly and directly related to the high resolution of the model. The interest in this paper is about computational features, parallel performance and scalability analysis on Earth Simulator (ES) supercomputer of GCRM (Global Cloud Resolving Model - regional version) meteorolgical model, currently developed at ESC by the Multiscale Climate Simulation Project Research Group.
The paper is organized as follows. In Section 1 the meteorological model is described, emphasizing above all numerical and computational features. In Section 2 the ES system and the types of parallelism available on it are presented. In Section 3 details related to computational experiments and valuation criteria are discussed. In Section 4 and 5, finally, computational results are shown and discussed, respectively.

1. THE GLOBAL/REGIONAL NON-HYDROSTATIC ATMOSPHERIC MODEL

The model is based on non-hydrostatic fully compressible three-dimensional Navier-Stokes equations in flux form; the prognostic variables are density perturbation, pressure perturbation, three components of momentum and temperature [9][10][11]. Equations are casted in spherical coordinates and they are discretized on an Arakawa C/Lorenz grid by using finite difference methods; the grid structure has a horizontal resolution of 5.5 km and 32 levels are used vertically in terrain-following σ-coordinate. A time splitting procedure is applied for fast/slow waves integration in order to improve computational efficiency [3][4][5] following Skamarock-Wicher formulation [6][7]; in particular, it employs Runge Kutta Gill method (2nd, 3rd and 4th order) for advection tendencies (using large time step) and forward/backward scheme for fast modes (using small time step).

The regional version is nested in a global/larger regional model (coarse grid) by 1-way interactive nesting. A sponge/relaxation boundary condition Davies-like [8] is used to allow the atmospheric interaction between the interior domain (regional domain on fine grid) and the external domain (global/larger regional model on coarse grid). Strictly speaking, the computational domain (fine grid) is composed by a halo area (outside the physical domain) and a sponge area (inside the physical domain); in the first one relaxation values of prognostic variables are defined by interpolation from the coarser grid or by a Newman condition, while in the latter prognostic variables are relaxed. Three types of sponge functions are implemented (hyperbolic, trigonometric and linear). More details about the model are available in [9] and a summary of the features discussed above is in Table 1.

Equation system		Non-hydrostatic fully compressible Navier-Stokes (flux form)
Prognostic variables		Density perturbation, pressure perturbation, three components of momentum and temperature
Grid system		Equations casted in spherical coordinates and discretized on an Arakawa-C/Lorenz grid by FDMs.
Horizontal Resolution and Vertical Levels		Horizontal resolution 5.5 km; 32 vertical levels in terrain-following coordinate
Time integration		HE-VI. Time splitting with Runge Kutta Gill 2nd/3rd/4th order for advection tendencies (large time step) and Forward-Backward for fast modes (small time step)
Boundary Conditions	**Lateral**	Sponge/relaxation boundary condition Davies-like
	Upper	Rayleigh friction

Table 1 Numerical features of the GCRM meteorological model (limited area version)

2. THE ES SYSTEM

The ES is a distributed-memory parallel system made up by 640 NEC SX-6 nodes and each node is a parallel/vectorial shared-memory MIMD (Multiple Instruction Multiple Data) computer having 8 CPUs. The peak performance is about 40 TFLOPS (64 GFLOPS per node), the total main memory is of 10 TB and currently it is one of the most powerful supercomputers in the world [12].
The operating system is SUPER-UX, which is based on Unix System V, suitable languages and libraries (FORTRAN90/ES, C++/ES, MPI/ES, OpenMP) are provided to achieve high performance parallel computing. In particular, two types of parallelism can be implemented: hybrid and flat. The first one means that parallelism among nodes (inter-node parallelism) is achieved by MPI (Message Passing Interface) or HPF (High Performance Fortran), while in each node (intra-node parallelism) by microtasking or OpenMP. The latter, instead, means that both intra- and inter-node parallelism is archived by HPF or MPI; consequently the ES is viewed as made up by 8x640 processors. Roughly speaking, in the first case a program is subdivided into *n* MPI-tasks (each one running on each node) and the intra-node parallelism is performed by OpenMP or microtasking. In the second case, instead, a program is subdivided into *n* MPI-tasks running on *n* processors. Detailed description of ES is reported in [1].

3. IMPLEMENTATION DETAILS AND COMPUTATIONAL EXPERIMENTS

The programming language is Fortran90 and the algorithm is based on a SPMD (Single Program Multiple Data) program. The parallel computation of the non-hydrostatic atmospheric model, which has been developed at ESC, is implemented by a hybrid parallelization model: intra- and inter-node parallelisms are performed by micro-tasking and by MPI, respectively.
The selected test refers to an intense precipitation event occurred during November 2002 on North-Western of Italy.
To study scalability of the code (performance changing varying problem size or number of processors) experiments have been carried out keeping constant the problem size and varying the number of processors. The sizes along *x*-, *y*- and *z*-direction of the considered computational domains are 600x1152x32 (small domain), 1200x1152x32 (medium domain, twice bigger) and 1200x2304x32 (large domain, four times bigger). In order to reduce the communication cost a block-wise processors grid is adopted in horizontal for all the performed tests; consequently each processor works on a sub-domain having 32 vertical levels. The number of used processors varies between 16 and 512 (that means 2 and 64 nodes on the ES, respectively); in all cases the step incrementing the number of processors is by powers of two.
To evaluate runtime performances suitable compilation options and environment variables have been set. In particular, the performance analysis tool *ftrace* has been used to collect detailed information on each called procedure (both subroutine and function); we emphasize that it is better to use this option just for tuning since it causes an overhead. Such collected information is related to elapsed/user/system times, vector

instruction execution time, MOPS (million of operations per second), MFLOPS (million of floating-point operations per second), VLEN (average vector length), Vector Operation Ratio (vector operation rate), MIPS (million of instructions per second) and so on. Criteria to achieve best performance are: vector time has to be close to user time; MFLOPS should be obviously as high as possible; VLEN, being the average vector length processed by CPUs, should be as much as possible close to the hardware vector length (256 on NEC SX-6) and, consequently, best performances are obtained on longer loops; vector operation rate (%) should be close to 100. Such information has been took into account for each performed experiment.

Scalability has been evaluated in terms of speed-up and efficiency. Since the program cannot be run sequentially, both speed-up and efficiency have been calculated using the elapsed time obtained running the parallel program with the minimum number of processors. In particular, speed-up has been evaluated by Amdhal's Law [13] in the following form:

$$S_p = \frac{T_2}{T_p}$$

where T_2 and T_p are the wall clock times on 2 nodes (16 processors) and p nodes ($p*8$ processors), respectively.

4. COMPUTATIONAL RESULTS

A feature verified in all performed tests is the good load balance among MPI processes: in fact, as showed in Table 2, the minimum value of the elapsed time is always very close to the maximum one.

domain size	# nodes	# processors	min (sec)	max (sec)
600x1152x32	2	16	1098,735	1098,845
	4	32	568,688	568,918
	8	64	292,654	293,462
	16	128	165,918	166,260
	32	256	91,712	92,698
1200x1152x32	4	32	1117,389	1117,858
	8	64	575,914	576,783
	16	128	339,519	341,198
	32	256	170,163	172,292
	64	512	100,435	101,677
1200x2304x32	8	64	1419,989	1420,469
	16	128	733,698	733,945
	32	256	387,153	387,458
	64	512	215,181	216,717

Table 2 Minimum and maximum real times

Computational results were analyzed according to criteria discussed in the previous Section. Results concerning elapsed times and speed-up are shown in Figure 2, Figure 3 and Figure 4 going on from the small (600x1152x32), to the medium (1200x1152x32) to the large (1200x2304x32) domain, respectively. Efficiency related to each run is in Table 3.

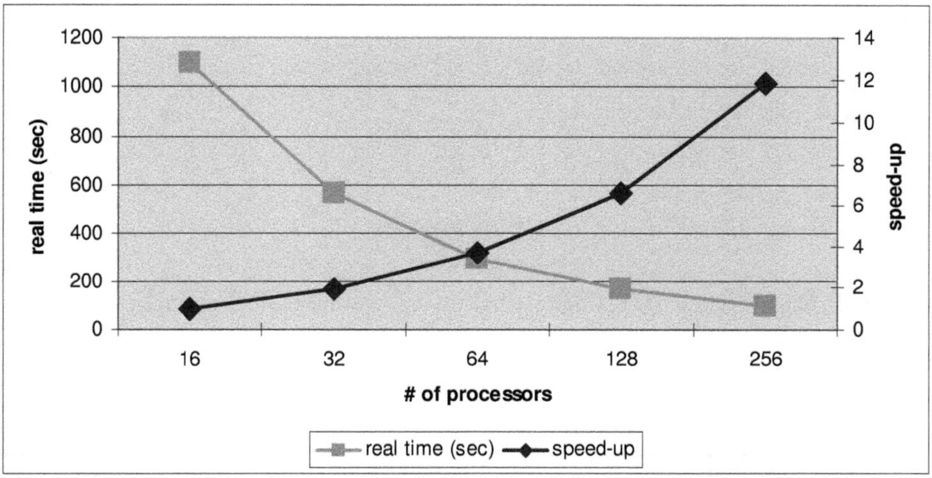

Figure 2 Elapsed time (sec.) and speed-up vs. number of processors on domain 600x1152x32

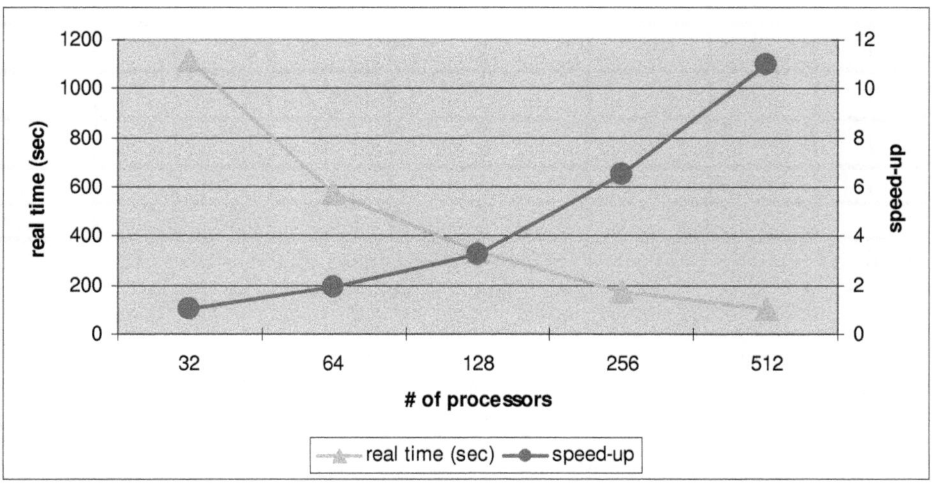

Figure 3 Elapsed time (sec.) and speed-up vs. number of processors on domain 1200x1152x32

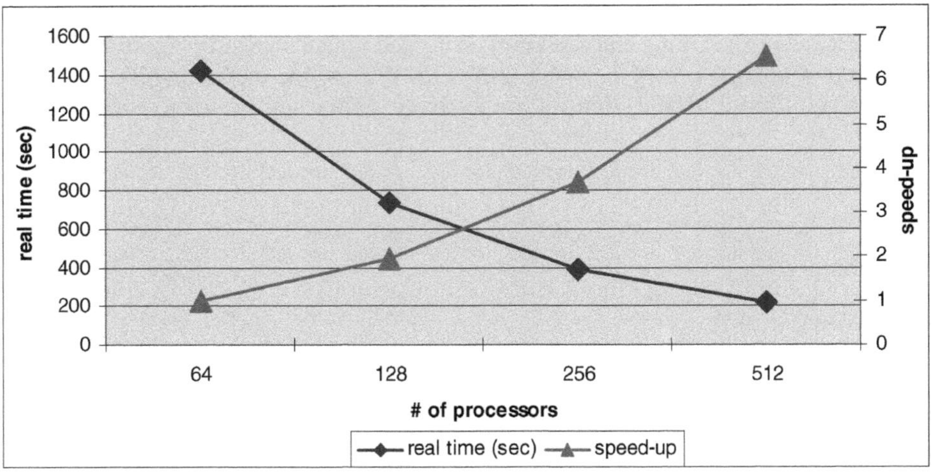

Figure 4 Elapsed time (sec.) and speed-up vs. number of processors on domain 1200x2304x32

Domain 600x1152x32			Domain 1200x1152x32			Domain 1200x2304x32		
Procs along *x*	Procs along *y*	Efficiency (%)	Procs along *x*	Procs along *y*	Efficiency (%)	Procs along *x*	Procs along *y*	Efficiency (%)
1	16	-						
1	32	96.5	2	16	-			
1	64	93.8	2	32	96.88	2	32	-
1	128	82.5	2	64	82.04	2	64	96.76
1	256	74.1	2	128	81.01	2	128	91.63
			2	256	68.68	2	256	81.79

Table 3 Efficiency (%) on each domain obtained changing the number of processors along *x*- and *y*-directions

Performance analysis concerns also performed GFLOPS: analogously, the sustained peak performance obtained on small, medium and large domain is shown in Figures 5 and a comparison of sustained peak performance versus the theoretical one is in Table 6.

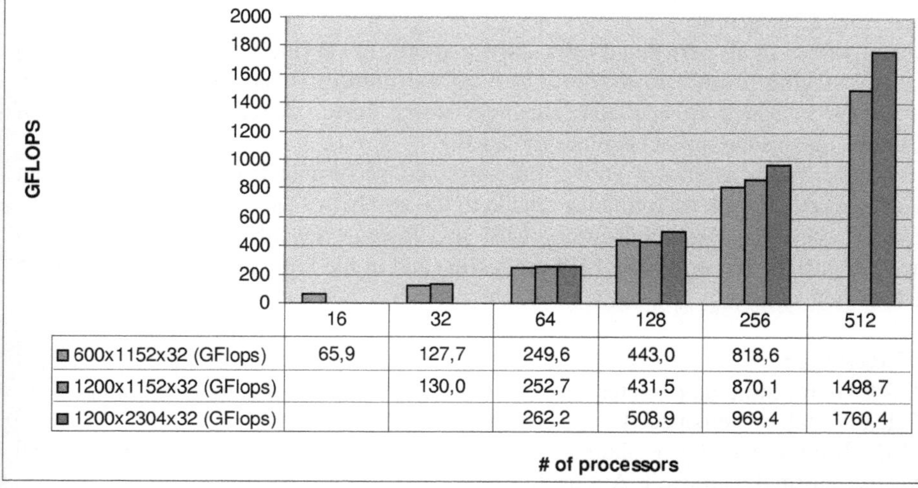

	16	32	64	128	256	512
600x1152x32 (GFlops)	65,9	127,7	249,6	443,0	818,6	
1200x1152x32 (GFlops)		130,0	252,7	431,5	870,1	1498,7
1200x2304x32 (GFlops)			262,2	508,9	969,4	1760,4

Figure 5 Sustained peak performance (GFLOPS) vs. number of processors

Domain 600x1152x32			Domain 1200x1152x32			Domain 1200x2304x32		
Procs along *x*	Procs along *y*	Sust/theor p.p. (%)	Procs along *x*	Procs along *y*	Sust/theor p.p. (%)	Procs along *x*	Procs along *y*	Sust/theor p.p. (%)
1	16	51.50						
1	32	49.89	2	16	50.79			
1	64	48.75	2	32	49.36	2	32	51.21
1	128	43.27	2	64	42.14	2	64	49.70
1	256	39.97	2	128	42.48	2	128	47.33
			2	256	36.59	2	256	42.98

Table 6 Comparison of sustained with theoretical peak performance (%) on each domain varying the number of processors

. CONCLUSIONS

Results shown in the Section 4 are very satisfactory and promising for the future developments.

In all cases, *ftrace* tool shows that MFLOPS achieved in each routine are more or less 0% of the theoretical peak performance and that Vector Operation Ratio is very close to 100%. The small gap between the minimum and maximum real time testifies a high

granularity and a good load balancing among processors: all times vary between 0.11 seconds and 2.13 seconds.
Obtained values in all performed tests show a speed-up varying nearby linearly and an efficiency staying always up to 70%. Varying the number of nodes on each domain, such values of speed-up and efficiency get worse since, increasing the number of processors, each subdomain becomes too small.
On the other hand, as shown above, the sustained performance is greater than about 40% of theoretical peak performance related to the number of used processors.
Moreover, the hybrid parallelization (MPI processes + OpenMP threads) has been compared with the flat one (only MPI processes) and in the first case an efficiency up to 25% has been achieved.

REFERENCES

[1] E.S.C. Earth Simulator Center http://www.es.jamstec.go.jp/esc/eng/index.html
[2] C.I.R.A. Italian Aerospace Research Center http://www.cira.it/
[3] J.B. Klemp and R. Wilhelmson (1978) The simulation of three-dimensional convective storm dynamics, J. Atmos. Sci., 35, 1070-1096.
[4] W.C. Skamarok and J.B. Klemp (1992) *The stability of time-split numerical methods for the hydrostatic and the nonhydrostatic elastic equation*, Mon. Wea. Rev. 120, 2109-2127.
[5] W.C. Skamarok and J.B. Klemp (1994) *Efficiency and accuracy of the Klemp-Wilhelmson time-splitting technique*, Mon. Wea. Rev. 122, 2623-2630
[6] W.C. Skamarock and L.J. Wicher (2001) Time-splitting methods for elastic models using forward time schemes, Monthly Weather Review, 130, 2088-2097.
[7] W.C. Skamarock and L.J. Wicher (1998) *A Time-splitting methods for elastic equations incorporating second-order Runge-Kutta time differencing* Monthly Weather Review, 126, 1992-1999.
[8] H.C. Davies (1976) *A Lateral boundary formulation for multi-level prediction models*, Quart. J. R. Met. Soc., 102, 405-408.
[9] K. Takahashi, X. Peng, K. Komine, M. Ohdaira, Y. Abe, T. Sugimura, K. Goto, H. Fuchigami, M. Yamada, K. Watanabe (2004) *Development of Non-hydrostatic Coupled Ocean-Atmosphere Simulation Code on the Earth Simulator* Proceeding of 7th International Conference on High Performance Computing and Grid in Asia Pacific Region, IEEE Computer Society, Ohmiya, 487-495.
[10] K. Takahashi, X. Peng, K. Komine, M. Ohdaira, K. Goto, M. Yamada, H. Fuchigami, T. Sugimura (2005) *Non-hydrostatic atmospheric GCM development and its computational performance*, in "Use of High Performance computing in meteorology", Walter Zwieflhofer and George Mozdzynski Eds., World Scientific, 50-62.
[11] K. Takahashi (2006) *Development of Holistic Climate Simulation codes for a non-hydrostatic atmosphere-ocean couplet system*, Annual Report of the Earth Simulator Center, available at http://www.es.jamstec.go.jp/esc/images/annual2004/index.html
[12] Top 500 web site: http://www.top500.org/
[13] G.M. Amdahl (1967) *Validity of the single-processor approach to achieving large scale computing capabilities* in *AFIPS Conference Proceedings*, 30 (Atlantic City, N.J., Apr. 18-20). AFIPS Press, Reston, Va., 483-485.

Parallel Computational Fluid Dynamics – Parallel Computing and Its Applications
J.H. Kwon, A. Ecer, J. Periaux, N. Satofuka and P. Fox (Editors)

A scalable high-order discontinuous Galerkin method for global atmospheric modeling*

Hae-Won Choi[a†], Ramachandran D. Nair[a‡] and Henry M. Tufo[a§]

[a]Scientific Computing Division, National Center for Atmospheric Research (NCAR), 1850 Table Mesa Drive, Boulder, CO 80305, USA

A conservative 3-D discontinuous Galerkin (DG) baroclinic model has been developed in the NCAR High-Order Method Modeling Environment (HOMME) to investigate global atmospheric flows. The computational domain is a cubed-sphere free from coordinate singularities. The DG discretization uses a high-order nodal basis set of orthogonal Lagrange-Legendre polynomials and fluxes of inter-element boundaries are approximated with Lax-Friedrichs numerical flux. The vertical discretization follows the 1-D vertical Lagrangian coordinates approach combined with the cell-integrated semi-Lagrangian method to preserve conservative remapping. Time integration follows the third-order SSP-RK scheme. To valid proposed 3-D DG model, the baroclinic instability test suite proposed by Jablonowski and Williamson is investigated. Parallel performance is evaluated on IBM Blue Gene/L and IBM POWER5 p575 supercomputers.

1. INTRODUCTION

The future evolution of the Community Climate System Model (CCSM) into an Earth system model will require a highly scalable and accurate flux-form formulation of atmospheric dynamics. Flux form is required in order to conserve tracer species in the atmosphere and accurate numerical schemes are essential to ensure high-fidelity simulations capable of capturing the convective dynamics in the atmosphere and their contribution to the global hydrological cycle. Scalable performance is necessary to exploit the massively-parallel petascale systems that will dominate high-performance computing (HPC) for the foreseeable future.

The High-Order Method Modeling Environment (HOMME) [5], developed by the Scientific Computing Section at the National Center for Atmospheric Research (NCAR), is a vehicle to investigate using high-order-element-based methods to build conservative and accurate dynamical cores. HOMME employs the spectral element (SE) methods on a cubed-sphere tiled with quadrilateral elements, can be configured to solve the shallow water or the dry/moist primitive equations, and has been shown to efficiently scale to 32,768 processors of an IBM Blue Gene/L (BG/L) system [16]. Nevertheless, a major disadvantage of the SE atmospheric model is that it is not inherently conservative. For climate and atmospheric applications, conservation of integral invariants such as mass and total energy is of significant importance. To resolve these issues, we recently included the DG atmospheric models to support HOMME framework.

In this paper we discuss our extension of the HOMME framework to include a 3-D DG option as a first step towards providing the atmospheric science community a new generation of atmospheric general circulation models (AGCMs). The DG method [2], which is a hybrid technique combining the finite element and finite volume methods, is inherently conservative and shares the same computational advantages as the SE method such as scalability, high-order accuracy, spectral convergence, and thus is an ideal candidate for climate modeling. The DG method is employed on a quadrilateral mesh of elements using a high-order nodal basis set of orthogonal Lagrange-Legendre polynomials with Gauss-Lobatto-Legendre (GLL) quadrature points. Time integration follows the strong stability-preserving Runge-Kutta (SSP-RK) scheme of Gottlieb *et al.* [4]. The globe is based on the singularity-free cubed-sphere geometry

*This paper is supported by the **DOE-SciDAC** program under award **#DE-FG02-04ER63870**.
†Corresponding Author. Email: `haewon@ucar.edu`.
‡Email: `rnair@ucar.edu`.
§Email: `tufo@ucar.edu`.

introduced by [13]. Parallelism is effected through a hybrid MPI/OpenMP design and domain decomposition through the space-filling curve approach described in [3]. Our work extends earlier efforts [3,10–12] in several important ways: first, we develop a scalable conservative 3-D DG-based dynamical core based on the hydrostatic primitive equations; second, we employ the vertical Lagrangian coordinate approach, developed by Starr [15] and later generalized by Lin [8]; and finally, we apply the 1-D cell-integrated semi-Lagrangian method [9] to preserve conservative remapping.

2. CONSERVATIVE DISCONTINUOUS GALERKIN MODEL

Inherently conservative numerical schemes are of fundamental importance in atmospheric and climate modelings in order to pertain conservation properties such as mass and total energy. Toward this effort, the 2-D DG shallow water model in the HOMME framework [10,11] has been recently extended to 3-D DG baroclinic models [12]. Main features of DG baroclinic model are the vertical discretization and the prognostic equations which are based on hyperbolic conservation laws where as the prognostic variables are pressure thickness δp, covariant wind vectors (u_1, u_2), potential temperature Θ, and moisture q.

2.1. Hydrostatic primitive equations on the cubed-sphere

The hydrostatic primitive equations in curvilinear coordinates employ the cubed-sphere geometry followed by [10,11]. A sphere is decomposed into '6 identical regions' by an equiangular central projection of

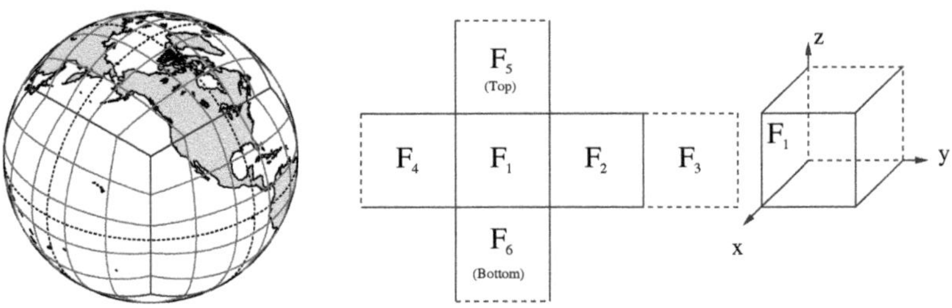

Figure 1. Cubed-sphere geometry for $N_{elem} = 6 \times 5 \times 5$ DG elements (left). Logical orientation of cube faces in HOMME (right).

the faces of an inscribed cube as displayed in Figure 1. This results in a nonorthogonal curvilinear (x^1, x^2) coordinate system free of singularities for each face of the cubed-sphere, such that $x^1, x^2 \in [-\pi/4, \pi/4]$. Each face of the cubed-sphere is partitioned into $N_e \times N_e$ rectangular non-overlapping elements (total number of element, $N_{elem} = 6 \times N_e^2$). The elements are further mapped onto the reference element bounded by $[-1,1] \otimes [-1,1]$ which has $N_v \times N_v$ (or $N_p \times N_p$) GLL grid points. Note that N_v and N_p denote the number of velocity and pressure points, respectively. The associated metric tensor, i.e., G_{ij}, in terms of longitude-latitude (λ, θ) is defined as follows:

$$G_{ij} = \mathbf{A}^T \mathbf{A}; \quad \mathbf{A} = \begin{bmatrix} R\cos\theta\, \partial\lambda/\partial x^1 & R\cos\theta\, \partial\lambda/\partial x^2 \\ R\,\partial\theta/\partial x^1 & R\,\partial\theta/\partial x^2 \end{bmatrix}. \tag{1}$$

The matrix $\mathbf{A}$ is used for transforming spherical velocity (u, v) to the 'covariant' (u_1, u_2) and 'contravariant' (u^1, u^2) 'cubed-sphere' velocity such that:

$$\begin{bmatrix} u \\ v \end{bmatrix} = \mathbf{A} \begin{bmatrix} u^1 \\ u^2 \end{bmatrix}; \quad u^i = G^{ij} u_j; \quad u_i = G_{ij} u^j; \quad G^{ij} = (G_{ij})^{-1}. \tag{2}$$

The hydrostatic primitive equations, consisting of the momentum, mass continuity, thermodynamic and moisture transport equations, can be expressed as a conservative form in curvilinear coordinates.

$$\frac{\partial u_1}{\partial t} + \nabla_c \cdot \vec{E}_1 = \sqrt{G} u^2 (f + \zeta) - RT \frac{\partial}{\partial x^1} (\ln p), \tag{3}$$

$$\frac{\partial u_2}{\partial t} + \nabla_c \cdot \vec{E}_2 = -\sqrt{G}u^1 (f + \zeta) - RT\frac{\partial}{\partial x^2} (\ln p), \tag{4}$$

$$\frac{\partial}{\partial t} (\Delta p) + \nabla_c \cdot \left(\mathbf{U}^j \Delta p\right) = 0, \tag{5}$$

$$\frac{\partial}{\partial t} (\Theta\Delta p) + \nabla_c \cdot \left(\mathbf{U}^j \Theta\Delta p\right) = 0, \tag{6}$$

$$\frac{\partial}{\partial t} (q\Delta p) + \nabla_c \cdot \left(\mathbf{U}^j q\Delta p\right) = 0, \tag{7}$$

where

$$\nabla_c \equiv \left(\frac{\partial}{\partial x^1}, \frac{\partial}{\partial x^2}\right), \ \vec{E}_1 = (E, 0), \ \vec{E}_2 = (0, E), \ E = \Phi + \frac{1}{2}\left(u_1 u^1 + u_2 u^2\right),$$

$$\mathbf{U}^j = \left(u^1, u^2\right), \ \Delta p = \sqrt{G}\delta p, \ \Theta = T\left(p_0/p\right)^\kappa, \ \kappa = R/C_p, \tag{8}$$

where E is the energy term, ζ is the relative vorticity, $\Phi = gh$ is the geopotential height and f is the Coriolis parameter.

2.2. Vertical discretization

The vertical discretization follows the 1-D vertical Lagrangian coordinates of Starr [15] based on an 'evolve and remap' approach developed by Lin [8]. A terrain following Lagrangian vertical coordinate, as shown in Figure 2 (left), can be constructed by treating any reference Eulerian coordinate as a material surface. The Lagrangian surface are subject to deform in the vertical direction during the integration, and need to be re-mapped onto a reference coordinate at regular intervals of time. By virtue of this approach, the hydrostatic atmosphere is vertically subdivided into a finite number of pressure intervals or pressure thicknesses. Moreover, the vertical coordinates and advection terms are absent thanks to the Lagrangian framework.

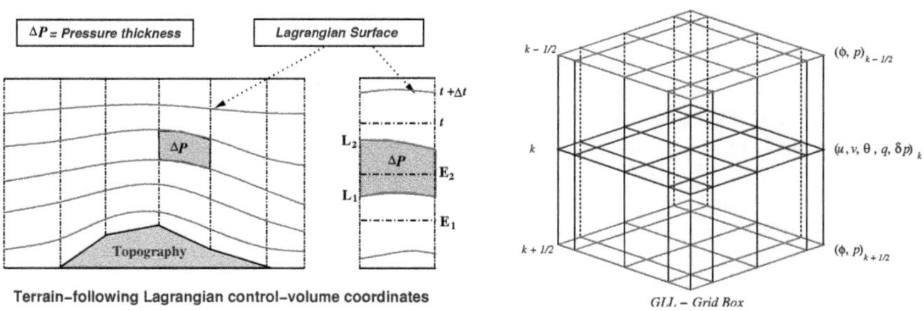

Figure 2. Lagrangian vertical coordinates system (left). The 3-D grid structure for the DG baroclinic model (right).

The entire 3-D system can be treated as a vertically stacked shallow water 2-D DG models, as demonstrated in Figure 2 (right), where the vertical levels are coupled only by the discretized hydrostatic relation. Therefore, vertical structures involve no parallel communications. Following Lin [8], at every time step δp is predicted at model levels and used to determine pressure at Lagrangian surfaces by summing the pressure thickness from top (p_∞) to bottom (p_s), $p_\ell = p_\infty + \sum_{k=1}^{\ell} \delta p_k$. The geopotential height at interfaces is obtained by using the hydrostatic relation, i.e., $\Delta\Phi = -C_p \Theta \Delta\Pi$ where $\Pi = (p/p_0)^\kappa$, and summing the geopotential height from bottom (Φ_s) to top, $\Phi_\ell = \Phi_s + \sum_{k=1}^{\ell} \Delta\Phi_k$. For the baroclinic model, the velocity fields (u_1, u_2), the moisture q, and total energy (Γ_E) are remapped onto the reference Eulerian coordinates using the 1-D conservative cell integrated semi-Lagrangian (CISL) method of Nair and Machenhauer [9]. The temperature field Θ is retrieved from the remapped total energy Γ_E.

2.3. DG discretization

The flux form of DG discretization can be formulated such that

$$\frac{\partial}{\partial t}\mathbf{U} + \nabla_c \cdot \vec{\mathcal{F}}(\mathbf{U}) = \mathbf{S}(\mathbf{U}), \tag{9}$$

where $\mathbf{U} = \left[u_1, u_2, \sqrt{G}\delta p, \Theta, q\right]^T$ denotes prognostic variables, $\vec{\mathcal{F}}(\mathbf{U})$ is flux function, and $\mathbf{S}(\mathbf{U})$ is source term. The corresponding weak Galerkin formulation of 3-D DG model can be written such that

$$\frac{\partial}{\partial t}\int_{\Omega^k} \mathbf{U}_h \cdot \vec{\varphi}_h \, d\Omega^k = \int_{\Omega^k} \vec{\mathcal{F}}(\mathbf{U}_h) \cdot \nabla_c \vec{\varphi}_h \, d\Omega^k - \oint_{\partial\Omega^k} \left(\vec{\mathcal{F}}(\mathbf{U}_h) \cdot \vec{\mathbf{n}}\right) \cdot \vec{\varphi}_h \, ds + \int_{\Omega^k} \mathbf{S}(\mathbf{U}_h) \cdot \vec{\varphi}_h \, d\Omega^k, \tag{10}$$

where the jump discontinuity at an element boundary requires the solution of a Riemann problem where the flux term $\vec{\mathcal{F}}(\mathbf{U}_h) \cdot \vec{\mathbf{n}}$ can be approximated by a Lax-Friedrichs numerical flux as shown in [11]. The resulting DG discretization leads to following ordinary differential equation (ODE):

$$\frac{d\mathbf{U}_h}{dt} = \mathbf{L}(\mathbf{U}_h), \quad \mathbf{U}_h \in (0, T) \times \Omega^k. \tag{11}$$

The above ODE can be solved by explicit time integration strategy such as the third-order Strong Stability Preserving Runge-Kutta (SSP-RK) scheme by Gottlib *et al.* [4].

$$\mathbf{U}_h^{(1)} = \mathbf{U}_h^{(n)} + \Delta t \mathbf{L}\left(\mathbf{U}_h^{(n)}\right), \tag{12}$$

$$\mathbf{U}_h^{(2)} = \frac{3}{4}\mathbf{U}_h^{(n)} + \frac{1}{4}\mathbf{U}_h^{(1)} + \frac{1}{4}\Delta t \mathbf{L}\left(\mathbf{U}_h^{(1)}\right), \tag{13}$$

$$\mathbf{U}_h^{(n+1)} = \frac{1}{3}\mathbf{U}_h^{(n)} + \frac{2}{3}\mathbf{U}_h^{(2)} + \frac{2}{3}\Delta t \mathbf{L}\left(\mathbf{U}_h^{(2)}\right). \tag{14}$$

3. NUMERICAL TEST

The baroclinic instability test proposed by Jablonowski and Williamson [6,7] is used to assess the evolution of an idealized baroclinic wave in the Northern Hemisphere. The baroclinic waves are triggered when overlaying the steady-state initial conditions with the zonal wind perturbation where as the initial conditions are given as quasi-realistic analytic expressions in [6,7]. Numerical computations are performed for the conservative 3-D DG model with 9th order polynomials (i.e., $N_v = N_p = 10$), horizontal resolution of 26 Lagrangian surfaces (i.e., $N_{lev} = 26$ where N_{lev} denotes the number of vertical levels) and total number of elements $N_{elem} = 216$. This case has 561,600 total degrees of freedoms (*d.o.f*). A Boyd-Vandeven filter in [1] is used for spatial filtering. Figure 3 demonstrates the triggering baroclinic waves and corresponding surface pressure P_s and temperature field T at 850 hPa from day 6 to day 10 . At day 6 the surface pressure shows few weak high and low pressure contours which leads to growth of very small-amplitude waves in the temperature field. At day 8 the baroclinic instability waves in surface pressure are well developed and the temperature waves can clearly be noticed. At day 10 the strong baroclinic pressure waves lead to two waves in the temperature field that have almost peaked and are beginning to wrap around the trailing fronts.

4. PARALLEL IMPLEMENTATION

The parallel implementation of HOMME is based on a hybrid MPI/OpenMP approach and domain decomposition is applied through Hilbert space-filling curve approach, Sagan [14] and Dennis *et al.* [3]. The approach generates best partitions when $N_e = 2^\ell 3^m 5^n$, where ℓ, m, and n are integers. The first step to partitioning the computing grid involves the mapping of the 2-D surface of the cubed-sphere into a linear array. Figure 4 illustrates the Hilbert space-filling curve and elements when $N_{elem} = 24$. Then the second step involves partitioning the linear array into P contiguous groups, where P is the number of MPI tasks. The space-filling curve partitioning creates contiguous groups of elements and load-balances

To perform parallel computing experiments, we uses the IBM Blue Gene/L (BG/L) and IBM POWER5 p575 systems at NCAR. The configuration of these systems is summarized in Table 1. A Message Passing Interface (MPI) job for IBM BG/L machine can be run in coprocessor mode (i.e., a single MPI task runs on each compute node) or in virtual-node mode (i.e., two MPI tasks are run on each compute

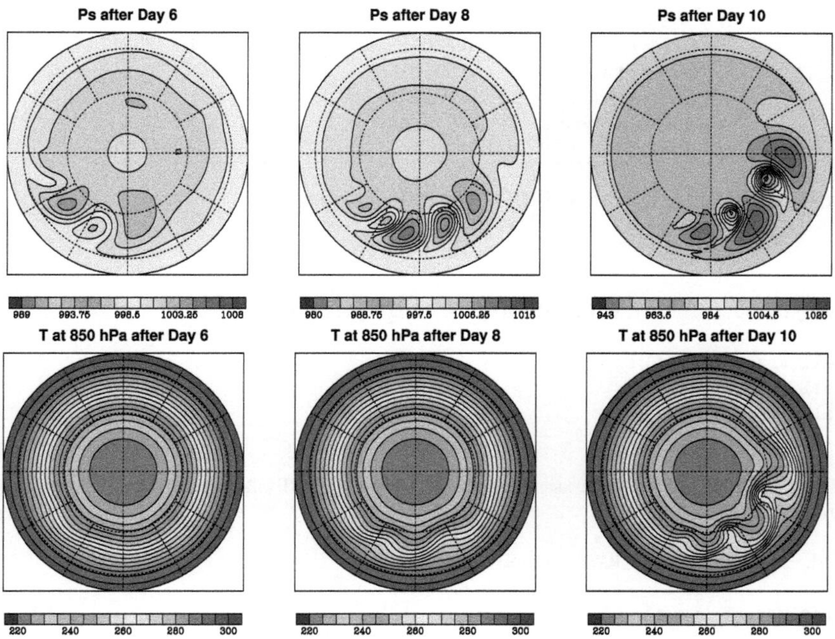

igure 3. Evolution of the baroclinic wave from integration day 6 to day 10: Surface pressure P_s [hPa] top) and Temperature field [K] at 850 hPa (bottom).

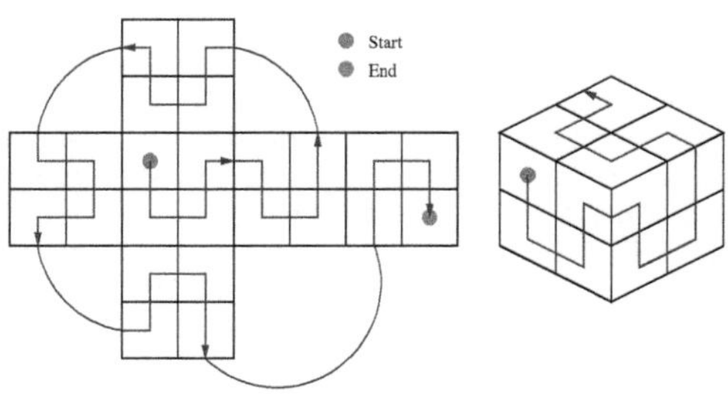

igure 4. A mapping of a Hilbert space-filling curve for $N_{elem} = 24$ cubed-sphere grid.

ode). On the other hand, 4 to 8 MPI tasks on each compute node are performed for IBM POWER5 achine. To determine sustained MFLOPS per processor, the number of floating-point operations per ime step was measured for the main DG time stepping loop using hardware performance counters for BM supercomputer.

- IBM Blue Gene/L system uses **libmpihpm** library and its link and code examples are given as follows:

Resource	IBM Blue Gene/L	IBM POWER5 p575
Clock cycle	0.7 GHz	1.9 GHz
Memory/proc	0.25 GB	2.0 GB
Total Processors	2048	624
Operating System	MK Linux	AIX 5.3
Compilers	IBM BGL XL	IBM AIX XL

Table 1
Comparison of IBM Blue Gene/L and IBM POWER5 p575 systems.

```
add -L$(BGL_LIBRARY_PATH) -lmpihpm_f -lbgl_perfctr.rts
...
call trace_start()
call dg3d_advance()
call trace_stop()
...
```

- IBM POWER5 p575 system uses **libhpm** library in HPM Toolkit and its link and code example are given as follows:

```
add -L$(HPM_LIBRARY_PATH) -lhpm -lpmapi -lm
...
#include "f_hpm.h"
...
call f_hpminit(taskid,'dg3d')
call f_hpmstart(5,'dg3d advance')
call dg3d_advance()
call f_hpmstop(5)
call f_hpm_terminate(taskid)
...
```

Note that all writing and printing functions are turned off during performance evaluations. Figure demonstrates IBM Blue Gene/L machine sustains between 253 to 266 MFLOPS per processor wit coprocessor mode and sustains between 238 to 261 MFLOPS per processor with virtual-node mod where as IBM POWER5 machine sustains between 715 to 732 MFLOPS per processor with 4 task per node mode, sustains between 706 to 731 MFLOPS per processor with 6 tasks per node mode, an sustains between 532 to 561 MFLOPS per processor with 8 tasks per node mode. Table 2 summarie the percentage of peak performance for strong scaling results for IBM Blue Gene/L and IBM POWER systems. IBM Blue Gene/L sustains 9.5% and 9.3% of peak performance for coprocessor and virtual-nod modes, respectively. However, IBM POWER5 sustains 9.6% of peak performance for 4 and 6 tasks pe node mode where as it sustains 7.4% of peak performance for 8 tasks per node mode. Note that th processors for IBM POWER5 system are grouped maximum 8 per node so that performance drops occu when full (i.e., 8) tasks per node have been used.

5. CONCLUSION

A conservative 3-D DG baroclinic model has been developed in the NCAR HOMME framework. Th 3-D DG model is formulated in conservative flux form. The computational domain is the singularity-fre cubed-sphere geometry. The DG discretization uses high-order nodal basis set of Lagrange-Legendr polynomials and fluxes of inter-element boundaries are approximated with Lax-Friedrichs numerical flu The vertical discretization follows the 1-D vertical Lagrangian coordinates approach. Time integratio follows the third-order SSP-RK scheme. To validate proposed 3-D DG model, the baroclinic instabilit test suite proposed by Jablonowski and Williamson is investigated. Currently, 3-D DG model perforn successfully upto 10-day simulation. Parallel experiments are tested on IBM Blue Gene/L and IBM

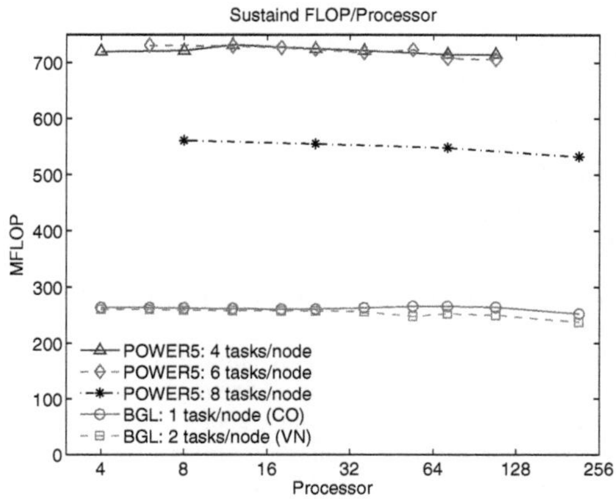

Figure 5. Parallel performance (i.e., strong scaling) results on **IBM BG/L** and **IBM POWER5 p575** systems.

Resource	Sustained MFLOPS	% of Peak Performance
POWER5: 4 tasks/node	732	9.6
POWER5: 6 tasks/node	731	9.6
POWER5: 8 tasks/node	561	7.4
BG/L: 1 task/node (CO)	266	9.5
BG/L: 2 tasks/node (VN)	261	9.3

Table 2
Summary of strong scaling results for IBM Blue Gene/L and IBM POWER5 p575 systems.

POWER5 p575 supercomputers. Conservative 3-D DG baroclinic model sustains 9.5% peak performance for IBM Blue Gene/L's coprocessor mode and sustains 9.6% peak performance for IBM POWER5's 4 and 6 tasks per node modes.

REFERENCES

1. J.P. Boyd, The erfc-log filter and the asymptotics of the Euler and Vandeven sequence accelerations, Proceedings of the Third International Conference on Spectral and High Order Methods, A. V. Ilin and L. R. Scott (eds), Houston J. Math. (1996) 267-276.
2. B. Cockburn and G.E. Karniadakis and C.-W. Shu, Discontinuous Gakerkin methods: Theory, Computation, and Applications, In Lecture Notes in Computational Science and Engineering, 11 (2000) 1-470, Springer.
3. J.M. Dennis and R.D. Nair and H.M. Tufo and M. Levy and T. Voran, Development of a scalable global discontinuous Galerkin atmospheric model, Int. J. Comput. Sci. Eng. (2006), In press.
4. S. Gottlieb and C.-W. Shu and E. Tadmor, Strong stability-preserving high-order time discretization methods, SIAM Review 43:1 (2001) 89-112.
5. HOMME: High-Order Methods Modeling Enviornment, National Center for Atmospheric Research (NCAR), 2006: `http://www.homme.ucar.edu`
6. C. Jablonowski and D.L. Williamson, A baroclinic wave test case for dynamical cores of general circulation models: model intercomparisons, TN-469+STR, National Center for Atmospheric Research

(NCAR), 2006.
7. C. Jablonowski and D.L. Williamson, A baroclinic wave test case for atmospheric model dynamical cores, Mon. Wea. Rev. (2006) In press.
8. S.-J. Lin, A "Vertically-Lagrangian" finite-volume dynamical core for global models, Mon. Wea. Rev. 132 (2004) 2293-2307.
9. R.D. Nair and B. Manchenhauer, The mass-conservative cell-integrated semi-Lagrangian advection scheme on the sphere, Mon. Wea. Rev. 130 (2002) 649-667.
10. R.D. Nair and S.J. Thomas and R.D. Loft, A discontinuous Galerkin transport scheme on the cubed sphere, Mon. Wea. Rev. 133 (2005) 814-828.
11. R.D. Nair and S.J. Thomas and R.D. Loft, A discontinuous Galerkin global shallow water model, Mon. Wea. Rev. 133 (2005) 876-888.
12. R.D. Nair and H.M. Tufo, A scalable high-order dynamical core for climate modeling, Proceedings of International Conference on Mesoscale Process in Atmosphere, Ocean and Environment Systems. IMPA 2006, IITD, New Delhi, India, 14-17 February, 2006.
13. R. Sadourny, Conservative finite-difference approximations of the primitive equations on quasi-uniform spherical grids, Mon. Wea. Rea. 100 (1972) 136-144.
14. H. Sagan, Space-Filling Curves, Springer-Verlag, 1994.
15. V.P. Starr, A quasi-Lagrangian system of hydrodynamical equations, J. Meterology, 2:1 (1945) 227-237.
16. A. St-Cyr, J.M. Dennis, R. Loft, S.J. Thomas, H.M. Tufo and T. Voran, Early experience with the 360 TF IBM Blue Gene/L Platform, Int. J. Comput. Meth. (2006), In press.

Parallel Computational Fluid Dynamics – Parallel Computing and Its Applications
J.H. Kwon, A. Ecer, J. Periaux, N. Satofuka and P. Fox (Editors)

Weather Prediction and computational aspects of Icosahedral-Hexagonal Gridpoint Model GME

H. S. Chaudhari, K. M. Lee, and J. H. Oh

Pukyoung National University, Busan, Republic of Korea

Key Words: Weather prediction; Icosahedral- hexagonal grid; GME

1. Introduction

Today most global atmospheric models use either the "spectral" method based on expansions in terms of triangularly truncated spherical harmonics [1] or finite differences implemented on regular grids in spherical (longitude-latitude) co-ordinates. Each of these methods has strengths and weaknesses. Global climate research requires faster and more accurate solutions of the geophysical dynamics on the spherical geometry. The spectral transform method based on spherical harmonics has been widely used to solve the dynamical part in the atmospheric GCM (General Circulation Model) because of its high accuracy. The computational cost of the spectral transform method, however, becomes extremely high as the resolution increases. This is because of the lack of an algorithm for fast Legendre transformation comparable to FFT (Fast Fourier Transformation), although alternative approaches such as the double FFT are developed [2]. In addition, on massively parallel computers with distributed memories, the spectral transform method has a serious problem: the spectral transform method requires extensive data transfer between computer nodes. For these reasons, it is widely believed that the next generation atmospheric GCM should employ the grid method rather than the spectral transform method [3]. One may employ the simple latitude–longitude grid system as the grid method. However, the latitude–longitude grid method also has another computational problem at high resolutions. In this grid system, the grid spacing near the poles becomes very small as the resolution becomes higher. This causes a very severe limitation on the time interval for advection or the wave-propagation problem to satisfy the CFL (Courant–Friedrich–Lewy) condition. This problem is called as "pole problem".

A new approach to weather prediction uses spherical geodesic grids generated from an icosahedron. Such grids are termed as geodesic because they resemble the geodesic domes designed by Buckminster Fuller. Geodesic grid models are the most suitable for running on massively parallel machines. The icosahedral grid method [4][5] is an attractive alternative to the spectral transform method and the simple latitude–longitude grid method. For the modeling of high-resolution atmospheric general circulation, an icosahedral grid has several advantages: the homogenous and isotropic distribution of gridpoints, the treatment of gridpoint data as two-dimensional structured data in program code, the easy parallelization on massively parallel computer with distributed memory and so on. A major advantage of the icosahedral-hexagonal grid is the avoidance of the so-called pole problem that exists in conventional latitude-longitude grids. The singularities at the poles lead to a variety of numerical difficulties including a severe limitation on the time step size unless special measures are undertaken. These difficulties simply vanish for grids not having such singularities. The grid point approach offers some advantages relative to spectral methods. One is elimination of ''spectral ringing'' in the vicinity of steep gradients. Another is the ability to ensure positivity in quantities such as cloud liquid water and turbulent kinetic energy. The grid point approach also avoids the large amount of global communication required by spectral transform techniques as well as the large number of arithmetic operations normally associated with Legendre transforms at high spatial resolution [8]. Icosahedral grid point model is more competitive at higher resolution than at lower resolution [6]. Tomita and Satoh [7] has demonstrated that the grid point method using a quasi-uniform grid system (e.g. Nonhydrostatic Icosahedral Atmospheric Model) has computational advantages over the spectral model at higher resolution (scale of which corresponds to 40km).

In the 1990s, several research groups have developed icosahedral gridpoint General circulation models using their own new techniques such as, operational global numerical weather prediction model GME [8] of Deutscher Wetterdienst (DWD), CSU AGCM (Atmospheric General Circulation Model) at Colorado State University [6][9], ICON (Icosahedral Nonhydrostatic) joint project of Max Plank Institute of Meteorology Hamburg (MPI) and DWD for the development of nonhydrostatic GCM [10] and Nonhydrostatic Icosahedral Atmospheric model of Frontier Research Center for Global Change [3][7][11].

The operational global weather forecast model of DWD (German Weather Service) employs a gridpoint approach with an almost uniform icosahedral–hexagonal grid. It has been named GME because it replaced operational global model (GM) and the regional model for central Europe (EM) [8].

In this study, we concentrate on new operational version of DWD numerical weather prediction model GME (GME2.9) and long-term simulation tests for seasonal prediction.

2. Brief Description of DWD GME Model

GME employs a grid point approach with quasi uniform icosahedral–hexagonal grid. Prognostic equations for wind components, temperature, surface pressure are solved by the semi-implicit Eulerian method. Only the two prognostic moisture equations (specific water vapor content and specific cloud liquid water content) use semi-Lagrangian advection in the horizontal direction to ensure monotonicity and positive definiteness [8]. GME constructs geodesic grid by starting with an ordinary icosahedron inscribed inside a unit sphere. The icosahedron has 12 vertices. As a first step in the construction of a spherical geodesic grid, each face of the icosahedron is subdivided into four new faces by bisecting the edges. This recursive bisecting process may be repeated until a grid of the desired resolution is obtained [8]. Such grids are quasi-homogeneous in the sense that the area of the largest cell is only a few percent greater than the area of the smallest cell. For example ni=192 has 40km resolution with 40 layers and it has 368,642 total number of grid points per layer. By combining the areas of pairs of the original adjacent icosahedral triangles, the global grid can logically also be viewed as comprising 10 rhombuses or diamonds, each of which has ni x ni unique grid points, where ni is the number of equal intervals into which each side of the original icosahedral triangles is divided. To facilitate the use of the model on parallel computer a diamond-wise domain decomposition is performed. For the 2-D domain decomposition the $(ni+1)^2$ grid points of each diamond are distributed to n1 x n2 processors [8]. Thus each processor computes the forecast for a sub-domain of each of the 10 diamonds. This is a simple yet effective strategy to achieve a good load balancing between processors.

3. Computational Performance of GME on Xeon Cluster (KISTI HAMEL)

Computational efficiency is necessary for long-term simulations. Model computational performance has been evaluated on KISTI (Korea Institute of Science and Technology Information) HAMEL cluster with Intel Xeon 2.8 GHz, 2 CPUs/node, 3GB memory/node and 32 bit REALs. The KISTI Supercomputing Center is the largest provider of supercomputing resources and high performance networks in Korea. As a public supercomputer provider, KISTI imposes limitations, regulations on user (in terms of usage of memory).

GME model run has been performed for 24-hr forecast starting with 17 Aug 2004 12UTC (MEGI Typhoon case). These runs show the full potential of high-resolution global modeling. The initial state for the model run was based on operational analysis dataset of ECMWF (T_L511, L91). Model performance test has been performed using 512 processors (Processing Elements-PE) of 256 nodes. Each processor (PE) computes the forecast in a sub-domain of all 10 diamonds. This approach improves the chance of achieving a better load balancing for the physical parameterizations. In the current version of GME, each computational sub-domain has a halo of just two rows and columns of grid points that have to be exchanged via MPI with those PEs that compute the forecast in the neighboring sub-domains. With 306 PEs on KISTI HAMEL cluster, the physical parameterization for a 24-h data forecast consume between 147s and 195s real time, the average real time is 170s. Performance of GME (40km/40 Layer, ni=192) on KISTI HAMEL cluster for a 24-h forecast has been presented in Fig. 1. GME

(40km/40 Layer, ni 192) 24-h forecast was performed using full physics with time step of 133s.

Fig. 1 depicts results of model performance for the use of 126 to 360 PEs. Between 126 to 208 processors (PEs), an almost linear speedup is obtained. Model reaches its saturation point between 324 to 360 processors. In short, optimum performance of model can be achieved with the use of 360 processors. It takes about 10 minutes to provide 24-hour forecast with the utilization of 306 processors.

For the use 360 to 512 PEs, expected speedup is not achieved (Fig. not shown here). In this case, increase in number of processors (more than 360) does not lead to efficient parallelization. The communication on system becomes too slow to support simultaneous processing of 360 (or more number of) processors. As number of processors increases, number of sub-domains also increases. And it need large amount of data to be exchanged. It also leads to increase in post-processing real time (this is happening in case of use of more than 360 processors). As a result model run time increases.

Early completion of forecasting job has vital importance in operational implementation. In this perspective, these results show possibility for the use of Linux-based cluster with tera-flops performance in operational implementation of GME. Since it is resources composed of PC clusters that meet the level of high computing power, it consumes minimal computing cost. In general, the computational performance depends on computer architecture and the degree of code optimization. The effort devoted to code optimization is limited [8]. Thus there is substantial potential for the further improvement of the performance of the model.

4. GME Results: NWP Approach

Severe weather phenomenon such as typhoon may be well suited for evaluating model's ability to predict weather. Typhoon Megi lashed South Korea's southern provinces with strong winds and heavy rains. GME model (40 km) simulation is performed with the initial condition of 17 Aug 2004 12UTC. Model simulated precipitation (during 27-51h) is compared with corresponding observed precipitation (Fig. 2). Model depicts heavy precipitation over East coast in the range of 250 mm-300mm. Model is able to capture the precipitation maxima over east coast of Korea. Overall, model simulated precipitation is in agreement with observation.

5. Seasonal Prediction Experiments with GME

Seasonal prediction experiments for the past cases are called as hindcast experiments. Each prediction suite included two experiments, hindcast run of model with observed SSTs and control run with climatological SSTs. ECMWF reanalysis dataset (ERA) is interpolated to icosahedral grid of GME and provided as an initial condition for GME runs.

For the simulations of 240km GME, model time step was reduced from 900 sec to 720 sec for the reason of numerical stability. For the simulations of 40km GME, model time step was 133.33 sec. Seasonal runs of 40 km (ni=192, L40) GME using 126 processors takes 40 hours of wallclock time on KISTI HAMEL Cluster. The selection criteria were based on dry/wet year of Korean monsoon. We selected following cases for the hindcast runs – Extreme cases of 1994, 2003.

5.1. Extreme Cases

5.1.1. Extreme case of 1994

Korean monsoon mean rainfall (June-August) is 648 mm and it has standard deviation of 167mm [12]. An extremely hot and dry summer of 1994 was reported in East Asian countries. It was a severe drought over Korean peninsula. Korean observed seasonal monsoon rainfall was 335 mm. This hindcast experiments of GME were performed using 240km resolution of GME (ni=32, L40). Precipitation difference maps (Experiment – Control) for the GME (left panel) is presented in Fig. 3. CMAP [13] precipitation difference patterns (JJA 1994- climatology) are also shown in Fig. 3 (right panel).

CMAP precipitation difference patterns depict strong negative precipitation anomaly over Korea and Japan sector. GME precipitation difference (Experiment – Control) patterns clearly show strong negative precipitation anomaly over Korean peninsula. It indicates dry summer over Korean Peninsula.

5.1.2. Extreme case of 2003

In 2003, Korea has experienced abnormal wet summer. Korean observed seasonal monsoon rainfall was 992 mm. This hindcast experiment was performed with 40 km GME. Precipitation difference patterns of GME are presented in Fig. 4 (left panel).

CMAP precipitation difference patterns depict positive anomaly over Korea-Japan region (Fig. 4, right panel). GME precipitation difference patterns also show positive precipitation anomaly over Korean peninsula (Fig. 4, left panel). It indicates abnormal wet summer over Korea.

6. Summary

In prediction viewpoint, hindcast experiment of GME was performed by specifying observed SSTs every day based on ECMWF Re-analysis (ERA) dataset. In brief GME incorporates many of the positive features of spectral models and finite difference models into a single framework. GME generally employs the same methods and procedures applied in other numerical weather prediction (NWP) grid schemes. However, the uniformity of the GME grid avoids unnecessary physics calculations in over resolved high-latitude zones that commonly occur in grids with polar singularities. Implementation of model on high performance computer based on Linux clusters can provide added advantage. High resolution of GME can play key role in the study of server weather phenomenon such as heavy rainfall, snowfall and typhoon cases. Seasonal prediction experiments (hindcast experiments) were conducted for selected cases. In dry and hot summer of 1994, GME model is able to capture negative anomalies over Korean peninsula. In wet summer of 2003, GME depicts stronger positive anomalies over entire Korean Peninsula.

We strongly believe that geodesic grids based on icosahedron hold a great deal of promise for the future of the atmospheric general circulation modeling.

Acknowledgement

We are thankful to DWD for providing model source codes, for the insightful discussions with Dr. Majewski of DWD and to ECMWF for providing dataset under the DWD-PKNU collaboration. This work was funded by the Korea Meteorological Administration R. & D. Program under Grant CATER 2006-1101. The authors would like to acknowledge the support from KISTI under "The Eighth Strategic Supercomputing Support Program" with Dr. Cho as the technical supporter. The use of the computing system of the Supercomputing Center is also greatly appreciated.

REFERENCES

1. Jarraud, M., and A. J. Simmons, (1983). The spectral technique. *Proc. ECMWF Seminar on Numerical Methods for Weather Prediction*, ECMWF, Reading, United Kingdom, pp. 1-59.
2. Cheong, H. B., (2000). Double fourier series on a sphere: Applications to elliptic and vorticity equations, *J. Comput. Phys.*, Vol. **157**, pp. 327-349.
3. Tomita, H., M. Satoh, and K. Goto, (2002). An Optimization of the icosahedral grid modified by spring dynamics. *J. Comput. Phys.*, Vol. **183**, pp. 307-331.
4. Sadourny, R., A. Arakawa, and Y. Mintz, (1968). Integration of the non-divergent barotropic vorticity equation with an icosahedral–hexagonal grid for the sphere. *Mon. Wea. Rev.*, Vol. **96**, pp. 351–356.
5. Williamson, D. L., (1968). Integration of the barotropic vorticity equation on a spherical geodesic grid. *Tellus*, Vol. **20**, pp. 642–653.
6. Ringler, T. D., R. P. Heikes, and D. A. Randall, (2000). Modeling the atmospheric general circulation using a spherical geodesic grid: A new class of dynamical cores. *Mon. Wea. Rev.*, Vol. **128**, pp. 2471–2490.
7. Tomita, H., and M. Satoh, (2004). The global nonhydrostatic model NICAM: numerics and performance of its dynamical core. *Proc. ECMWF seminar on Recent developments in numerical methods for atmosphere and ocean modelling* ,pp. 71.
8. Majewski, D., D. Liermann, P. Prohl, B. Ritter, M. Buchhold, T. Hanisch, G. Paul, W. Wergen, and J. Baumgardner, (2002). The operational global icosahedral–hexagonal gridpoint model GME: Description and high-Resolution tests. *Mon. Wea. Rev.*, Vol. **130**, pp. 319–338.
9. Randall, D. A., T. D. Ringler, R. P. Heikes, P. Jones, and J. Baumgardner, (2002). Climate modeling with spherical geodesic grids. *Computing in Science and Engr.*, Vol. **4**, 32-41.
10. Bonaventura, L., (2003). Development of the ICON dynamical core: Modelling strategies and preliminary results. *Proc. ECMWF/SPARC Workshop on Modelling and Assimilation for the Stratosphere and Tropopause*, Reading, United Kingdom, ECMWF, pp. 197–214.
11. Tomita, H., M. Tsugawa, M. Satoh, and K. Goto, (2001). Shallow Water Model on a modified icosahedral geodesic grid by using spring dynamics. *J. Comput. Phys.*, Vol. **174**, pp. 579-613.

12. Kim, B. J., R. H. Kripalani, J. H. Oh, and S. E. Moon, (2002). Summer monsoon rainfall patterns over South Korea and associated circulation features. *Theoretical and Applied Climatology*, Vol. **72**, pp. 437-471.
13. Xie, P., and P. A. Arkin, (1997). Global precipitation: a 17-year monthly analysis based on gauge observation, satellite estimates and numerical model outputs. *Bull. Amer. Meteor. Soc.*, Vol. **78**, pp. 2539-2558.

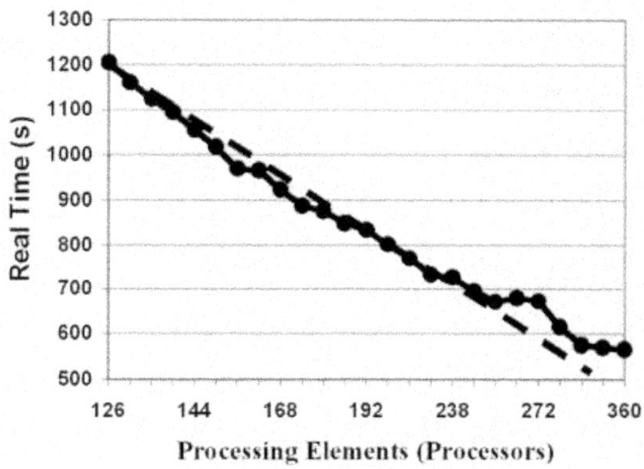

Figure 1. Speed up of GME on KISTI HAMEL cluster, linux-based cluster with tera-flops performance (Intel Pentium Xeon, 2 CPUs/node). No special optimization for HAMEL cluster taken so far (ni=192, 40km, 40 layers, dt=133s, full physics, 24-h forecast).

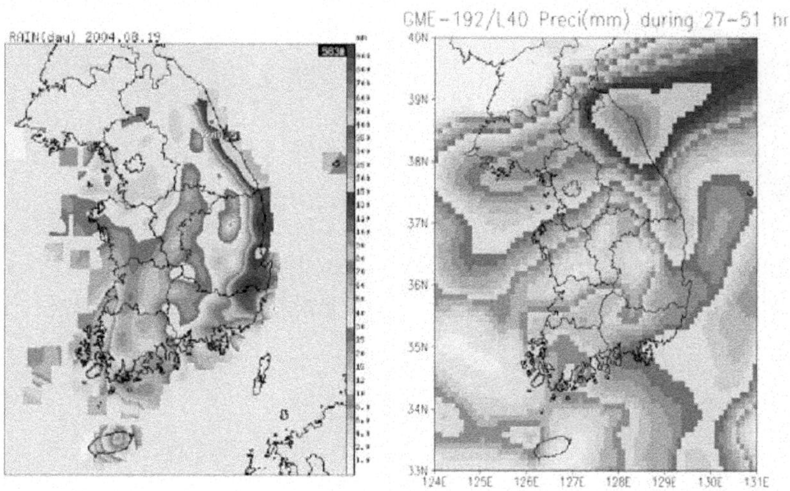

Figure 2. Observed Precipitation (mm/day) on 2004-8-19 (left panel) and GME Precipitation forecast during 27-51hr (starting with 17 Aug 2004, 12 UTC).

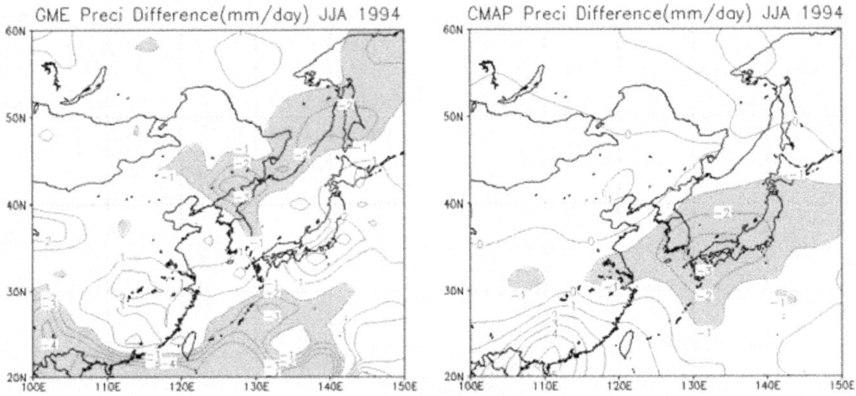

Figure 3. Precipitation difference (Experiment – Control) patterns for GME (left panel) for JJA 1994. CMAP precipitation difference patterns (JJA 1994 – climatology) are shown in right panel.

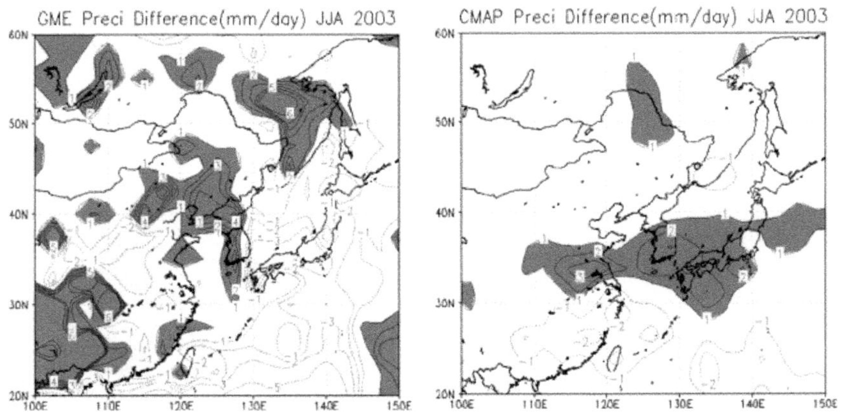

Figure 4. GME precipitation difference (Experiment – Control) patterns for JJA 2003 (left panel). CMAP precipitation difference patterns (JJA 2003- climatology) are shown in right panel.

Parallel Computational Fluid Dynamics – Parallel Computing and Its Applications
J.H. Kwon, A. Ecer, J. Periaux, N. Satofuka and P. Fox (Editors)

Numerical Investigation of Flow Control by a Virtual Flap on Unstructured Meshes

Sungyoon Choi* and Oh Joon Kwon

Department of Aerospace Engineering
Korea Advanced Institute of Science and Technology(KAIST)
373-1 Guseong-dong, Yuseong-gu, Daejeon 305-701, Republic of Korea

Key Words: Flow Control, Unstructured Meshes, Virtual Flap, Numerical Method

Introduction

Recently, numerous researches have been conducted about the improvement of the aerodynamic characteristics of existing airfoils using flow control. Flow control methods are divided into two large catagories depending on whether it requires extra energy for control or not. Passive control uses mechanical devices such as flap or slat, while active control utilizes pneumatic control devices such as blowing, suction, or synthetic jet actuator.

Gurney flap is a simple flat plate, generally 1-2% of the airfoil chord in height, attached perpendicular to the airfoil surface at the bottom of the airfoil in the vicinity of the trailing edge. In spite of the simple control device, Gurney flaps dramatically increase lift without degradation of the airfoil performance when the height is less than the boundary layer thickness. Until now, the flow control effect of Gurney flap has been investigated by several researchers[1, 2]. The results showd that compared to a clean airfoil, Gurney flap reduces the pressure on the suction side and increases the pressure on the pressure side. These pressure variations are induced by the enhancement of circulation related to the extra downward turning of the flow generated behind the Gurney flap. As a result of the pressure variations, additional pressure difference between the upper and lower surfaces occurres, and this causes the increment of the maximum lift, the decrement of the zero-lift angle of attack, and the increment of the nose-down pitching moment, similar to the effect of increasing the camber of the airfoil[2]. The amount of lift increment is proportional to the height of the Gurney flap. But when the height exceeds the boundary layer thickness, the drag also increases. Therefore the flap height should be determined in a way that the maximum lift-to-drag ratio can be achieved.

In the present study, numerical simulations of virtual Gurney flaps were conducted by imposing continuous or periodic jet in the direction normal to the lower surface of airfoil at the trailing edge, and the effects of virtual flap on the aerodynamic characteristics of an airfoil were investigated.

Numerical Method

The Governing equations are the two-dimensional, compressible, Reynolds-averaged Navier-Stokes equations. The Spalart and Allmaras one-equation turbulence model was employed for turbulent closure.

A vertex-centered second-order accurate finite-volume method was used for the spacial discretization of the convective fluxes on unstructured meshes. The viscous terms were computed using a second-order accurate central differences scheme. The time advancement was achieved using second-order accurate implicit time integration based on a point Gauss-Seidel relaxation method and dual time-step sub-iteration. The characteristic boundary condition was imposed at the far-field boundary, and the no-slip iso-thermal boundary condition was applied at the solid surface.

To simulate the flow control jet, the jet velocity was imposed along the control slot on the airfoil surface:

$$u = V_j \cos(\theta + \Phi), \quad v = V_j \sin(\theta + \Phi) \tag{1}$$

where θ is the boundary surface angle; Φ is the control jet angle; and V_j is the control jet velocity. The control jet velocity was obtained from the non-dimensional control momentum coefficient, $C_\mu (C_\mu = <J> / cq_\infty)$. Here, $<J> = \rho V_j^2 W_j$ is the control momentum; q_∞ is the freestream dynamic pressure; W_j is the control slot width; and c is the chord length.

To reduce the large computational time required for the unsteady time-accurate solution, a parallel computation algorithm was adopted by splitting the computational domain into local sub-domains using the MeTis library. In the present node-based solver, to avoid excessive communication, information about all vertices needed for flux integration was retained in each sub-domain, at the cost of slightly larger storage resource. Then, only two communication phase were required at each stage of the time integration, one for the evaluation of the solution gradients for second-order reconstruction and the other for updating the information of the vertices included in the communication boundary at the end of the time integration. Fig. 1 shows the communication boundary between parallel sub-domains for the current vertex centered scheme. In the figure, *a* represents the number of the added node in the neighbor processor matching with the node *i* for data communication. Mesh partitioning, assignment of the vertices for each sub-domain, and generation of the information for data communication are completed in pre-processing.

To simulate the boundary layer accurately and to avoid the numerical difficulties involved in using highly stretched high aspect-ratio cells, hybrid unstructured meshes were employed in the present study. In Fig. 2, a hybrid mesh around NACA 0015 airfoil with quadrilateral cells inside the viscous boundary layer and isotropic triangular cells for the rest of the computation domain is presented.

Results and Discussion

In order to check the efficiency of the parallel computation, a transonic inviscid flow around an NACA 0012 airfoil was calculated using three different meshes. The coarse mesh, the medium mesh, and the fine mesh were composed of 5,079 cells, 20,394 cells, and 87,157 cells, respectively. The calculations were made at a freestream Mach number of 0.8 and the incidence angle of 1.25 deg.

In Fig. 3, comparison of residual convergence characteristics and the corresponding lift variation between single and multiple processors is presented as a function of iteration for the fine mesh. It shows that when the data communication was executed at every iteration of the Gause-seidal relaxation, the convergence characteristics were almost identical with each other, independent to the number of processors. But when the communication was performed only at the first iteration of the Gauss-Seidal iterations, the convergence characteristics were deteriorated and this degradation of convergence was further amplified as the number of processors increased. However, the convergence characteristic of the lift coefficient was not affected significantly by the number of communications in the Gause-seidal relaxation. Fig. 4 depicts the relative speedups achieved as a function of the number of processors for the three different meshes. The results show that good scalability was obtained for the fine mesh up to eight processors, independent to the number of the communication in the Gause-Seidal relaxation. However, as the number of processors further increased, the parallel performance became poorer in the case of the full communication, while the opposite trend was obtained for partial communication. As the mesh size decreased, the relative speedups became lower on the same nubmer of processors.

In order to check the accuracy of the parallel computation, convergence characteristics of the aerodynamic force and moment, and resultant velocity vectors near the trailing edge were compared for the flow control case in Fig. 5. The calculation was made at a freestream Mach number of 0.1, the incidence angle of 0 deg, and the Reynolds number of 3.6×10^5. For the parallel computation, 32 processors were used. The imposed blowing jet parameters values were $V_j=0.2$, $L_j=99.5\%c$, $W_j=0.5\%c$, where L_j is the control slot location on the airfoil. The results show all convergence hystories of the parallel computation are identical to those of single processor. The velocity vectors from both cases shown to be the same. In the vicinity of the trailing edge, the velocity vectors show the traces of two distict counter-rotating vortices, similar to those of the Gurney flap. These results verifing that perpendicular blowing effectively works as a virtual flap, similar to the Gurney flap.

In Fig. 6, comparison of the aerodynamic force and moment between the flows without and with continuous blowing. Three different jet strengths were employed for the continuous jet applied through 0.5% witdth control slot at 99.5% of the chord.

At a given angle of attack, virtual flap increased lift without changing the lift curve slope. The amount of lift increment became larger as the blowing jet strengh increased. Compared to the case without blowing, the lift increased 1.3, 1.54, and 2.05 times higher for $V_j=0.05$, $V_j=0.1$, and $V_j=0.2$ at the incidence of 4 deg, respectively. While the virtual flap was not as effective to suppress airfoil stall as Gurney flap, the stall angle of attack did not change much, despite the increment of the jet strength. This characteristic is different from that of Gurney flap whose stall angle of attack is inversely proportional to the height of the flap. When the jet strength increased, the drag coefficient was reduced at a given angle of attack. This tendency is different from that of Gurney flap which shows drag increment as the flap height increases. Compared to the case without blowing, drag coefficients were lower by 0.99, 0.96, and 0.79 times for $V_j=0.05$, $V_j=0.1$, and $V_j=0.2$ at the incidence of 4 deg, respectively. This drag reduction contributed to the increment of the lift-to-drag ratio as shown in Fig. 6(d). Fig. 6(c) shows the effect of continuous blowing on the moment coefficient about the quarter of the chord. Similar to the Gurney flap, there was an increment in the nose-down pitching moment when the continuous blowing was applied, mostly due to the circulation induced by the downward turning flow. The amount of the moment increment became larger as the jet strength increased.

The effects of the periodic jet on the aerodynamic characteristics of an NACA 0015 airfoil are presented in Fig. 7. It is shown that periodic jets with various non-dimensional control jet frequency(F_j) for $V_j=2.0$, $L_j=99.5\%c$, $W_j=0.5\%c$ are less effective than continuous jet at all angles of attack considered. While the amount of the lift increment was lower than the continuous jet, the result of periodic jet showed similar flow control patterns even at less energy consumption. As the frequency increased, the flow control effect was slightly enhanced, but not significantly. Therefore, it can be stated that high frequency periodic jets are more efficient for flow control. Unlike the continuous jet, the drag coefficient of periodic jet was higher than that of without control from the incidence angle of 4 deg, and the amount became higher as the jet frequency became lower. As a result, the lift-to-drag ratio was slightly lower than that of a clean airfoil from the incidence angle of 10 deg for $F_j=0.57$ in spite of the increment in lift. The moment coefficient became more negative than without blowing, similar to the case of the continuous jet. This result shows that periodic jet control is also useful for the moment control of an airfoil

Conclusions

The flow control effect of virtual Gurney flap on an NACA 0015 airfoil was numerically investigated using a parallelized unstructured Navier-Stokes flow solver. The virtual Gurney flap simulated by imposing jet showed flow control characteristics much similar to the solid Gurney flap, such as the increment in lift and the nose-down pitching moment. It also relieved the defects of the solid Gurney flap, such as the reduction of the lift-to-drag ratio and the decrement of stall angle of attack. When the virtual Gurney flap was simulated by a periodic jet, it showed flow control patterns similar to the continuous jet, but with less effectiveness. As the frequency of the periodic jet increased, the efficiency of the flow control also increased.

REFERENCES

[1] C. S. Jang, J. C. Ross, and R. M. Cummings, (1992). "Computational Evaluation of an Airfoil with a Gurney Flap", AIAA 92-2708.

[2] R. Myose, M. Papadakis, and I. Heron, (1998). "Gurney Flap Experiments on Airfoils, Wings, and Reflection Plane Model", Journal of Aircraft, Vol. 35, No.2, pp. 206-211.

[3] V. Venkatakrishnan, (1994). "Parallel Implicit Unstructured Grid Euler Solvers", AIAA Journal, Vol. 32, No. 10, pp. 1985-1991.

[4] D. J. Mavriplis, (2000). "Viscous Flow Analysis Using a Parallel Unstructured Multigrid Solver", AIAA Journal, Vol. 38, No.11, pp. 2067-2076.

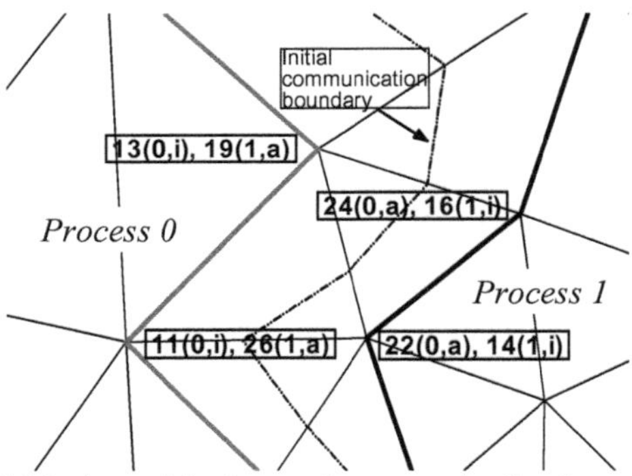

Fig. 1 Mesh partitioning and communication boundary.

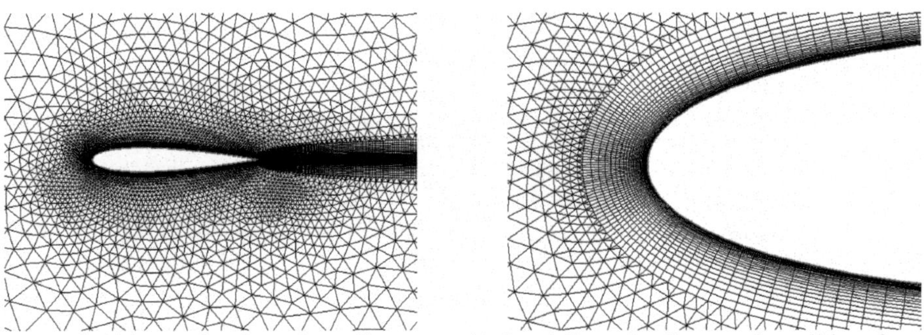

a) Far-field isotropic triangular cells b) Quadrilateral cells near the solid surface

Fig. 2 Hybrid unstructured mesh around the NACA 0015 airfoil.

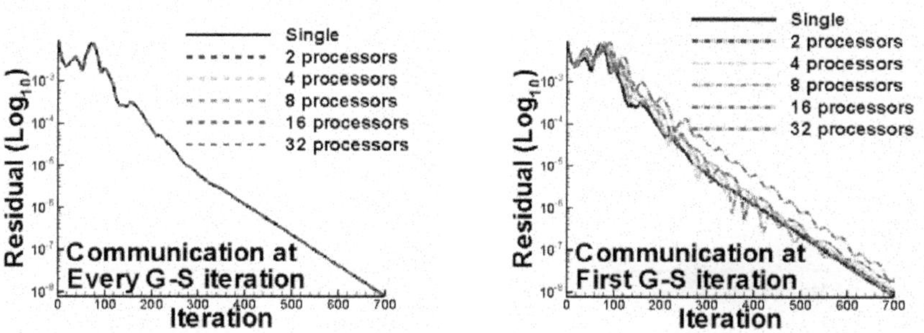

(a) Residual convergence

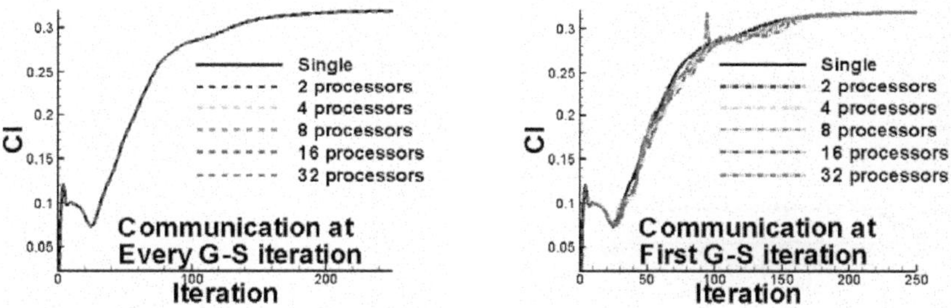

(b) Lift coefficient convergence

Fig. 3 Comparison of convergence characteristics between single processor and multiple processors.

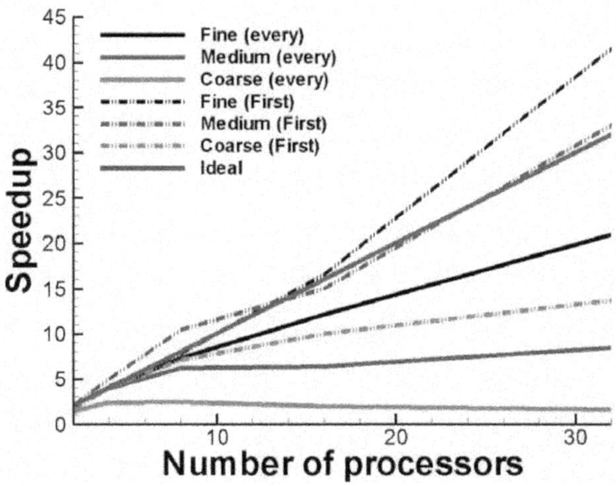

Fig. 4 Relative speedups for different number of processors.

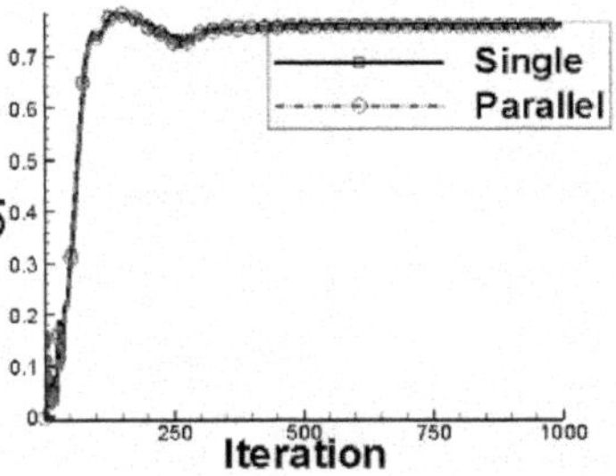

(a) Lift coefficient convergence

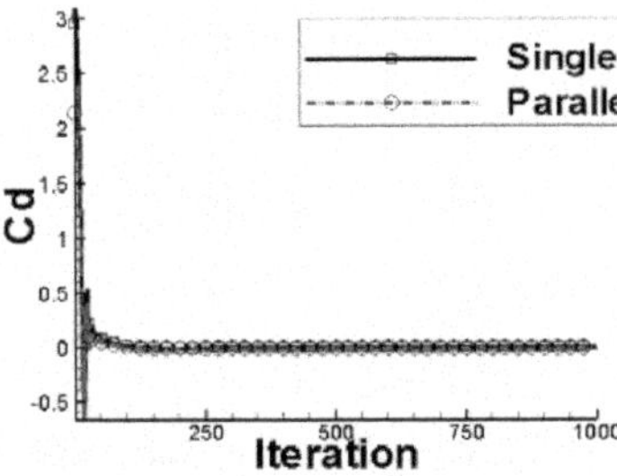

(b) Drag coefficient convergence

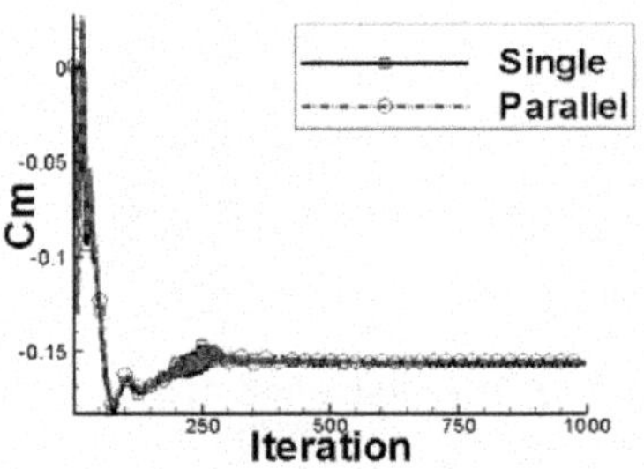

(c) Moment coefficient convergence

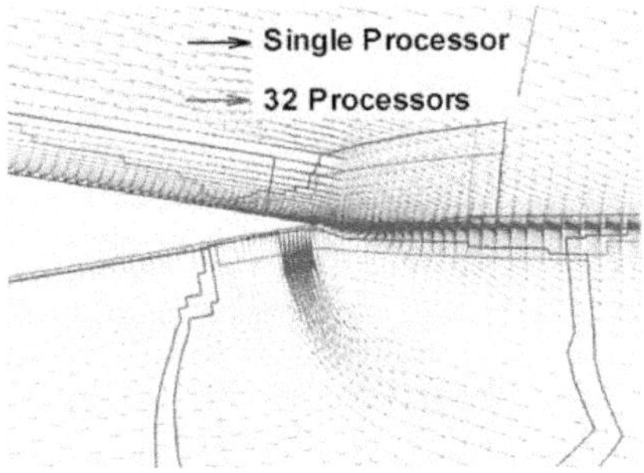

(d) Velocity vectors

Fig. 5 Comparison of aerodynamic force and moment convergence characteristics, and resultant velocity vectors between single processor and multiple processors.

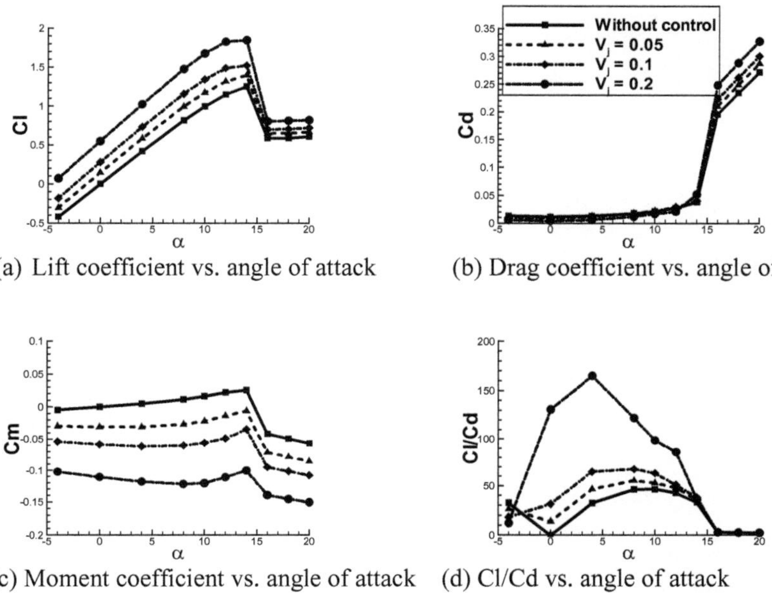

(a) Lift coefficient vs. angle of attack (b) Drag coefficient vs. angle of attack

(c) Moment coefficient vs. angle of attack (d) Cl/Cd vs. angle of attack

Fig. 6 Comparison of aerodynamic force and moment coefficients between flows with and without continuous blowing.

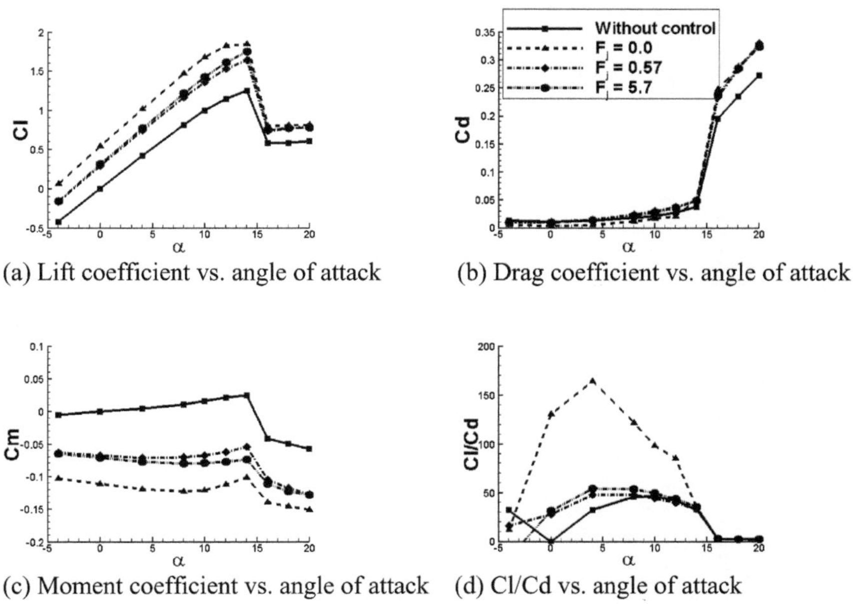

(a) Lift coefficient vs. angle of attack (b) Drag coefficient vs. angle of attack

(c) Moment coefficient vs. angle of attack (d) Cl/Cd vs. angle of attack

Fig. 7 Comparison of aerodynamic force and moment coefficients between flows with and without periodic blowing.

Parallel Computational Fluid Dynamics – Parallel Computing and Its Applications
J.H. Kwon, A. Ecer, J. Periaux, N. Satofuka and P. Fox (Editors)

Implicit Kinetic Schemes for the Ideal MHD Equations and their Parallel Implementation

Ramesh K. Agarwal* and Heru S.R. Reksoprodjo**

*Department of Mechanical and Aerospace engineering, Washington University, One Brookings Drive, St. Louis, MO 63130
** National Institute for Aviation Research, Wichita State University, Wichita, KS 67208

e-mail: rka@wustl.edu, heru_reksoprodjo@yahoo.com

The implicit kinetic schemes, namely the Kinetic Flux-Vector-Split (KFVS) and the Kinetc Wave/Particle Split (KWPS) schemes are derived for the Euler and ideal MHD equations. The schemes are applied to compute the 2D flow fields due to a cylindrical blast wave, and the supersonic flow over a planar blunt body, both with and without the magnetic field. For the ideal 2D MHD equations, the homogeneity of the flux vector is achieved by employing the approach due to MacCormack, and a Poisson solver is used at every time-step to enforce the solenoidal condition on the magnetic field. The computations are performed on a SGI Origin 2000, 64 R12000 MIPS processors supercomputer. Almost a linear speed up is obtained.

Key words: Kinetic Scheme, Implicit Scheme, Euler Equations, Magnetohydrodynamics

NOMENCLATURE

$\mathbf{A}$	Jacobian matrix
$\hat{a}$	a "dummy" variable
BGK	Bhatnagar-Gross-Krook
$\vec{B}$	magnetic field
$\vec{c}$	thermal/peculiar velocity $(\equiv \vec{v} - \vec{u})$
D	non-translational (internal) d.o.f.
e_t	specific total energy $\left(\equiv \frac{1}{2}u_k^2 + \frac{D+3}{2\rho}p + \frac{1}{2}w_k^2\right)$
$\mathbf{F}$	flux vector
f	probability density distribution function
$\mathbf{I}$	identity matrix
$\mathrm{J}(f,f)$	collision integral
KFVS	Kinetic Flux-Vector Split
KWPS	Kinetic Wave/Particle Split
MHD	magnetohydrodynamics
m	molecular mass
P_0	total pressure $(\equiv p + \frac{1}{2}B_k^2)$
p	thermal (hydrodynamic) pressure
$\mathbf{Q}$	field vector
$\vec{u}$	fluid velocity
$\mathbf{V}$	primitive variable vector
$\vec{v}$	molecular velocity
$\vec{w}$	Alfvén wave velocity $\left(\equiv \frac{1}{\sqrt{\rho}}\vec{B}\right)$
β	$\equiv \frac{\rho}{2P_0}$
γ	ratio of specific heats $\left(\equiv \frac{D+5}{D+3}\right)$
δ	Kronecker delta
ϵ	non-translational velocity
ρ	fluid density
$\mathbf{\Psi}$	collision invariant vector

INTRODUCTION

In recent years, there has been a considerable interest in the kinetic schemes for solving the Euler [1,8], Navier-Stokes [3], and ideal MHD [10,13] equations. However, to date, all the papers on kinetic schemes reported in the literature employ an explicit formulation due to its simplicity. The major drawback of the explicit formulation is the restriction placed on the time-step allowed for the stability of the scheme.

Kinetic schemes are based on the fact that the set of equations governing the motion of fluid flows at the continuum level, i.e., Euler, Navier-Stokes, and Burnett equations, can be obtained by taking the moments of the Boltzmann equation, which can be expressed as

$$\frac{\partial f}{\partial t} + v_i \frac{\partial f}{\partial x_i} = \mathrm{J}(f,f) \tag{1}$$

at the molecular level with respect to the collision invariants $\mathbf{\Psi}$, which represent quantities conserved during a collision. This is often referred to as the *"moment method strategy"*. This operation is defined as the following mapping operation:

$$\langle f \rangle = \int_{\Re^D} d^D\epsilon \int_{\Re^3} d^3v\,(\mathbf{\Psi} f) \tag{2}$$

where

$\Psi = \left[\begin{array}{ccc} m & mv_l & \frac{1}{2}m\epsilon^2 + \frac{1}{2}mv_k^2 \end{array}\right]^T$

In the Euler limit, the gas is thought to relax instantaneously to a state of collisional equilibrium, at which the solution to the Boltzmann equation is simply the Maxwellian velocity distribution function:

$$f = \frac{\rho}{m}\left(\frac{\beta}{\pi}\right)^{\frac{D+3}{2}} \exp\left[-\beta\epsilon^2 - \beta\left(v_k - u_k\right)^2\right] \tag{3}$$

One of the well-known and extensively used kinetic schemes, the Kinetic Flux-Vector Splitting (KFVS) scheme proposed by Mandal & Deshpande [8], splits the flux term in the Boltzmann equation into positive and negative parts based on the sign of the molecular velocity ($\vec{v}$). Taking the moments of this split-flux with respect to the collision invariant vector results in the KFVS algorithm. This scheme, however, requires the evaluation of the computationally expensive error functions.

Agarwal & Acheson [1] have proposed a new flux splitting at Boltzmann level, which they call Kinetic Wave/Particle Splitting (KWPS). They have shown that this new splitting does not require the evaluation of error functions, increasing computational speed. By recognizing that the molecular velocity of an individual gas particles can be expressed as the sum of the average fluid velocity of the gas ($\vec{u}$) and each particle's thermal (peculiar) velocity ($\vec{c}$), the Boltzmann flux can be split into two components: the convective part and the acoustic part. These two parts of the Boltzmann flux are then upwind discretized and the moments of this discretized equation with the collision invariants then result in the KWPS scheme for the Euler equations.

Recently, for the first time, implicit formulations of two kinetic schemes, namely the Kinetic Flux-Vector Split (KFVS) and the Kinetic Wave/Particle Split (KWPS) schemes were systematically derived by Reksoprodjo & Agarwal [11]. They showed that for steady state calculations, the implicit kinetic schemes were significantly more efficient than the explicit schemes while retaining the accuracy and robustness of explicit schemes.

The above-mentioned schemes have also been extended to obtain numerical solutions for the ideal magnetohydrodynamics (MHD) equations. For the ideal 1-D MHD equations, the KFVS scheme was first formulated by Croisille et al [4] to solve the *7-wave* model. However, their scheme was not directly obtained from the Boltzmann equation due to the lack of appropriate distribution function. This methodology was also employed by Reksoprodjo & Agarwal [10] in developing the KWPS scheme for the ideal MHD equations and by Xu [13] for his BGK-scheme. Later, Tang & Xu [12] derived the *8-wave* multidimensional BGK-scheme for the ideal MHD equations. It is interesting to note here that recently Huba & Lyon [5] presented a method of obtaining the fluid portion of the *8-wave* ideal MHD equations directly from the molecular level. They succeeded in modifying the Boltzmann equation and the Maxwellian distribution function to include an acceleration term due to the magnetic field.

Implicit upwind-split schemes for the 8-wave ideal MHD equations are impossible to derive directly due to the non-homogeneity of the flux vectors with respect to the state vector. By introducing a *"dummy"* equation ($\partial_t \hat{a} = 0$), MacCormack [7] was successful in recovering the homogeneity of the flux vectors. Using this new expanded system of equations, he derived an implicit scheme based on the modified Steger-Warming flux-splitting algorithm. It should be noted that Powell's *"magnetic divergence wave"* [9] is inherently included in this formulation. This methodology has been applied in this paper to derive the implicit formulation of the kinetic schemes for the ideal MHD equations. A Poisson solver is used at every time-step to enforce solenoidal condition requirement on the magnetic field.

In this paper, the derivations of the implicit kinetic schemes for the Euler and the ideal MHD equations are presented and validated by computing several test cases in 2-D, namely the cylindrical blast wave, and the supersonic flow over a planar blunt-body with and without the magnetic field.

IMPLICIT KINETIC SCHEMES FOR THE EULER EQUATIONS

To obtain the implicit kinetic scheme for the Euler equations, an implicit algorithm for the Boltzmann equation is formulated. The flux terms are linearized as follows:

$$(v_i f)^{n+1} \approx (v_i f)^n + \Delta t \left[\frac{\partial (v_i f)}{\partial t}\right]^n \approx v_i (f)^n + v_i \left(\frac{\partial f}{\partial \mathbf{U}} \Delta \mathbf{U}\right)^n \tag{4}$$

and the flux vectors and Jacobians are then obtained as

$$\mathbf{F}_i = \langle v_i f \rangle \quad \text{and} \quad \mathbf{A}_i = \left\langle v_i \frac{\partial f}{\partial \mathbf{U}} \right\rangle \tag{5}$$

The KFVS methodology first splits the molecular velocity ($\vec{v}$) into positive and negative spaces, and then takes the moments with respect to the collision invariant vector. On the other hand, the KWPS methodology first expresses the molecular velocity as the sum of the average fluid velocity ($\vec{u}$) and the particle's thermal velocity ($\vec{c}$) prior to splitting them into positive and negative spaces. Subsequently, the split-flux vectors and Jacobians are obtained. Due to the lack of space, the resulting split-flux vectors and Jacobians are not given here explicitly, but can be obtained from the formulations for the ideal MHD equations simply by setting the magnetic field $\left(\vec{B}\right)$ to zero.

IMPLICIT KINETIC SCHEMES FOR THE IDEAL MHD EQUATIONS

Employing the methodology of MacCormack [7], the implicit kinetic schemes for the ideal MHD equations can now be obtained directly from the explicit formulations. For the KFVS scheme this process is very straightforward. Its split flux-vectors can be written as

$$\mathbf{F}_i^{\pm} = \frac{1+\mathrm{erf}\left(u_i\sqrt{\beta}\right)}{2}\begin{bmatrix} \rho u_i \\ \rho u_i u_j + P_0\delta_{ij} - B_iB_j \\ \rho u_i e_t + P_0 u_i - B_i u_k B_k \\ u_iB_j - B_iu_j \\ 0 \end{bmatrix} \pm \frac{\mathrm{erf}\left(-u_i^2\beta\right)}{2\sqrt{\pi\beta}}\begin{bmatrix} \rho \\ \rho u_j \\ \rho e_t + \frac{1}{2}\left(P_0 - B_i^2\right) \\ B_j - B_i\delta_{ij} \\ 0 \end{bmatrix} \tag{6}$$

The expressions obtained for the split-flux Jacobians for the KFVS scheme are

$$\mathbf{A}_i^{\pm} = \frac{1+\mathrm{erf}\left(u_i\sqrt{\beta}\right)}{2}\left[u_i\mathbf{I} + \mathbf{B}_i^{(1)}\mathbf{C}^{-1}\right] \pm \frac{\mathrm{erf}\left(-u_i^2\beta\right)}{2\sqrt{\pi\beta}}\left[\mathbf{I} + \mathbf{B}_i^{(2)}\mathbf{C}^{-1}\right] \tag{7}$$

where

$$\mathbf{B}_i^{(1)} = \begin{bmatrix} 0 & \rho\delta_{il} & 0 & 0 & 0 \\ 0 & \rho u_j\delta_{il} & \delta_{ij} & B_l\delta_{ij} - B_i\mathbf{I} & -\frac{1}{2}B_k^2\delta_{ij} \\ 0 & \rho e_t\delta_{il} + P_0\delta_{il} - B_iB_l & u_i & B_lu_i - B_iu_l & -\frac{1}{2}B_k^2u_i \\ 0 & B_j\delta_{il} - B_i\mathbf{I} & 0 & 0 & -B_iu_j \\ 0 & 0 & 0 & 0 & -u_i \end{bmatrix}$$

and

$$\mathbf{B}_i^{(2)} = \begin{bmatrix} 0 & 0 & 0 & 0 & 0 \\ 0 & \rho\delta_{ij}\delta_{il} - 2B_iB_j\beta\delta_{il} & 0 & 0 & 0 \\ 0 & \frac{1}{2}\rho u_i\delta_{il} - B_i\left(2u_kB_k - u_iB_i\right)\beta\delta_{il} & \frac{1}{2} & \frac{1}{2}\left(B_l - B_i\delta_{il}\right) & -\frac{1}{4}B_k^2u_i \\ 0 & -2B_i\left(u_j - u_i\delta_{ij}\right)\beta\delta_{il} & 0 & 0 & -B_i\delta_{ij} \\ 0 & 0 & 0 & 0 & -1 \end{bmatrix}$$
$$+ \left[\begin{bmatrix} 1 \\ u_j \\ e_t + \frac{1}{2}\left(\frac{1}{2\beta} - w_i^2\right) \\ \frac{1}{\rho}\left(B_j - B_i\delta_{ij}\right) \\ 0 \end{bmatrix} - \left(u_i\right)\begin{bmatrix} 0 \\ \delta_{ij} - \frac{1}{P_0}B_iB_j \\ \frac{1}{2}u_i - \frac{1}{2P_0}B_i\left(2u_kB_k - u_iB_i\right) \\ -\frac{1}{P_0}B_i\left(u_j - u_i\delta_{ij}\right) \\ 0 \end{bmatrix}\right]\begin{bmatrix} -\frac{1}{2} \\ 0 \\ \beta \\ B_j\beta \\ -\frac{1}{2}B_k^2\beta \end{bmatrix}^T$$

Note that the numerical value of the dummy variable ($\hat{a} \equiv 1$) has been substituted in all formulations.

For the KWPS scheme, however, the derivation is not as simple. The split-flux Jacobians are obtained by comparison with those of the KFVS scheme using the split-flux Jacobians for the Euler equations as guidelines. The split-flux vectors for the KWPS scheme are

$$\mathbf{F}_i^{\pm} = \frac{u_i+|u_i|}{2}\begin{bmatrix} \rho \\ \rho u_j \\ \rho e_t \\ B_j - B_i\delta_{ij} \\ 0 \end{bmatrix} + \frac{1}{2}\begin{bmatrix} 0 \\ P_0\delta_{ij} - B_iB_j \\ P_0u_i - B_iu_kB_k \\ -B_i\left(u_j - u_i\delta_{ij}\right) \\ 0 \end{bmatrix} \pm \frac{1}{2\sqrt{\pi\beta}}\begin{bmatrix} \rho \\ \rho u_j \\ \rho e_t + \frac{1}{2}\left(P_0 - B_i^2\right) \\ B_j - B_i\delta_{ij} \\ 0 \end{bmatrix} \tag{8}$$

By comparison with equation (7), the following expression is obtained for the split-flux Jacobians for the KWPS scheme:

$$\mathbf{A}_i^{\pm} = \frac{u_i+|u_i|}{2}\left[\mathbf{A}_i^{(0)}\right] + \frac{1}{2}\left[\mathbf{B}_i^{(1)}\mathbf{C}^{-1}\right] \pm \frac{1}{2\sqrt{\pi\beta}}\left[\mathbf{I} + \mathbf{B}_i^{(2)}\mathbf{C}^{-1}\right] \tag{9}$$

where

$$\mathbf{A}_i^{(0)} = \begin{bmatrix} 1 & 0 & 0 & 0 & 0 \\ 0 & \mathbf{I} & 0 & 0 & 0 \\ 0 & 0 & 1 & 0 & 0 \\ 0 & 0 & 0 & \mathbf{I} & -B_i\delta_{ij} \\ 0 & 0 & 0 & 0 & 0 \end{bmatrix}, \quad \mathbf{B}_i^{(1)} = \begin{bmatrix} 0 & \rho\delta_{il} & 0 & 0 & 0 \\ 0 & \rho u_j\delta_{il} & \delta_{ij} & B_l\delta_{ij} - B_i\mathbf{I} & -\frac{1}{2}B_k^2\delta_{ij} \\ 0 & \rho e_t\delta_{il} + P_0\delta_{il} - B_iB_l & u_i & B_lu_i - B_iu_l & -\frac{1}{2}B_k^2u_i \\ 0 & B_j\delta_{il} - B_i\mathbf{I} & 0 & 0 & -B_i\left(u_j - u_i\delta_{ij}\right) \\ 0 & 0 & 0 & 0 & 0 \end{bmatrix}$$

and

$$\mathbf{B}_i^{(2)} = \begin{bmatrix} 0 & 0 & 0 & 0 & 0 \\ 0 & \rho\delta_{ij}\delta_{il} - 2B_iB_j\beta\delta_{il} & 0 & 0 & 0 \\ 0 & \rho u_i\delta_{il} - 2B_i\left(2u_kB_k - u_iB_i\right)\beta\delta_{il} & \frac{1}{2} & \frac{1}{2}\left(B_l - B_i\delta_{il}\right) & -\frac{1}{4}B_k^2u_i \\ 0 & -2B_i\left(u_j - u_i\delta_{ij}\right)\beta\delta_{il} & 0 & 0 & -B_i\delta_{ij} \\ 0 & 0 & 0 & 0 & -1 \end{bmatrix} + \begin{bmatrix} 1 \\ u_j \\ e_t + \frac{1}{2}\left(\frac{1}{2\beta} - w_i^2\right) \\ \frac{1}{\rho}\left(B_j - B_i\delta_{ij}\right) \\ 0 \end{bmatrix}\begin{bmatrix} -\frac{1}{2} \\ 0 \\ \beta \\ B_j\beta \\ -\frac{1}{2}B_k^2\beta \end{bmatrix}^T$$

Finally, the matrix **C** used in the preceding formulations is expressed as

$$\mathbf{C} = \frac{\partial \mathbf{U}}{\partial \mathbf{V}} = \frac{\partial([\ \rho \ \ \rho\vec{u} \ \ \rho e_t \ \ \vec{B} \ \ \hat{a}\ ])}{\partial([\ \rho \ \ \vec{u} \ \ p \ \ \vec{B} \ \ \hat{a}\ ])} = \begin{bmatrix} 1 & 0 & 0 & 0 & 0 \\ u_j & \rho\mathbf{I} & 0 & 0 & 0 \\ \frac{1}{2}u_k^2 & \rho u_l & \frac{D+3}{2} & B_l & -\frac{1}{2}B_k^2 \\ 0 & 0 & 0 & \mathbf{I} & 0 \\ 0 & 0 & 0 & 0 & 1 \end{bmatrix}$$

NUMERICAL RESULTS AND DISCUSSIONS

The performance of implicit kinetic schemes is evaluated by computing a number of test cases in 2-D, namely the cylindrical blast wave [12] and the supersonic flow over a planar blunt-body. The blast wave test case is computed with a vertically directed magnetic field. This necessitates the use of a procedure to enforce zero $\nabla \cdot \vec{B}$. A projection scheme based on the Poisson solver is used. Another widely used scheme is to use a source term proportional to the numerical value of $\nabla \cdot \vec{B}$ [9,6]. On the other hand, the blunt-body computations use an out-of-plane magnetic field, thus automatically satisfying the solenoidal condition requirement.

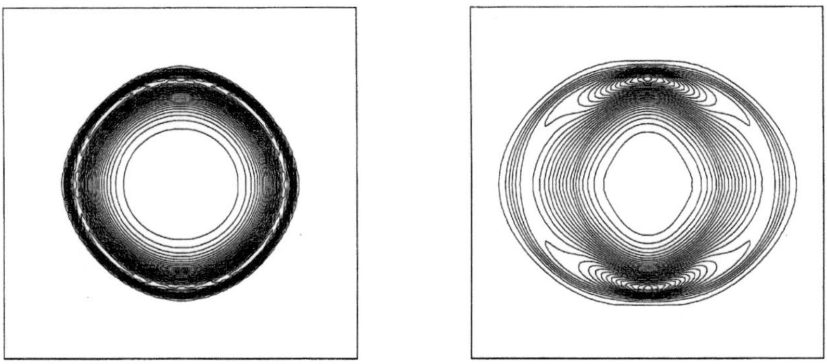

(a) Euler equations (b) Ideal MHD equations

Fig. 1. Density contour plots of the Euler and the ideal MHD computations

Fig. 1 shows the density contours for both the Euler and the ideal MHD calculations. As expected, the plot for the Euler case shows symmetrical shock wave, while in the ideal MHD case the shock along the horizontal centerline is diffused and weakened considerably compared to the shock along the vertical centerline. This is further confirmed by the centerline profiles of both the hydrodynamic and magnetic pressures, shown in Fig. 2. This may suggest that some of the dissipated energy from the shock wave is transferred to the magnetic field instead of being converted into thermal energy.

The supersonic flow over a blunt-body is computed on a 78×51 grid with a leading edge radius of 0.1 m. The freestream conditions for the Mach 5.85 flow are 510 Pa and 55 K. The CFL numbers are 0.4 for the explicit scheme, and 2.0 for the implicit scheme. Solution is assumed to attain convergence when the L_2-norm of the density change is approximately 10^{-7}. For the MHD calculation, an out-of-plane magnetic field 0.05 T is applied. To accelerate convergence of the MHD computations, the initial magnetic field is applied to the converged Euler solution.

Fig. 3 compares the density contour plots for both the Euler and the ideal MHD calculations. It clearly shows how the application of the magnetic field affects the flow, the most obvious is that the shock wave is pushed away from the body. Additionally, it can be shown that the shock wave has decreased in strength, as can be seen in the density and pressure stagnation line profiles in Fig. 4.

It has been argued that the main strength of the implicit schemes is the significant reduction in the cost of computation especially in the case of steady-state calculations, when compared to the explicit formulation due to the ability to utilize larger time-step. To show that this is indeed the case, the convergence history of the blunt-body test case, which is chosen to be the L_2-norm of the density change at each time-step, is computed and presented in Fig. 5 for both the Euler and the ideal MHD computations. It is clear that the implicit formulation does require less number of time-step to reach the convergence criterion of 10^{-7} when compared to the explicit formulation.

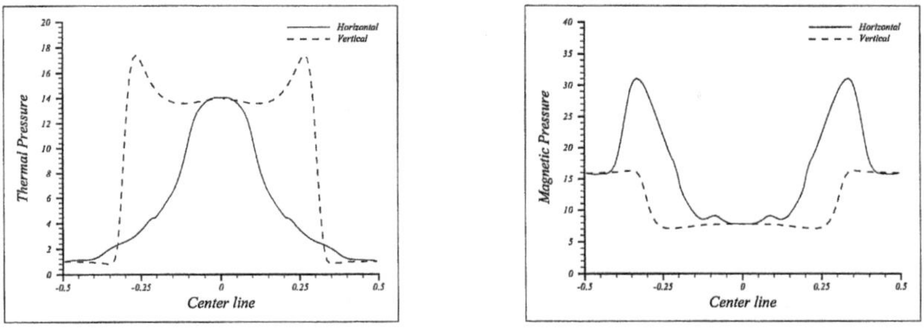

(a) Hydrodynamic pressure (b) Magnetic pressure

Fig. 2. Centerline profiles of the hydrodynamic pressure ($= p$) and magnetic pressure ($= \frac{1}{2}B_k^2$)

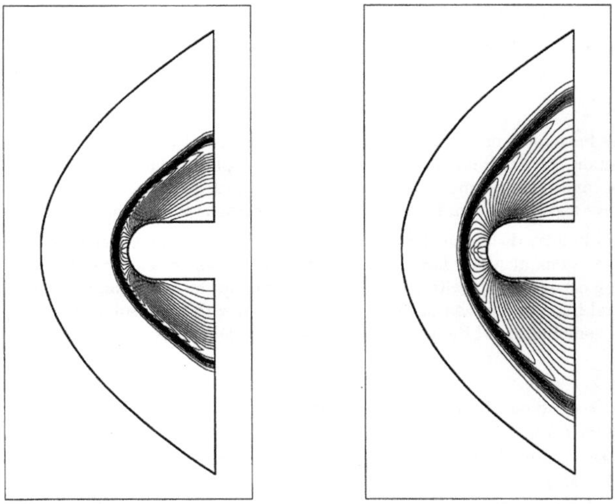

(a) Euler equations (b) Ideal MHD equations

Fig. 3. Density contour plots for the Euler and the ideal MHD computations

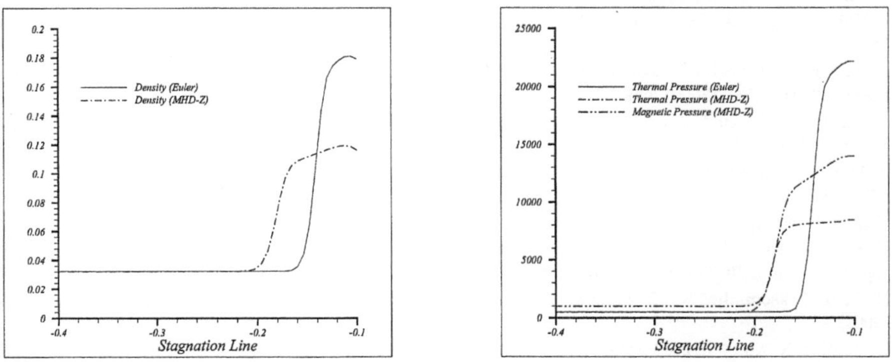

(a) Stagnation density profiles (b) Stagnation pressure profiles

Fig. 4. Stagnation line density and pressure profiles for the blunt-body computations

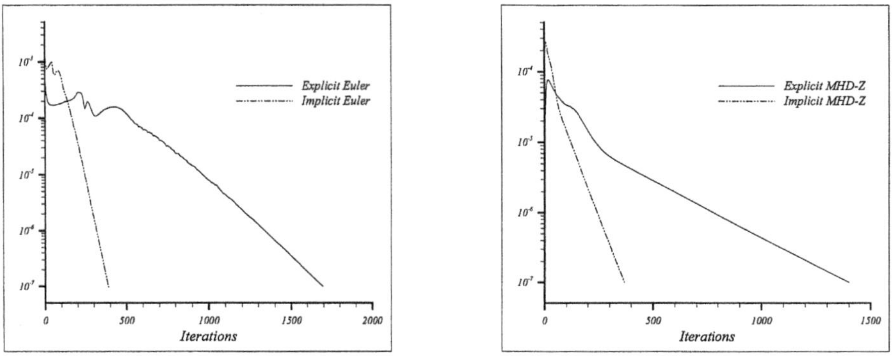

(a) Euler equations (b) Ideal MHD equations

Fig. 5. Plots of convergence history for the Euler and the ideal MHD computations

CONCLUSIONS

Implicit kinetic schemes have been developed for the Euler and the ideal MHD equations. Numerical tests show tha the implicit schemes retain the accuracy and robustness of the explicit schemes while increasing their efficiency by reducing the number of time-steps required in obtaining the numerical solution. Excellent results are obtained for th 2-D cylindrical blast wave and flow past a blunt-body with and without the magnetic field.

Furthermore, it is clearly indicated that an effective and efficient method of divergence control is of utmost importanc in obtaining accurate solutions of MHD flow problems. The projection scheme has been found to be an expensiv approach numerically, especially in generalized coordinate system. On the other hand, the use of source terms does no really address the central issue of creation of the divergence errors within the solution, it only helps in advecting th errors away. Perhaps, a combination of the two methods may provide a better alternative.

REFERENCES

[1] R. K. Agarwal, K. E. Acheson, *A Kinetic Theory Based Wave/Particle Flux Splitting Scheme for the Euler Equa tions*, AIAA Paper 95-2178, (1995).

[2] M. Brio, C. Wu, *An Upwind Differencing Scheme for the Equations of Ideal Magnetohydrodynamics*, J. Comp Physics, 75, (1988), 400–422.

[3] S. Chou, D. Baganoff, *Kinetic Flux-Vector Splitting for the Navier-Stokes Equations*, J. Comp. Physics, 130, (1997) 217–230.

[4] J.-P. Croisille, R. Khanfir, G. Chanteur, *Numerical Simulation of the MHD Equations by a Kinetic-Type Metho* J. Sci. Comput., 10, (1995), 481–492.

[5] J. Huba, J. Lyon, *A New 3D MHD Algorithm: The Distribution Function Method*, J. Plasma Physics, 61, (1999) 391–405.

[6] P. Janhunen, *A Positive Conservative Method for Magnetohydrodynamics Based on HLL and Roe Methods*, Comp. Physics, 160, (2000), 649–661.

[7] R. MacCormack, *An Upwind Conservation Form Method for the Ideal Magnetohydrodynamics Equations*, AIA Paper 99-3609, (1999).

[8] J. Mandal, S. Deshpande, *Kinetic Flux-Vector Splitting for Euler Equations*, Computers and Fluids, 23, (1994) 447–478.

[9] K. Powell, *An Approximate Riemann Solver for Magnetohydrodynamics*, ICASE Report 94-24, (1994).

[10] H. S. R. Reksoprodjo, R. K. Agarwal, *A Kinetic Scheme for Numerical Solution of Ideal Magnetohydrodynami Equations with a Bi-Temperature Model*, AIAA Paper 2000-0448, (2000).

[11] H. S. R. Reksoprodjo, R. K. Agarwal, *Implicit Kinetic Schemes for Euler Equations*, AIAA Paper 2001-262 (2001).

[12] H.-Z. Tang, K. Xu, *A High-Order Gas-Kinetic Method for Multidimensional Ideal Magnetohydrodynamics*, J. Com Physics, 165, (2000), 69–88.

[13] K. Xu, *Gas-Kinetic Theory Based Flux Splitting Method for Ideal Magnetohydrodynamics*, ICASE Report 98-5 (1998).

Parallel Computational Fluid Dynamics – Parallel Computing and Its Applications
J.H. Kwon, A. Ecer, J. Periaux, N. Satofuka and P. Fox (Editors)

Parallel simulation of turbulent flow in a 3-D lid-driven cavity

J. M. McDonough[a]

[a]Departments of Mechanical Engineering and Mathematics
University of Kentucky, Lexington, KY 40506-0503 USA
E-mail: jmmcd@uky.edu   Website URL: http://www.engr.uky.edu/~acfd

In this paper we present a new version of the standard CFD benchmark lid-driven cavity problem in which now all six sides of the 3-D cavity are in motion to promote more efficient mixing. We solve this problem with a new form of large-eddy simulation at moderate Reynolds number to demonstrate the ability of this procedure to automatically predict transition to turbulence. We report parallel speedups observed on a general-purpose symmetric multiprocessor employing MPI for parallelization.

Keywords: CFD benchmarks, 3-D flows, lid-driven cavity, LES, turbulence.

1. INTRODUCTION

The lid-driven cavity (LDC) problem has served as a benchmark for validation of CFD codes essentially from the beginning of CFD itself. The carefully-performed study of Burggraf [1], although not the first investigation of this problem, provided early accurate 2-D results that still are cited today. Considerably later, with the advent of supercomputers, Ghia *et al.* [2] published landmark computations of demonstrated accuracy that have long provided a quantitative benchmark against which to test CFD codes. Three-dimensional solutions to this problem began to appear by the mid to late 1980s, *e.g.*, Freitas *et al.* [3], and work has continued to the present (for a review, see Shankar and Deshpande [4] and references therein). But only recently have studies of the calibur of [2] in 2-D begun to appear for the 3-D problem (see Albensoeder and Kuhlmann [5]). Clearly, the main impediment has been the same as it was in the early 2-D simulations—very long run times for 3-D problems. This highlights the importance of parallel computation and the utility of employing overall algorithms and numerical methods that are readily parallelizable.

Recent work on the 3-D LDC problem has been restricted to application of motion on at most top and bottom boundaries, and usually only on the top boundary. Although the resulting flow fields contain a large-scale vortical structure, essentially no matter what the value of Reynolds number, *Re*, this structure does not necessarily lead to effective mixing. In the present study we introduce motion by sliding <u>all</u> bounding faces of the cubical box containing the fluid and demonstrate in a qualitative way the more complicated stream-lines compared with sliding a single plane. Of importance in the context of parallelization is analysis of parallel efficiency of the present 3-D code, which will ultimately be needed to enable thorough studies with guaranteed accuracy of results.

The remainder of the paper consists of an analysis section presenting details of the problem being solved and solution procedures employed. This is followed by a section containing results for both the physical problem and for parallelization attempts. In a final section we summarize the work and draw conclusions.

2. ANALYSIS

Figure 1 displays the geometry of the problem considered here: a 3-D unit cube with unit velocity components applied on each of the faces, in opposite directions on opposing faces. It is apparent from simulations that this choice of velocity directions to some extent mitigates well-known pressure singularities in corners of the cube. The equations describing this flow situation are well known—the 3-D incompressible Navier–Stokes (N.–S.) equations expressed here as

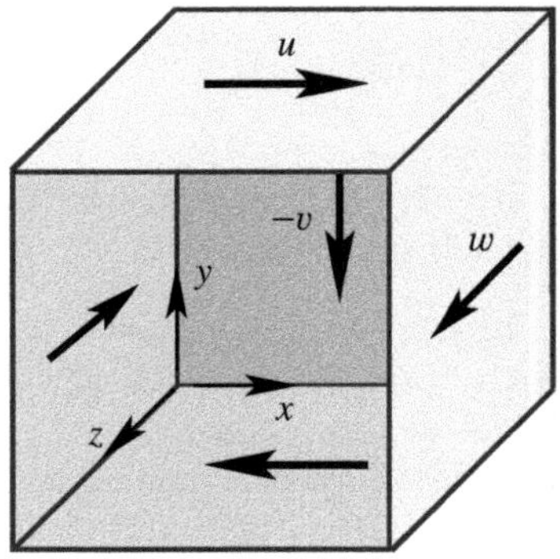

Figure 1. Basic problem geometry.

$$\nabla \cdot \boldsymbol{U} = 0 \tag{1a}$$

$$\frac{D\boldsymbol{U}}{Dt} = -\nabla P + \frac{1}{Re}\Delta \boldsymbol{U}\,, \tag{1b}$$

where $\boldsymbol{U} = (u, v, w)^T$ is the dimensionless velocity vector; P is pressure scaled with twice the dynamic pressure, and $Re \equiv UL/\nu$ with U and L being reference velocity and length, respectively (both set to unity), and ν is kinematic viscosity. D/Dt is the usual substantial (material) derivative; ∇ is the gradient operator, and Δ is the Laplacian. No-slip boundary conditions are imposed with components of $\boldsymbol{U}$ set to ± 1 on the faces of the cube, as indicated in the figure.

The code employed for this study has been reported previously (McDonough and Dong [6] and McDonough *et al.* [7]) and embodies a new form of large-eddy simulation (LES) utilizing the following concepts: *i*) filter solutions rather than equations; *ii*) model subgrid-scale (SGS) dependent variables, not their statistics; *iii*) directly employ subgrid-scale results as part of a multi-scale formalism instead of discarding them. This represents a specific instance of the class of LES methods sometimes termed "synthetic-velocity" approaches. These take on several different forms, the best known of which are probably the linear-eddy models (LEMs) and one-dimensional turbulence (ODT) models of Kerstein and various coworkers (see, *e.g.*, [8] and [9]) and the "estimation" approaches of Domaradzski and coauthors (*e.g.*, [10]). The method described here was first presented by McDonough *et al.* [11], analyzed in detail by Hylin and McDonough [12] and summarized in the monograph by Sagaut [13]. The specific form of the SGS model has been modified rather significantly in recent years due to work by McDonough and Huang [14], McDonough [15] and Holloway and McDonough [16]. In particular, the SGS model now takes the simple form

$$q_i^* = A_i M_i\,, \qquad i = 1, 2, \ldots, N_v\,, \tag{2}$$

for N_v dependent variables corresponding to the typical LES decomposition into large- and small-scale parts:

$$\boldsymbol{q}(\boldsymbol{x},t) = \widetilde{\boldsymbol{q}}(\boldsymbol{x},t) + \boldsymbol{q}^*(\boldsymbol{x},t)\,, \tag{3}$$

with $\boldsymbol{q} \equiv (q_1, q_2, \ldots, q_{N_v})^T$.

In (2) q_i^* is the small-scale part of the ith dependent variable; A_i is its amplitude, and M_i represents its temporal behavior. Equations (2) are evaluated in all grid cells of the large-scale discretization of the problem domain at every discrete time step. The amplitudes are calculated using deconvolution (see, *e.g.*, Adams and Stolz [17]) to obtain estimates of fluctuating quantities from large-scale results, calculating the corresponding second-order structure functions and employing a generalization of Kolmogorov's K41 theory (see Frisch [18]) to obtain energy as a function of wavenumber, and then invoking the Parseval identity to equate Fourier and physical L^2 norms.

The M_is appearing in Eq. (2) are components of the "poor-man's Navier–Stokes (PMNS) equation" first presented by McDonough and Huang [19] and analyzed in considerable detail in [14]. This epithet seems to have first been used in [18] to draw an analogy between the N.–S. equations and a simple quadratic map. But such maps do not carry much of the N.–S. structure (they are scalar!) and so are not generally suitable for use as SGS models. In [19] and [14] a Galerkin procedure was employed to obtain a 2-D version of the following discrete dynamical system (DDS) given now in 3D:

$$a^{(n+1)} = \beta_1 a^{(n)}(1-a^{(n)}) - \gamma_{12}a^{(n)}b^{(n)} - \gamma_{13}a^{(n)}c^{(n)}\,, \tag{4a}$$
$$b^{(n+1)} = \beta_2 b^{(n)}(1-b^{(n)}) - \gamma_{21}a^{(n)}b^{(n)} - \gamma_{23}b^{(n)}c^{(n)}\,, \tag{4b}$$
$$c^{(n+1)} = \beta_3 c^{(n)}(1-c^{(n)}) - \gamma_{31}a^{(n)}c^{(n)} - \gamma_{32}b^{(n)}c^{(n)}\,. \tag{4c}$$

When only incompressible flow is treated (as is the case herein), we identify $a \equiv M_1$, $b \equiv M_2$, $c \equiv M_3$. Additional components of the DDS can be derived to account for further physics (*e.g.*, heat transfer, chemical kinetics) as was done in [19].

It is worthwhile to provide a few remarks concerning this approach to construction of SGS models. First, with the A_is calculated as indicated above, it is theoretically possible to obtain relatively accurate SGS amplitudes even on very coarse grids (as will be demonstrated below). At the same time, use of structure functions introduces multi-point statistical information into the model—lack of which has always been a general criticism of usual turbulence models. Second, it can be shown that the PMNS equation is related to the symbol of a pseudodifferential operator of the N.–S. equations (McDonough *et al.* [20]). As such, it locally (in space and time) carries most of the N.–S. physics as has been previously evidenced by the ability to "curve fit" turbulent experimental flow data to these equations (see Yang *et al.* [21]). Third, the numerous bifurcation parameters (β_is, γ_{ij}s) appearing in these equations can be evaluated from high-pass filtered large-scale results (deconvolution again) because they are directly related to specific terms and parameters of the N.–S. equations; hence, they do not pose a "closure problem." Finally, we observe that Eqs. (4) (and any additional ones accounting for further physics) are not only inexpensive to evaluate (and embarassingly parallelizable), but they also permit account to be taken of interactions of all represented physics on the sub-grid scales—something not well accomplished by most other SGS models. Among other things, this leads to an ability to automatically predict transition to and from turbulence, as results presented below demonstrate.

The numerical analysis employed for the code is rather standard, but highly efficient and robust; it utilizes the following techniques implemented on a staggered, structured grid: centered differencing in space, trapezoidal integration in time with Douglas–Gunn [22] time splitting, quasilinearization of nonlinear terms, (spatial) filtering of under-resolved large-scale results, all applied in the context of Gresho's projection 1 [23] to handle velocity-pressure coupling. The code is parallelized via MPI as reported by McDonough and Yang [24]. We observe that the discretizations employed imply formal second-order accuracy in both space and time, but use of the highly-efficient, robust projection method reduces this to only first order, temporally. But turbulent solutions to the N.–S. equations are known to be sufficiently nonsmooth as to invalidate typical Taylor expansion based truncation error estimates, so only first-order accuracy can be guaranteed in any case.

3. RESULTS

The results reported in this section consist of two different types, and a subsection is devoted to each. In the first subsection we present flow field results computed on a coarse grid of 26^3 points to demonstrate effectiveness of the turbulence modeling procedure. In the second subsection we describe the outcome of parallelization efforts and compare findings for several different grids.

3.1. Coarse-grid flow results

Figures 2 (a) and (b) display streamlines of the large-scale part of the solution at one instant in time during the initial transient. We remark that this part of the solution is all that one would be able to obtain using typical LES approaches; *i.e.*, it corresponds to components of only $\widetilde{\boldsymbol{q}}$ in Eq. (3). The streamlines are shaded with vorticity magnitude using a scale reduction factor of 10 to better highlight regions of high vorticity. Part (a)

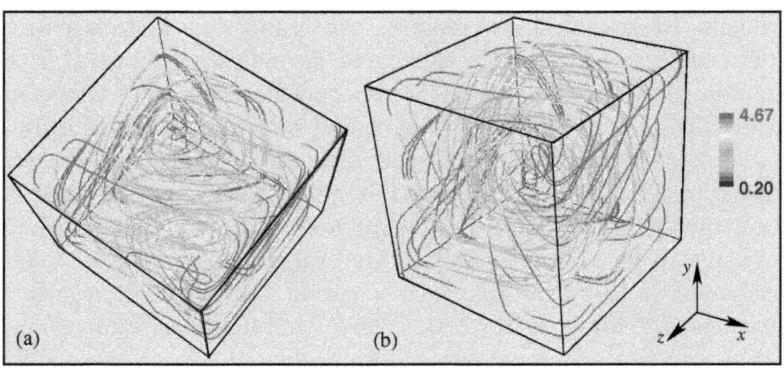

Figure 2. Large-scale velocity streamlines shaded with vorticity magnitude: (a) top view; (b) corner view indicating single vortex.

of the figure provides a view through the top x-z plane of the cavity, displaying what appear to be two main vortical structures. These exhibit fairly smooth, but complicated trajectories except in a bottom corner of the box appearing toward the top of Fig. 2(a), and in the diagonally opposite upper corner (near the bottom of part (a)). At these two locations the streamlines are more contorted.

But part (b) of the figure suggests these two structures are connected and comprise one large, quite complicated vortex whose core extends diagonally across the cavity from a lower corner to an upper one. In addition to these observations on overall structure, it is important to note that highest vorticity magnitudes occur near the moving walls of the cavity, especially along edges and in corners, as would be expected.

Figure 3 presents several depictions of small-scale streamlines, which we emphasize are not available in usual LES results. Part (a) displays such streamlines, shaded with small-scale vorticity magnitude (again rescaled to emphasize regions of high vorticity). Seeds for producing these have been placed throughout the box, with some concentration near edges and corners. Although it is difficult to detect from this single time slice (the same

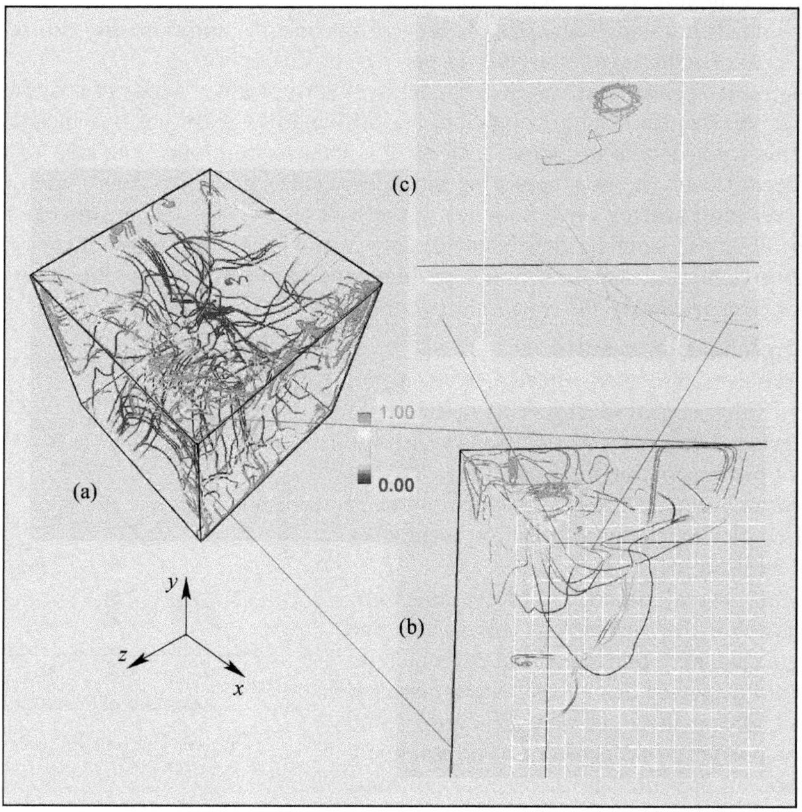

Figure 3. Small-scale streamlines. (a) complete small-scale flow field; (b) zoom-in near upper-forward corner; (c) second zoom-in showing size of vortex.

one used for large-scale results in Fig. 2), at this relatively late time in the initial transient there is only a moderate amount of small-scale turbulent activity in the middle region of the cavity, but still significant generation of vortices near edges and corners. Indeed, this persists well beyond the time that the large-scale portion of the flow has achieved steady state. On the other hand, peak small-scale vorticity magnitudes are decaying by this time, so it is likely that a completely laminar steady state would ultimately be attained.

Parts (b) and (c) of this figure are successive zoom-ins intended to show the range of sizes of spatial structures generated by the SGS model. In part (b) we present a few of the streamlines started with seeds only in the upper corner of the box nearest the viewer. We have included in this figure a plane from the computational 26^3 grid situated behind all of the streamlines shown. From this it can be inferred that all vortical structures shown in part (b) are no larger in absolute size, relative to the grid size, than they appear. Part (c) displays a second zoom-in including little more than a single cell and a vortex that is immediately in front of the plane of grid cells. It is clear from this that small-scale velocity components generated by the SGS model permit representation of vortical structures that are only a small fraction of the size of the grid spacing. In addition, one can see a wide range of sizes of structures, especially in part (b) of this figure.

In general, it is clear that the SGS model is able to produce scales of structures both larger and smaller than the grid spacing, and although we have not specifically demonstrated this here, details in figures such as 3(a) change from one time step to the next with some vortical structures persisting for milliseconds (of physical time) and others appearing and disappearing on time scales a tenth of this or less. The consequence of all this is the ability to simulate flow behavior deep in the inertial subrange of the turbulence energy spectrum even using very coarse gridding and to locally in space and time generate the backscatter necessary for transition to turbulence.

3.2. Parallelization results

The problem discussed above has been run with three different grid spacings corresponding to uniform grids of 26^3, 51^3 and 101^3 points with the number of processors ranging from two to 32. Results presented in Fig. 4 show speedups obtained via MPI employing the parallelization techniques given in detail in [24].

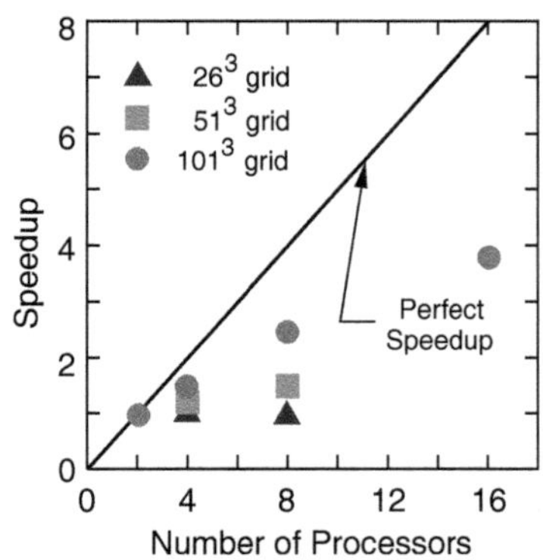

Figure 4. Parallel speedups.

There are several important observations to be made regarding the contents of this figure, and other tests that were performed but for which results are not displayed here. First, it was observed that no significant speedups were achieved beyond those obtained with 16 processors, so no results for those cases are presented here. Second, all computations were performed both with, and without, activation of the SGS turbulence model. On the two coarser grids the run time penalty for invoking the turbulence model could be as great as a factor of six, or as little as three, depending on the grid and the number of processors being employed. On the other hand, on the fine grid employing the turbulence model resulted in slightly less than a factor of two increase in run time, uniformly for all numbers of processors. At the same time, parallel speedups, *per se* (or lack thereof in the 26^3 case) were unaffected by the turbulence model. In particular, we deduce from this that our parallelization of the turbulence model was at essentially the same level of effectiveness as that for the large-scale algorithm. As a consequence, we have displayed only parallel speedups for turbulence calculations in the figure.

We next remark that speedups shown in Fig. 4 are significantly inferior to those reported in [24] using precisely the same code, but for a rather different problem—and executed on different processors. Indeed, it is clear from the figure that on coarse grids there is no speedup (in fact, independent of the number of processors), while on the more moderate grid there is limited speedup, but not at a level to justify the required machine resources. We have performed extensive tests to isolate this seeming discrepancy, and the data presented here immediately suggest its source. It is fairly obvious that for the coarse and moderate grid sizes, communication times are evidently overwhelming computation times: there is no other explanation for the fact that no improvement with number of processors is seen for the 26^3 grid, and very little for the 51^3 grid. The current results were computed on a HP SuperDome SMP, just as were those presented in [24], but the system has been upgraded with new processors that are between $1\frac{1}{2}$ and 3 times faster than the earlier ones—but with no upgrade to communication bus bandwidth. Hence, compute times aare significantly reduced, but communication times remain the same as for the earlier version of the machine. It is only with the 101^3 grid that processors are kept busy long enough for overall run times to not be dominated by communication. Clearly, for any machine configured in this way, small simulations should not be parallelized.

4. SUMMARY/CONCLUSIONS

We have introduced a new version of the 3-D lid-driven cavity problem that leads to more complicated fluid parcel trajectories and thus, enhanced mixing, but at the same time weakens corner singularities. We employed an advanced form of LES to solve this problem and presented preliminary results that show very complicated streamline structures on both large and small scales, despite a relatively low Reynolds number. Finally, we demonstrated moderate speedups via parallelization.

We conclude from these results that the SGS modeling approach described here should be further studied in other more complicated flow situations, and at a more quantitative level for the LDC problem. Furthermore, our parallelization efforts emphasize that with modern high-speed processors use of parallelization should probably be reserved for very large problems to more effectively utilize machine resources.

REFERENCES

1. O. R. Burggraf, J. Fluid Mech. 24 (1966) 113.
2. U. Ghia, K. T. Ghia and C. T. Shin, J. Comput. Phys. 48 (1982) 387.
3. C. J. Freitas, R. L. Street, A. N. Findikakis and J. R. Koseff, Int. J. Numer. Meth. Fluids 5 (1985) 561.
4. P. N. Shankar and M. D. Deshpande, Annu. Rev. Fluid Mech. 32 (2000) 93.
5. S. Albensoeder and H. C. Kuhlmann, J. Comput. Phys. 206 (2005) 536.
6. J. M. McDonough and S.-J. Dong, PARALLEL COMPUTATIONAL FLUID DYNAMICS

Trends and Applications, Jenssen *et al.* (Eds.), North-Holland Elsevier Pub., Amsterdam, 173, 2001.
7. J. M. McDonough, T. Yang and M. Sheetz, PARALLEL COMPUTATIONAL FLUID DYNAMICS Advanced Numerical Methods, Software and Applications, Chetverushkin *et al.* (Eds.), North Holland -Elsevier Pub., Amsterdam, 473, 2004.
8. A. R. Kerstein, Comb. Sci. and Tech. 60 (1988) 391.
9. T. Echekki, A. R. Kerstein, and T. D. Dreeben, Combust. and Flame 125 (2001) 1083.
10. J. A. Domaradzski and E. M. Saiki, Phys. Fluids 9 (1997) 2148.
11. J. M. McDonough, Y. Yang and E. C. Hylin, Proc. 1st Asian Computational Fluid Dynamics Conference, Hui *et al.* (Eds.), Hong Kong, 747, 1995.
12. E. C. Hylin and J. M. McDonough, Int. J. Fluid Mech. Res. 26 (1999) 539.
13. P. Sagaut, Large Eddy Simulation for Incompressible Flows, Springer-Verlag, Berlin, 2001.
14. J. M. McDonough and M. T. Huang, Int. J. Numer. Meth. Fluids 44 (2004) 545.
15. J. M. McDonough, paper c34, Proc. 2002 Spring Tech. Meeting, Central States Sec., Combustion Institute, Knoxville, TN, 2002.
16. J. C. Holloway and J. M. McDonough, paper E33, Proc. 3rd Jt. Mtg. US Secs., Combustion Institute, Chicago, IL, 2003.
17. N. A. Adams and S. Stolz, in Modern Simulation Strategies for Turbulent Flow, B. J. Geurts (Ed.), R. T. Edwards, Inc., Philadelphia, 2001.
18. U. Frisch, TURBULENCE the Legacy of A. N. Kolmogorov, Cambridge University Press, Cambridge, 1995.
19. J. M. McDonough and M. T. Huang, paper ISSM3-E8, Proc. 3rd Int. Symp. on Scale Modeling, Nagoya, 2000.
20. J. M. McDonough, C. B. Velkur and J. Endean, in preparation for submission to Physica D, 2006.
21. T. Yang, J. M. McDonough and J. D. Jacob, AIAA J. 41 (2003) 1690.
22. J. Douglas, Jr. and J. E. Gunn, Numer. Math. 6 (1964) 428.
23. P. M. Gresho, Int. J. Numer. Meth. Fluids 11 (1990) 587.
24. J. M. McDonough and T. Yang, PARALLEL COMPUTATIONAL FLUID DYNAMICS Multidisciplinary and Applications, Winter *et al.* (Eds.), North Holland-Elsevier Pub., Amsterdam, 45, 2005.

Parallel Computational Fluid Dynamics – Parallel Computing and Its Applications
J.H. Kwon, A. Ecer, J. Periaux, N. Satofuka and P. Fox (Editors)

Numerical Analysis of Supersonic Jet Flow from Vertical Landing Rocket Vehicle in Landing Phase

Toshiyuki Suzuki,[a] Satoshi Nonaka,[b] Yoshifumi Inatani[b]

[a]*Institute of Aerospace Technology, JAXA*
7-44-1 Jindaiji-Higashi-machi, Chofu, Tokyo 182-8522, Japan

[b]*Institute of Space and Astronautical Science, JAXA*
3-1-1 Yoshinodai, Sagamihara, Kanagawa 229-8510, Japan

Keywords: Large Eddy Simulation, Opposing Jet;

1. INTRODUCTION

As a future space transportation system, a working group at Institute of Space and Astronautical Science (ISAS) of Japan Aerospace Exploration Agency (JAXA) proposed a concept of a fully reusable launch vehicle.[1] The proposed vehicle is a single stage vertical taking off and landing (VTOL) type one. For the purpose of verification of this concept, a small-scale test vehicle shown in Fig. 1 was developed. The vehicle was named Reusable Vehicle Testing (RVT). Using the RVT, a series of flight tests has been conducted to explore the capability of the vertical landing and to acquire ground operational techniques for repeated flight. One of the most considerable issues regarding the aerodynamics of the RVT is an influence of jet flow on attitude control of the RVT. In the landing phase, the RVT ejects supersonic jet toward subsonic freestream so as to obtain enough drag force for vertical landing. However, because the jet ejection significantly disturbs the flowfield around the RVT, the flowfield is expected to be highly unsteady which characterized by massive separations and formation of large scale vortical structures. This unsteady nature of flowfield becomes a cause of oscillation of the RVT. Therefore, it is important to understand the unsteadiness of the flowfield in order to conduct attitude control, guidance and navigation while descending. For this purpose, several wind tunnel experiments have been carried out in JAXA.

In the recent work of Nonaka et al.,[2] a 1/12-scale model of the RVT which can produce an opposing supersonic jet was built and a subsonic wind tunnel testing was made using the model to experimentally simulate a interaction flowfield between the opposing jet and freestream. The flowfield around the model was visualized using the particle image velocimetry (PIV) technique[3]. From the instantaneous velocity vectors deduced from the image, it was found that the opposing jet from the model produced a vortex ring around the vehicle, and that the vortex ring moved unsteadily in another moment. Due to this unsteady flow motion, aerodynamic forces measured on the model were fluctuated with time. Unfortunately, because the wind tunnel experiment could not reproduce some parameters in the flight environment of the RVT such as Reynolds number and an engine nozzle exit diameter, etc., it is uncertain whether the obtained experimental data can be reliably extrapolated to the flight environment. Therefore, a computer code is needed to be able to simulate the opposing jet flow from the RVT accurately. Such a computer code will be helpful to correlate the experimental data to a realistic flight environment of the RVT, and of a full scale reusable vehicle in the future.

Fig. 1 Flight test of small-scale test vehicle RVT. (Oct. 2003)

Large Eddy Simulation (LES) is expected to be a powerful and efficient technique to analyze such an unsteady flowfield. It is the purpose of the present study to simulate the flowfield around the RVT by using LES to examine how the qualitative nature of the flowfield affects the aerodynamic forces on the RVT. However, there have been no prior attempts to calculate the interaction flowfield between the jet and freestream of this kind. For this reason, we first calculate the flow environment for the wind tunnel conditions in this study. Calculated results are compared with the existing experimental data[2,3] to validate our LES modeling. Simulation of the actual flight environment of the RVT will be made in future work.

2. EXPERIMENTAL CONFIGURATION

Experimental results discussed in this paper were obtained by two independent series of experiment: (1) aerodynamic forces and surface pressure measurement[2] and (2) measurement of velocity distributions by using the PIV technique[3]. The aerodynamic forces and surface pressure measurements were conducted in the planetary environmental wind tunnel facility at ISAS of JAXA. The test section diameter and length are 1.6m and 2.0m, respectively. The PIV measurements were carried out in the 2m x 2m low speed wind tunnel facility at Institute of Aerospace Technology (IAT) of JAXA. Though the details of the measurements can be found in Refs. 2 and 3, the experimental configurations are described briefly here.

In order to investigate the aerodynamic characteristics of the RVT with jet ejection, aerodynamic forces and surface pressure were measured. The size of the wind tunnel model is about 1/12 of the actual RVT. The total length and base diameter are 252mm and 185mm, respectively. The model has a nozzle at the center of base region to produce supersonic jet toward freestream. The nozzle throat and exit diameter are 2mm and 3mm, respectively. Nitrogen gas for the jet is supplied by flexible tubes connected to a chamber, and is then ejected toward freestream from the nozzle exit. The force measurements were carried out by using an inertial balance which can measure six components of aerodynamic forces. For surface pressure measurements, differential pressure gauges are installed at the body surface of the model. In this study, nine pressure gauges were embedded along the entire model surface.

In order to understand the flow structure in detail, the velocity distributions around the model were measured by using the PIV technique. The oil particles with a diameter of about 1μm were introduced in the freestream and also in the jet flow. The particles around the model were illuminated by a laser light sheet produced by a double pulse Nd-YAG laser. Time separation between laser pulses was set between 2 and 30μs according to the flow velocity. By using two cameras, a wide range of flowfield can be observed simultaneously. From the obtained image pair, a particle displacement during the interval of pulses was determined, and then velocity vectors in the flowfield were calculated. In this study, more than 300 instantaneous image pairs were obtained for one test condition. Each of image pairs was obtained every 0.5s. The size of model and the mechanism of jet ejection for the PIV measurement were the same as those for the surface pressure measurement.

Wind tunnel test conditions are summarized in Table 1. In order to simulate the flight environment in the wind tunnel, following three parameters are corresponded with those of the RVT flight environment: Mach number of jet at the nozzle exit M_j, ratio of nozzle exit pressure to atmospheric pressure p_j/p_∞ and momentum flux ratio of freestream to jet f_∞/f_j. In these parameters, nozzle exit pressure p_j is controlled by a chamber pressure p_{ch}, which is determined by following equation,

Table 1. RVT flight and wind tunnel test conditions.

	RVT flight	Wind tunnel test
Freestream parameters		
Pressure, p_∞ [Pa]	101330	101330
Temperature, T_∞ [K]	291.3	291.3
Velocity, u_∞ [m/s]	70.0	26.4
Mach number, M_∞	0.206	0.077
Reynolds number, R_e	9.96×10^6	3.23×10^5
Jet parameters		
Mach number, M_j	2.41	2.41
Pressure ratio, p_j/p_∞	1.38	1.38
Flux ratio, f_∞/f_j	1.3	1.3

$$\frac{p_{ch}}{p_j}=\left(1+\frac{\gamma-1}{2}M_j^2\right)^{\frac{\gamma}{\gamma-1}}. \tag{1}$$

Test conditions shown in Table 1 correspond with the flight environment when the RVT vehicle descends at a velocity of 70m/s, and throttles with 100% thrust at sea level.

3. NUMERICAL METHODS

Numerical methods used in the present calculation are briefly described here. In this study, we use a flow solver called Unified Platform for Aerospace Computational Simulation (UPACS), a standard CFD code developed in IAT of JAXA.[4] The UPACS is a compressible Navier-Stokes flow solver based on a cell-centered finite volume method on multi-block structured grids. The code is parallelized by a flexible domain decomposition concept and Message Passing Interface (MPI).

Governing equations are dimensionless form unsteady filtered Navier-Stokes equations. For modeling a non-resolvable sub-grid scale (SGS) stress, Smagorinsky model with a model constant of $C_s = 0.1$ is used. In near wall regions, C_s is multiplied by the van Driest type wall damping factor to represent molecular viscosity effect. The convection terms are discretized by utilizing AUSM-DV scheme and MUSCL approach for maintaining 2nd-order spatial accuracy. The viscous terms are discretized using 2nd-order central scheme. Time integration is performed implicitly by Matrix-Free Gauss-Seidel (MFGS) scheme with 3 sub-iterations. The time step is set to $dt = 1.0 \times 10^{-4}$ in order to obtain power spectral density of the pressure coefficient fluctuations in reasonable CPU time. The computations are accomplished using 66 processors of Fujitsu PRIMEPOWER HPC2500, which is the central machine of Numerical Simulator III system in JAXA.

The computational domain extends 40 times as large as base diameter of the model. Note that only half of physical domain is used for computation because of symmetry. The grid is designed to provide an adequate resolution of the dominant mean flow structures near the interaction region between the jet and freestream, and contains 14.1 million points distributed over 66 blocks. The computational grid uses viscous grid spacing suitable for turbulent boundary layer computations at body surface.

The freestream properties shown in Table 1 are imposed at the outer boundary. At the body surface except for the nozzle exit, no-slip boundary condition is assumed. The body surface is assumed to be adiabatic. In this study, calculation of flow in nozzle section is not included. Instead, the boundary conditions at the nozzle exit are given by following: The pressure of the jet flow at the nozzle exit p_j is determined from the pressure ratio p_j / p_∞ shown in Table. 1. The temperature of the jet flow T_j is given by following equation,

$$\frac{T_{ch}}{T_j} = 1 + \frac{\gamma - 1}{2} M_j^2 . \qquad (2)$$

The velocity u_j is determined by assuming Mach number of jet flow at the nozzle exit.

4. RESULTS AND DISCUSSIONS

4.1. Overall Features of Flowfield

LES of unsteady flowfield under wind tunnel test condition is carried out for the cases with and without jet ejection. In this study, we assume the angle of attack to be 0 deg. Figure 2(a) shows a calculated time-averaged velocity contours and streamlines of the flowfield at cross sectional plane for the case without jet ejection. Corresponding result obtained by PIV measurement is shown in Fig. 2(b) for the purpose of comparison. The color contour describes the velocity component along freestream direction. The velocity is normalized by the freestream velocity, and the same scale is used for both of figures. As shown in Fig. 2(a), the velocity at the model base region becomes lower because of the stagnation flow. The flow is then separated around the corner of the model, and makes a large recirculation region at the side of the model. From the comparison between Figs. 2(a) and (b), qualitative agreement between the calculation and the experiment is fairly good.

Figures 3(a) and (b) show the calculated and measured velocity contours and streamlines around the model for the case with jet ejection, respectively. From the calculated result, one can see that there appears a large interaction region between the freestream and the opposing jet flow near the model base region. In this region, a stagnation point is formed ahead of the jet flow, and the freestream is biased by the jet flow. At the side surface of the model, the flow separation which is observed in the case without jet ejection can not be recognized for the case with jet ejection. As a result, the flow is attached to the model. These trends correspond with those of obtained by the PIV measurement.

As shown in Fig. 3(b), one can see that there appears a flow recirculation in the interaction region in the PIV measurement. However, such the flow recirculation can not be recognized from the calculated result shown in Fig. 3(a). The reason for this difference is still unknown. One possible

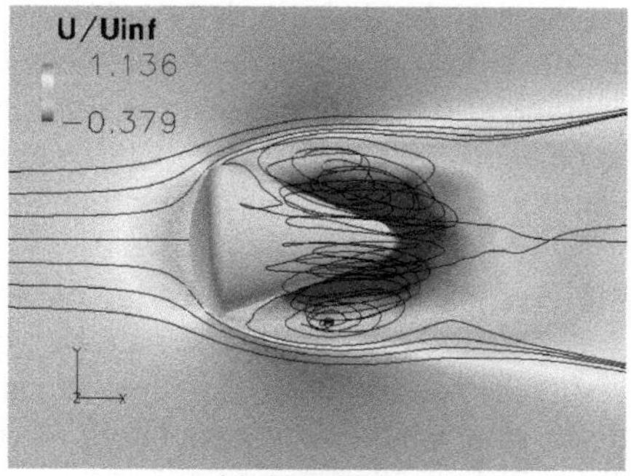

(a) Calculation

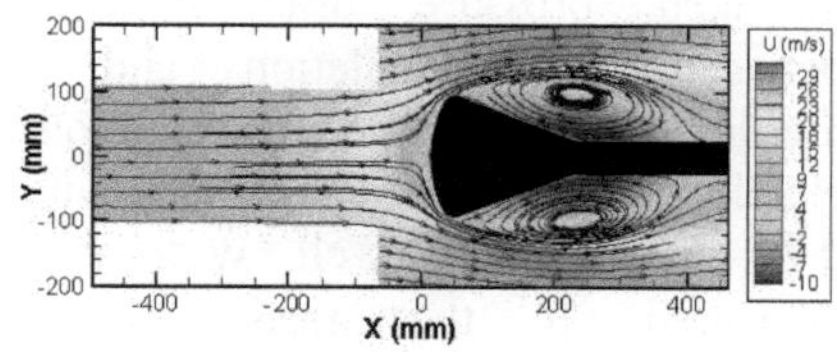

(b) PIV measurement

Fig. 2 Time-averaged velocity contours and streamlines of flowfield for the case without jet ejection.

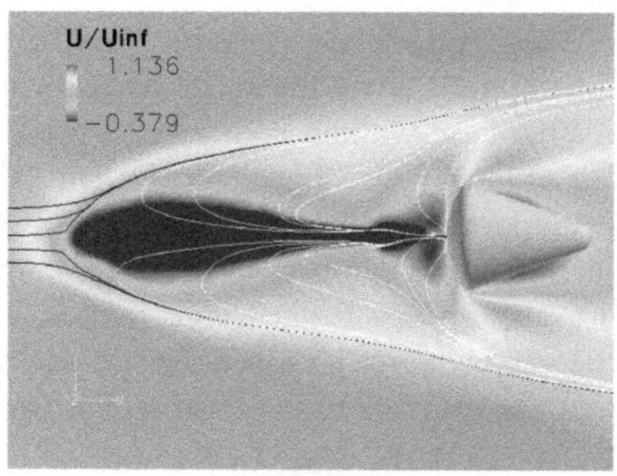

(a) Calculation

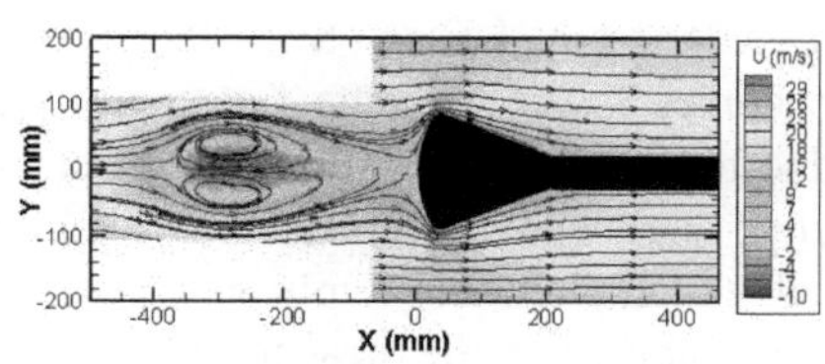

(b) PIV measurement

Fig. 3 Time-averaged velocity contours and streamlines of flowfield for the case with jet ejection.

reason may be that the time scale computed in this study is not enough to obtain the detailed flow structure in the interaction region. The calculated result represents the average of 0.3s, while measured result represents the average of 150s. This point needs to be confirmed in future work. However, the dominant mean flow around the model is nearly reproduced in this calculation.

4.2. Aerodynamic Forces Acting on Wind Tunnel Model

Because the angle of attack is assumed to be 0deg in this study, discussion is made only about drag force. Table 2 shows the comparison of a time-averaged pressure drag coefficient between the calculation and the measurement. For the case without jet ejection, the calculated coefficient value is 0.74, and agrees well with that of measured. In the measurement, the pressure drag coefficient decreases for the case with jet ejection. This tendency is well reproduced in the present calculation. However, the quantitative agreement is yet to be achieved. These points will be discussed later.

Table 2. Comparison of pressure drag coefficient.

	Calculation	Measurement
W/o jet ejection	0.74	0.75
With jet ejection	0.63	0.20

The calculated pressure coefficient distributions along the entire body surface for the case without and with jet ejection are shown in Figs. 4(a) and (b), respectively. Measured values with identification numbers of pressure gauges are also shown in the same figures. For the case without jet ejection, one can see that the pressure at the base region (from S1 to S4) is higher because of the stagnation flow. At the side surface of the model (from S5 to S9), the pressure becomes lower than the freestream value due to the flow separation shown in Figs. 2(a) and (b). As can be seen, the calculated solution duplicates well with that obtained by the measurement.

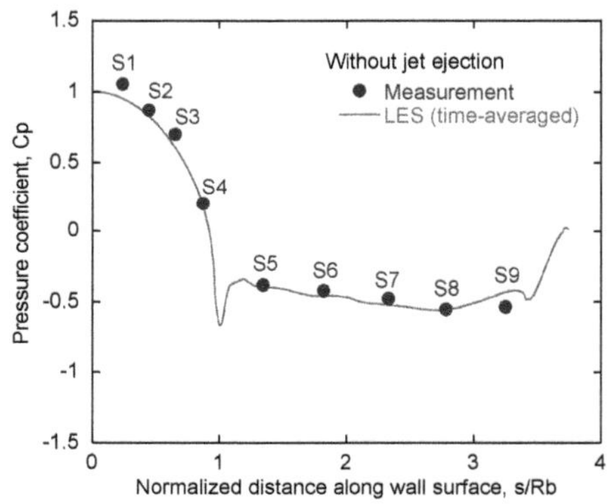

(a)Without jet ejection

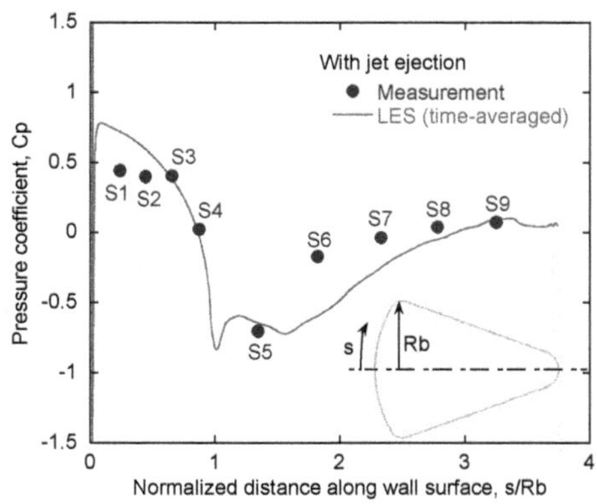

(b) With jet ejection

Fig. 4 Time-averaged pressure coefficient distribution along the body surface.

For the case with jet ejection, in both the calculation and the measurement, the pressure coefficient at the base region (from S1 to S4) for the case with jet ejection is lower than that for the case without jet ejection. This is because the freestream toward the base region is biased by the jet flow, and goes further downstream rather smoothly. As a result, dynamic pressure acting on the model base region is reduced. On the other hand, the pressure along the side surface (from S6 to S9) for the case with jet ejection is higher than that for the case without jet ejection. This is because the separated flow region that was seen in the case without jet ejection disappears with jet ejection. Consequently, not only the reduced base pressure but also the increased pressure along the side surface results in the reduction of the pressure drag force. This is the reason why the pressure drag coefficient decreases due to the jet ejection.

From the comparison between the calculation and the measurement for the case with jet ejection, one can see that the calculated solution overestimates the measured values at the base region (from S1 to S2), and underestimates at the side surface (from S6 to S8). Due to these differences, the reduction of the pressure drag force becomes smaller than that of measurement. This is the reason why the calculated pressure drag coefficient becomes larger than that of experiment.

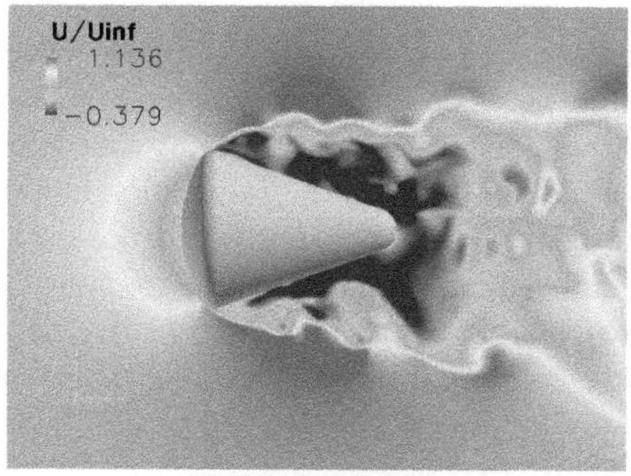

(a) W/o jet ejection

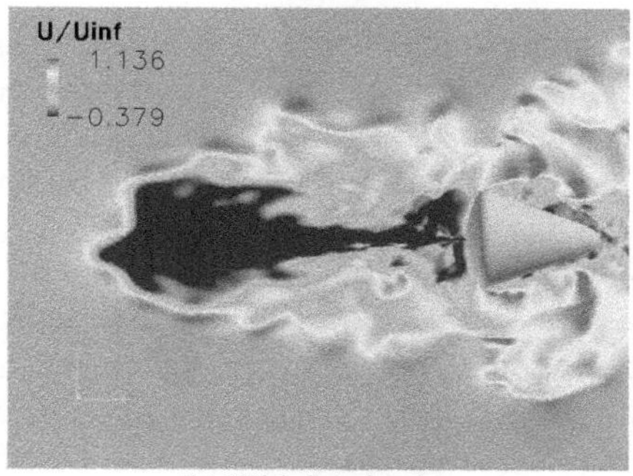

(b) With jet ejection

Fig. 5 Instantaneous velocity contours at cross sectional plane.

4.3. Unsteady Nature of Flowfield

Figures 5(a) and (b) show the instantaneous velocity contours for the cases without and with jet ejection, respectively. The color contours shown in these figures describe the velocity component along freestream direction. By comparing with the time-averaged results shown in Figs. 2(a) and 3(a), unsteady nature of the flowfield is clearly observed for both cases.

Figure 6 shows the power spectrum density of pressure coefficient at the position of the pressure gauge designated by S4 (see Fig. 4). The calculated results

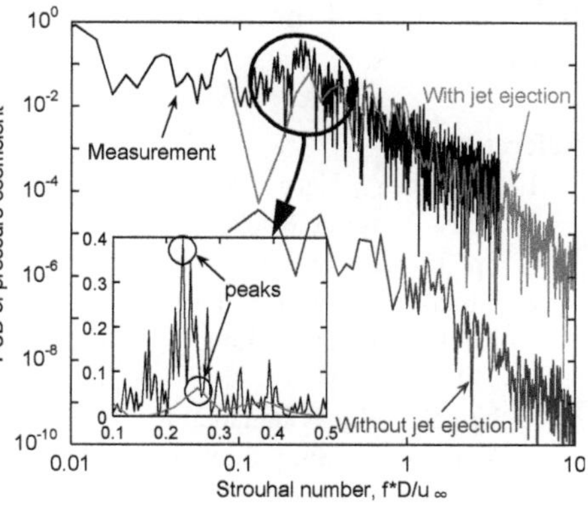

Fig. 6 Power spectrum density of pressure coefficient at the location of pressure transducer designated by S4.

are presented for the cases without and with jet ejection. The result obtained by the measurement is also shown in the same figure. From the comparison of calculated results between the cases without and with jet ejection, it can be seen that the amplitude of the power spectrum density becomes higher due to the jet ejection. From the comparison between the calculation and the measurement for the case with jet ejection, one can see that the characteristics of the power spectrum density obtained by the present calculation are similar to those of measured. In the present calculation, the dominant Strouhal number for the case with jet ejection is about 0.25, and the corresponding frequency is 36Hz. This tendency corresponds with that of measured.

5. SUMMARY

Numerical simulation of flowfield around the vertical landing rocket vehicle under wind tunnel test condition is performed by using LES technique. Calculated results are presented for the cases with and without jet ejection, and are compared with existing experimental data for the purpose of code validation.

For the case without jet ejection, it is shown that dominant mean flow structures around the model are nearly reproduced in this calculation. Calculated time-averaged pressure coefficient distribution duplicates well with that of measured. For the case with jet ejection, although a quantitative agreement of measured pressure coefficient values with those given by the present calculation is yet to be accomplished, the general trend in the measured pressure coefficient distributions was reproduced well in the present calculation. It is also found that calculated characteristics of power spectrum density of pressure coefficient are similar to those of measured for the case with jet ejection.

ACKNOWLEDGEMENT

Authors would like to thank Dr. S. Ogawa, Dr. S. Enomoto and Dr. H. Abe of IAT of JAXA for fruitful discussions and appreciate various helps given by K. Watanabe of ISAS of JAXA.

References

1. Inatani, Y., Naruo, Y. and Yonemoto, K., "Concept and Preliminary Flight Testing of a Fully Reusable Rocket Vehicle," *Journal of Spacecraft and Rockets*, Vol. 38, No. 1, 2001, pp. 36-42.
2. Nonaka, S., Osako, Y., Ogawa, H. and Inatani, Y., "Aerodynamics of Vertical Landing Rocket Vehicle in Landing Phase," Advances in the Astronautical Sciences, Vol. 117, pp. 791-803, 2004.
3. Nonaka, S., Watanabe, K., Ogawa, H., Kato, H., and Inatani, Y., "Aerodynamics of Vertical Landing Rocket Vehicle with Engine Thrust in Landing Phase," AIAA Paper 2006-256, 2006.
4. Yamane, T., Yamamoto, K., Enomoto, S., Yamazaki, H., Takaki, R. and Iwamiya, T, "Development of a Common CFD Platform - UPACS -," Proceedings of International Conference on Parallel Computational Fluid Dynamics, Trondheim, Norway, 2001, pp. 257-264.

Parallel Computational Fluid Dynamics – Parallel Computing and Its Applications
J.H. Kwon, A. Ecer, J. Periaux, N. Satofuka and P. Fox (Editors)

Parallel Computation of a Large Number of Lagrangian Droplets in the LES of a Cumulus Cloud

I. S. Kang[a], Y. Noh[a], and S. Raasch[b]

[a]*Department of Atmospheric Science, Yonsei University, 134 Sinchon-dong, Seodaemun-gu, Seoul 120-749, Korea*

[b]*Institute of Meteorology and Climatology, University of Hannover, 30419, Hannover, Germany*

Key Words: LES; cloud; suspended particles; particle settling; turbulence

A LES model was developed to simulate the motion of a large number of Lagrangian particles, or droplets, in turbulent flows with an aim to the application to a cumulus cloud. A highly efficient parallel algorithm was developed in order to calculate the motion of a large number of Lagrangian particles whose number and size keep changing with time through condensation and collision. Simulation results provide various important information of the cloud process such as entrainment and internal mixing, the distribution of droplets within a cloud, and the evolution of particle size spectrum and the probability density function of particle velocity, which was not available earlier.

1. INTRODUCTION

Turbulence plays a critical role in cloud dynamics. It determines how dry air is entrained into a cloud and mixes within it, how droplets collide with each other to increase the droplet size by coalescence, and how droplets are distributed within a cloud and ultimately precipitate.

Although there have been a few attempts to simulate a cloud by using LES, they are limited to the case of a non-precipitating cloud [1] neglecting cloud microsphysics or to the case of calculating the droplet distribution by Eulerian approach [2], mainly because of the difficulty in dealing with complex thermodynamics and microphysics relevant to cloud dynamics. There has been no LES model of a cloud with Lagrangian droplets so far.

Meanwhile, the simulation of Lagrangian droplets in a cloud is not only more natural to the cloud process, but also it is able to provide the information directly, which is critical to understand and model the cloud process, such as the distribution of droplets within a cloud, the evolution of droplet size spectrum, and the probability density function of droplet velocity. In this paper, we illustrated the newly developed LES model simulating the motion of a large number of Lagrangian droplets in turbulent flows of a cumulus cloud, in which droplets undergo condensation and collision. In particular, we introduce a new parallel algorithm developed to handle a large number of Lagrangian droplets. We also showed LES results that are in good agreement with known cloud physical process.

2. MODEL

The LES model we used in this study was developed based on the PALM (PArallelized LES Model), which has been applied extensively to the atmospheric boundary layer [3], the ocean deep convection [4], and the ocean mixed layer [5]. Subgrid-scale turbulence is modeled according to Deardorff [6], which is widely used in the LES of geophysical turbulence. Recently, PALM was extended successfully to simulate the motion of Lagrangian particles in turbulent flows [7]. For the simulation of a cloud, thermodynamics was incorporated into the model, and the growth of an individual particle by condensation/evaporation was treated using the standard formula based on water vapour pressure and saturation pressure [8].

Meanwhile, the most important expansion of the model was made for the simulation of a large number of Lagrangian particles whose size and number keep changing with time. Since it is not realistic to simulate all droplets in a cloud, we introduced the concept of weighting factor A_p which represents the ratio of real and simulated volumes of droplets, i.e.,

$$q_l = A_p \frac{1}{\rho}\left[\frac{\rho_L}{\Delta V}\sum_{i=1}^{N_p}\frac{4}{3}\pi r_i^3\right] \tag{1}$$

where N_p is the number of particles in a grid per unit volume ΔV, q_l is the total liquid water content and ρ_L and r_i are the density and radius of a droplet. After coalescence and advection, A_p is modulated to conserve the total liquid water content. Collision between particles is not simulated explicitly, but the growth of an individual particle is calculated based on the mean particle concentration and size within a radius around a particle, using the formula [8]

$$\frac{dr}{dt} = \frac{EM}{4\rho_L}u(r) \tag{2}$$

where E is an effective average value of collection efficiency, $u(r)$ is terminal falling speed of droplet.

Each droplet experiences the increase of its radius by condensation and coalescence, while it is advected by the simulated turbulent field simultaneously as

$$\frac{dV_i}{dt} = \frac{1}{\tau_p}(u_i - V_i - w_s\delta_{i3}) \tag{3}$$

where V_i and u_i are the velocity of a droplet and fluid, respectively, τ_p is the inertial response time, and w_s is the settling velocity in the rest fluid.

The model domain was 6 x 1 x 6 km, and the grid size was 20 m in all directions. Experiments started with a two-dimensional thermal near the bottom with 2.4 x 10^4 droplets with $r_i = 0.1\mu m$. The relative humidity was given as 99% in the whole domain, and the total buoyancy was given by $Q_0 = 0.3$ and 2.4 m^3s^{-3}.

3. PARALLELIZATION TECHNIQUE

Parallelization of the PALM code, which is written in FORTRAN 95, is achieved by two-dimensional domain decomposition along both horizontal directions x and y. Communication is realized by MPI. A more detailed description is found in Raasch and Schröter [9]. However, we currently prefer to use a 1d-decomposition (along x or y) because this significantly reduces the communication overhead. Fig. 1 shows the PALM performance for up to 256 processors of an IBM-Regatta p690 series using a 1d-decomposition and a quite large model domain with 1536 * 768 * 242 grid points. Even then the communication overhead (caused by calls of `MPI_SENDRECV` and `MPI_ALLTOALL`) is less than 20% of the total CPU time. About 25% (`MPI_SENDRECV`) of the communication time is needed for the exchange of ghost points (2d-data arrays) and the remaining time is required for the transposition of a 3d-array (`MPI_ALLTOALL`) in order to solve a Poisson equation for the perturbation pressure using an FFT-method. Both actions are required to be carried out after each time step.

The newly developed particle model follows the domain-decomposition concept. Particles are handled by and stored in the memory of the processors assigned to those subdomains where the particles have been released. A particle is defined by a set of features (position, velocity components, diameter, etc.). For a better handling, these features are not stored as single variables but as components of a derived data type variable. One element of this data type defines a complete particle and the whole set of particles is stored in a 1d-array of this derived data type. As soon as a particle crosses a subdomain boundary, it is transferred (i.e. its data are transferred) to the respective neighbouring subdomain (processor) by using `MPI_SENDRECV`. Transfers are done after each timestep. The respective particles are stored in send buffers. Before sending them, the number of particles to be transferred is sent to the receiving processor and it is checked, if the particle array of this processor has enough free elements to store them. If not, the upper bound of the particle array is automatically increased by deallocating and subsequently allocating this array with the newly required number of elements. The particles received are directly stored in the particle array beginning with the index of the first free position. As a next step the particle array elements are resorted by shifting the array elements in order to eliminate the “holes” caused by those particles which left the subdomain.

Although this parallelization method does not guarantee a good load balancing by itself, because e.g. for some reason all particles may gather in the same subdomain causing a high CPU load on the respective processor while all others idle, the nature of our simulations and domain decomposition provides that all processors always handle about the same number of particles.

From a performance view, the most difficult part of the code development was the part to calculate the interaction of particles, i.e. the particle growth by collision, which requires to determine the average diameter of all particles in the vicinity of the particle under consideration which are smaller than the diameter of this particle. So far we define the vicinity of a particle as the grid box in which it currently resides and so the problem is reduced to determine all the particles belonging to the same box.

There is no problem at the beginning of a model run since particles are stored in a 1d-array `P(N)`, where `N` is the number of particles in use, and sorted following the sequence of the grid point data as stored in the memory by FORTRAN. For example, `P(1:10)` contains data of those particles located in grid box `(1,1,1)`, `P(11:20)` those of particles in `(2,1,1)`, etc. If the number of particles in each box is known, the particles of this box (i.e. their respective array index) can easily be accessed. The second advantage of storing particles in this order is that if e.g. the new particle coordinate x for the next time step is calculated in a loop like

```
DO  n = 1, N
   P%x(n)  = P%x(n)  + delta_t  *  u_int
ENDDO
```

where `u_int` is the u-component of the velocity field of the fluid as interpolated from the neighbouring grid points and `delta_t` is the timestep, then the u-velocity grid point data are accessed in this loop in the order as they are stored in memory. This is computationally very efficient because it takes advantage of the cache mechanism avoiding time-consuming cache misses.

However, during the simulation, due to the turbulent mixing, the correlation between the particles and the grid boxes soon gets lost. Although the data of the first particle are still stored in `P(1)`, the particle now probably stays in a box far away from that where it was initially started. Carrying out the above loop now requires a more or less random access to the memory in order to get the velocity data needed for interpolation. This causes a large number of cache misses which significantly slows down the computational speed, especially for large grids whose data does not fit into the cache. An even bigger problem is to determine the particles in the vicinity of a particle under consideration, which would need a nested loop search like

```
DO  n = 1, N
   DO  k = 1, N
      IF ( k /= n ) THEN
         IF  ( ABS( P%x(k)-P%x(n)  )  <  threshold  )  THEN
            ...
         ENDIF
      ENDIF
   ENDDO
ENDDO
```

requiring an order of N^2 operations.

The main idea of our optimization is a re-sorting of the particles after each time step in order to avoid the cache misses and the loop search. The sorting is carried out in a way that the original correlation between the sequence of the particle data and the sequence the grid point data as stored in memory is reestablished. If particles are always sorted according to the grid box order, the particles in the vicinity of a particle (i.e. in the same grid box) can be accessed very fast using a loop like

```
DO n = 1, N
   DO  k = lower_ijk, upper_ijk
     ...
   ENDDO
ENDDO
```

where `lower_ijk` and `upper_ijk` are the lower and upper index of those particles within grid box `(i,j,k)`. This method only requires operations which are much more in the order of `N`.

Using the sorting method, the CPU-time for the particle calculation with a total number of 250000 particles needs about 12% of the total integration time for a computational domain with about 4.5 million grid points (with only 0.3% for sorting) , compared with 74% in the case of using no sorting.

4. RESULTS

Fig. 2 shows the evolution of distributions of potential temperature and droplets during the development of a cumulus cloud. Contrary to the typical vortex-pair shape of a dry cloud, a strong core of buoyancy maximum is maintained with stronger upward motion. Distribution of the variance of temperature reveals that entrainment occurs mostly at the top of a cloud initially, and subsequently occurs at the side as well at the later stage.

The entrainment of dry air into a cloud near the top increases the variability in the droplet spectra, and it also causes the droplet size there by enhanced collision rate. The evolution of the droplet spectra also shows the slow growth by condensation at the initial stage, followed by the rapid growth by coalescence, which is in good agreement with the observational evidence (Fig. 3). Experiments with larger Q_0 shows much larger enhancement of buoyancy and much faster droplet size growth at the initial stage. Furthermore, the spatial distributions of droplet spectrum and concentration within a cloud could be obtained from LES results.

Fig. 4 shows that the most droplets slant to the downward after 900 seconds, signifying the initiation of precipitation. It also shows the decrease of total number of particles with time following precipitation. We can also obtain the information of the spatial and temporal variations of precipitation in association with the cloud condition.

Furthermore, history of the position and velocity of an individual particle could be traced from LES results.

5. CONCLUSION

In the present work, we calculated for the first time the motion of a large number of Lagrangian droplets in turbulent flows of a cumulus clouds by LES model. A highly efficient parallel algorithm was developed in order to calculate the motion of a large number of Lagrangian particles whose number and size keep changing with time through condensation and collision. Simulation results provide various important information of the cloud process, which was not available earlier, without introducing parameterizations.

More realistic simulation of a cumulus cloud is expected in the future with more elaborate treatment of collision process, initial condition of particles and consideration of the effects of background atmosphere such as wind shear.

REFERENCES

1. Siebesma, A. P. and Coauthors (2003). A large-eddy simulation intercomparison study of shallow cumulus convection. J. Atmos. Sci., 60, 1201-1219.
2. Kogan Y.L., Khairoutdinov, M.P., Lilly, d.K., Kogan, Z.N. and Liu, Qingfu (1995). Modeling of stratocumulus cloud layers in a large eddy simulation model ith explicit microphysics. Am. Meteorol. Soc., 52, 2923-2940.
3. Raasch, S. and Harbusch, G. (2001). An analysis of secondary circulations and their effects causes by small-scale surface inhomogeneities using large-eddy simulation. Bound. Layer Meteorol., 101, 31-59.
4. Noh, Y., Cheon, W.G. and Raasch, S., (2003[a]). The role of preconditioning on the onset of open-ocean deep convection. J. Phys. Oceanorg., 33, 1145-1166.
5. Noh, Y, Min, H.S. and Raasch, S. (2004). Large eddy simulation of the ocean mixed layer: the effect of wave breaking and Langmuir circulation. J. Phys. Oceanogr., 34, 720-735.
6. Deardorff, J. W. (1980). Stratocumulus-capped mixed laters derived from a three-dimensional model. Boun. Layer Meteorol., 18, 495-572.
7. Noh, Y., Kang, I.S., Herold, M. and Raasch, S. (2006). Large eddy simulation of particle settling in the ocean mixed layer. Phys. Fluids 18, (accepted).
8. Roger, R. R. and M. K. Yau (1989). 'A Short Course in Cloud Physics' (Elsevier).
9. Raasch, S. and M. Schröder (2001). A large eddy simulation model performing on massively parallel computers. Z. Meteor. 10, 363-372.

List of Figures

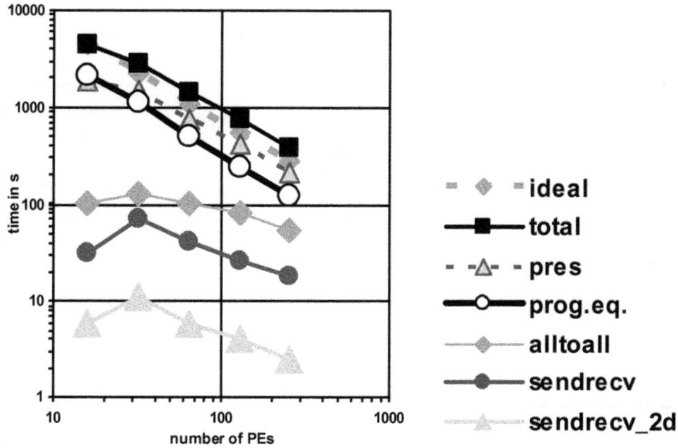

Fig. 1. PALM performance on the IBM-Regatta p690 series for a model domain with 1536 * 768 * 242 gridpoints

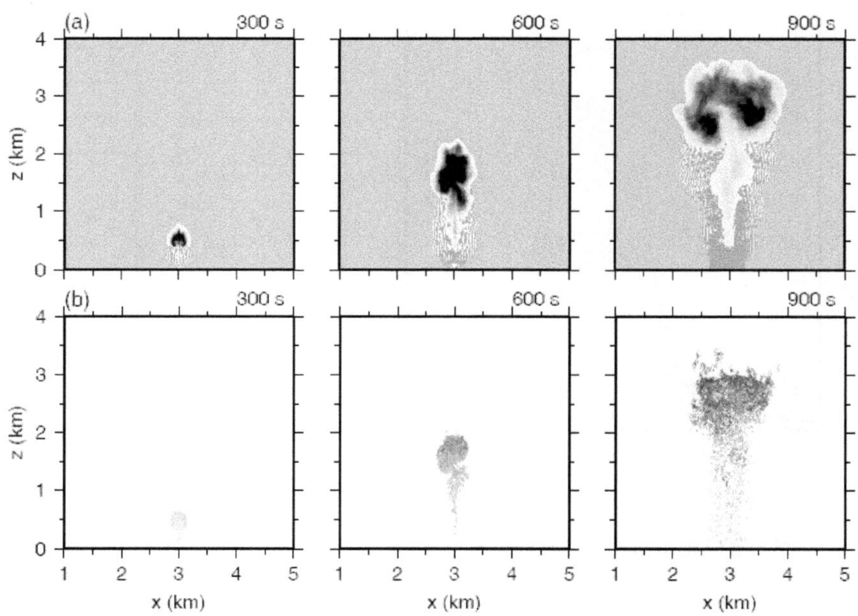

Fig. 2. Evolution of the distribution of (a) potential temperature and (b) droplets at the vertical cross-section ($Q_0 = 0.3\ m^3s^{-2}$).

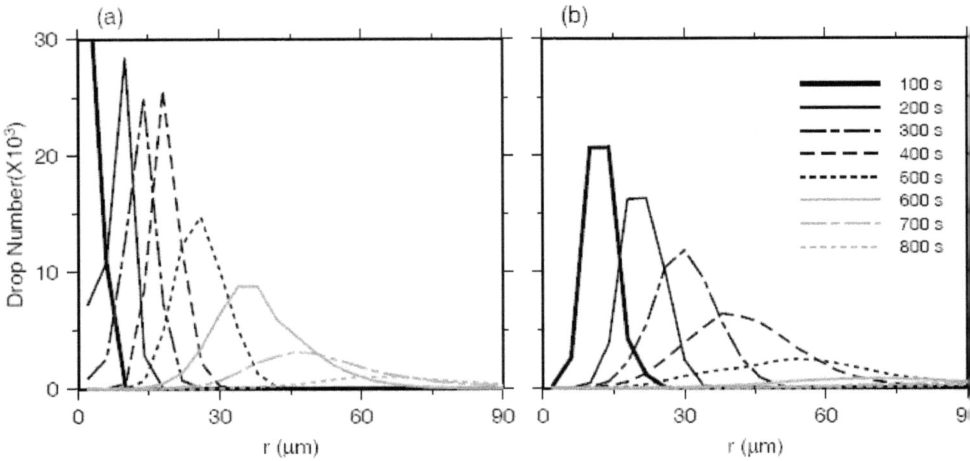

Fig. 3. Evolution of a droplet spectrum with time : (a) $Q_0 = 0.3\ m^3s^{-2}$, (b) $Q_0 = 2.4\ m^3s^{-2}$

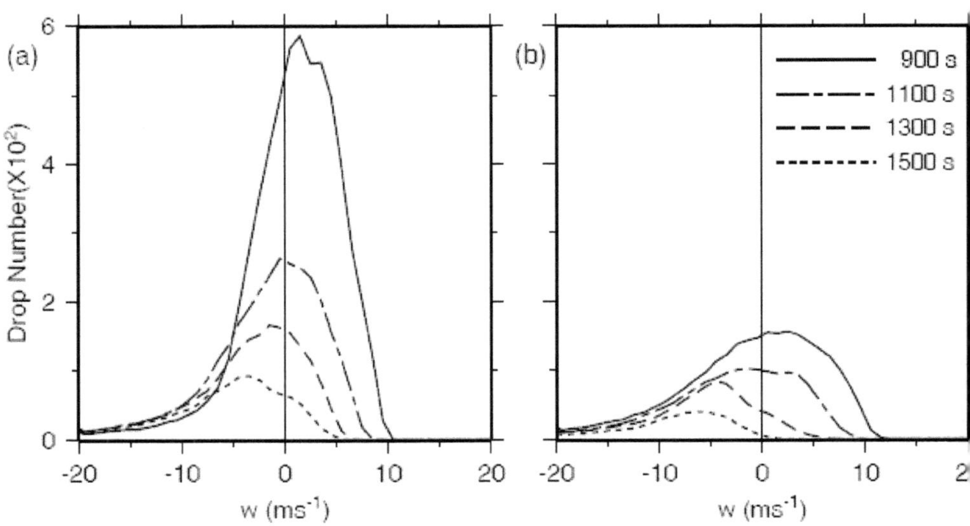

Fig. 4. Distribution of vertical velocity of droplets : (a) $Q_0 = 0.3\ m^3s^{-2}$, (b) $Q_0 = 2.4\ m^3s^{-2}$

Parallel Computational Fluid Dynamics – Parallel Computing and Its Applications
J.H. Kwon, A. Ecer, J. Periaux, N. Satofuka and P. Fox (Editors)

A Fluid-Structure Interaction Problem on Unstructured Moving-Grid using OpenMP Parallelization

Masashi Yamakawa[a] and Kenichi Matsuno[a]

[a]*Department of Mechanical and System Engineering, Kyoto Institute of Technology, Matsugasaki, Sakyo-ku, Kyoto 606-8585, Japan*

In this paper, to calculate a fluid-structure interaction problem in three-dimensional system, we proposed an extended unstructured moving-grid finite-volume method. In this case, it is important that physical and geometric conservation laws are satisfied on the moving grid, and the method is able to satisfy these laws perfectly. In this method, flux and other variables are estimated on control volumes in space-time unified domain. So, in the case of three-dimensional system, they are treated as four-dimensional domain which has a three-dimensional space term and a one-dimensional time term. In this paper, a three dimensional gun-tunnel problem as a large-scale fluid-structure interaction problem is dealt with and estimated using OpenMP parallelization.

Keyword: Moving-Grid Unstructured Grid

1. INTRODUCTION

Fluid-structure interaction problems are seen in various field of industry, for example in aeronautics, civil engineering, turbomachinery, and they are also researched in computational fluid dynamics. To compute these problem in body-fitted coordinate system, the grid moves as time advances, and it is important to treat flux and conservative variables on moving grid. So, the change of grid must not affect flow variables. In other words, a geometric conservation law should be satisfied on the moving grid. To satisfy the geometric conservation laws, we have proposed the moving-grid finite-volume method in unstructured grid system[1]. The method adopts a control volume in the space-time unified domain, and it is solved iteratively at every time-step in order to assure physical and geometric conservation laws and numerical accuracy. The method is suitable to solve the fluid-structure interaction problem. In this paper, the problem is dealt with in three-dimensional coordinate system. In this case, estimated control volumes are treated as four-dimensional domain which has a three-dimensional space term and a one-dimensional time term. In four-dimensional domain, however it is difficult to estimate metric of each planes of the control volume, they make possible to calculate the problem accuracy. In this paper, the method is applied to a gun-tunnel problem, in which a piston is moved by air pressure. And, to apply a large scale calculation, we introduced OpenMP parallelization. Then, the efficiencies of the method and application in parallel environment are discussed.

2. THREE-DIMENSIONAL UNSTRUCTURED MOVING-GRID FINITE-VOLUME METHOD

2.1. Governing Equation

In this paper, we deal with Euler equation in three-dimensional coordinate system. Then the equation can be written in the conservation law form as follow:

$$\frac{\partial q}{\partial t}+\frac{\partial E}{\partial x}+\frac{\partial F}{\partial y}+\frac{\partial G}{\partial z}=0, \tag{1}$$

where

$$q=\begin{pmatrix}\rho\\ \rho u\\ \rho v\\ \rho w\\ e\end{pmatrix},\quad E=\begin{pmatrix}\rho u\\ \rho u^2+p\\ \rho uv\\ \rho uw\\ u(e+p)\end{pmatrix},\quad F=\begin{pmatrix}\rho v\\ \rho uv\\ \rho v^2+p\\ \rho vw\\ v(e+p)\end{pmatrix},\quad G=\begin{pmatrix}\rho w\\ \rho uw\\ \rho vw\\ \rho w^2+p\\ w(e+p)\end{pmatrix}. \tag{2}$$

The unknown variables ρ, u, v, w, p and e represent the gas density, velocity components in the x, y and z directions, pressure, and total energy per unit volume, respectively. The working fluid is assumed to be perfect, and the total energy e per unit volume is defined by

$$e=\frac{p}{\gamma-1}+\frac{1}{2}\rho\left(u^2+v^2+w^2\right), \tag{3}$$

where the ratio of specific heats γ is typically taken as being 1.4.

2.2. Three-Dimensional Unstructured Moving-Grid Finite-Volume Method

To deal with a fluid-structure interaction problem, a grid system must dynamically change and deform its shape according to the movement of boundary between fluid and moving body in body-fitted grid system. Then, one of the most important issue in the grid system is how to satisfy a geometric conservation law on the moving grid. To solve this problem, in this section, the unstructured moving-grid finite-volume method in three-dimensional coordinate system is explained. The method is able to satisfy the geometric conservation law as well as a physical conservation law.

At first, the Euler equation in conservation form Eq.(1) can be written in divergence form:

$$\tilde{\nabla}\tilde{\mathbf{F}} = 0 \,, \tag{4}$$

where

$$\tilde{\mathbf{F}} = E\vec{\mathbf{e}}_x + F\vec{\mathbf{e}}_y + G\vec{\mathbf{e}}_z + q\vec{\mathbf{e}}_t \,, \tag{5}$$

$$\tilde{\nabla} = \vec{\mathbf{e}}_x\frac{\partial}{\partial x} + \vec{\mathbf{e}}_y\frac{\partial}{\partial y} + \vec{\mathbf{e}}_z\frac{\partial}{\partial z} + \vec{\mathbf{e}}_t\frac{\partial}{\partial t}. \tag{6}$$

Here $\vec{\mathbf{e}}_x$, $\vec{\mathbf{e}}_y$, $\vec{\mathbf{e}}_z$ and $\vec{\mathbf{e}}_t$ are unit vectors in *x*, *y*, *z* and *t* directions respectively.

In the three-dimensional coordinate system, we deal with the control-volume as four-dimensional space-time unified domain (*x, y, z, t*). Then, the method is based on a cell-centered finite-volume method, thus, flow variables are defined at the center of cell in unstructured mesh. The unified domain is shown in Fig.1.

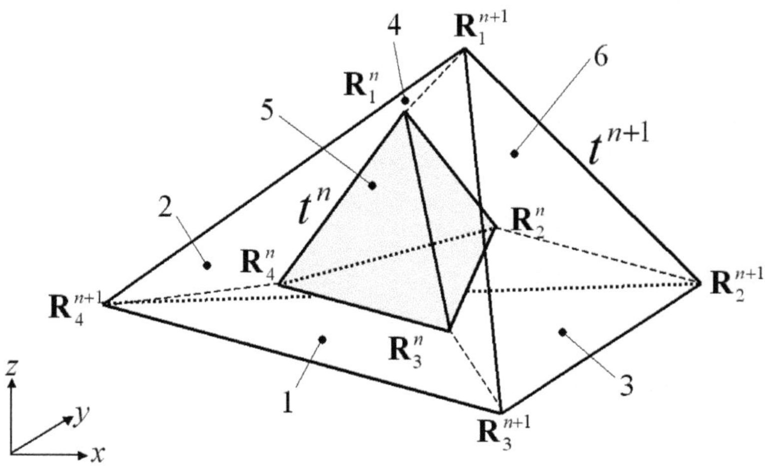

Figure 1. Control-volume in space-time unified domain

N-time step tetrahedron is made by four-grid points($\mathbf{R}_1^n$, $\mathbf{R}_2^n$, $\mathbf{R}_3^n$, $\mathbf{R}_4^n$). And n+1-time step tetrahedron is made by other grid points($\mathbf{R}_1^{n+1}$, $\mathbf{R}_2^{n+1}$, $\mathbf{R}_3^{n+1}$, $\mathbf{R}_4^{n+1}$). Then, the space-time unified control-volume is formed by moving of tetrahedron from n-time step to n+1 time step. So, the control-volume is constructed by eight grid points($\mathbf{R}_1^n$, $\mathbf{R}_2^n$, $\mathbf{R}_3^n$, $\mathbf{R}_4^n$, $\mathbf{R}_1^{n+1}$, $\mathbf{R}_2^{n+1}$, $\mathbf{R}_3^{n+1}$, $\mathbf{R}_4^{n+1}$).
We apply volume integration to Eq.(4) with respect to this control volume. Then Eq.(4) can be written in surface integral form, using the Gauss theorem:

$$\int_V \tilde{\nabla}\tilde{\mathbf{F}}dV = \oint_S \tilde{\mathbf{F}} \cdot \mathbf{n}dS = \sum_{l=1}^{6}\left(En_x + Fn_y + Gn_z + qn_t\right)_l = 0 \tag{7}$$

where $\mathbf{n}$ is a outward unit normal vector of control volume boundary. V is the polyhedron control volume and S is its boundary. $\mathbf{n}_l$ = ($\mathbf{n}_x$, $\mathbf{n}_y$, $\mathbf{n}_z$, $\mathbf{n}_t$)$_l$ (l = 1,2,...6) is the normal vector of control volume boundary, and the length of the vector equals to the volume of the boundary. For example at l = 1, the volume of the boundary is constructed by six grid points ($\mathbf{R}_1^n$, $\mathbf{R}_2^n$, $\mathbf{R}_3^n$, $\mathbf{R}_1^{n+1}$, $\mathbf{R}_2^{n+1}$, $\mathbf{R}_3^{n+1}$) as shown in Fig.2. This is formed by sweeping a triangle ($\mathbf{R}_1^n$, $\mathbf{R}_2^n$, $\mathbf{R}_3^n$) to next time-step triangle ($\mathbf{R}_1^{n+1}$, $\mathbf{R}_2^{n+1}$, $\mathbf{R}_3^{n+1}$). The components $\mathbf{n}_x$, $\mathbf{n}_y$, $\mathbf{n}_z$, and $\mathbf{n}_t$ are unit normal vectors in x, y, z and t directions respectively, and these components are expressed by the combination of a suitable inner product and a suitable outer product. Assuming $\mathbf{R}_1^n = (x_1^n, y_1^n, z_1^n)$, component n_x for example at l = 1 is written as:

$$\begin{aligned}
n_x = {} & \Delta t/3\{(y_3^{n+1} - y_3^n)(z_1^{n+1} - z_3^n) - (z_3^{n+1} - z_3^n)(y_1^{n+1} - y_3^n)\} \\
& + \Delta t/6\{(y_1^{n+1} - y_3^n)(z_1^n - z_3^n) - (z_1^{n+1} - z_3^n)(y_1^n - y_3^n)\} \\
& + \Delta t/3\{(y_1^{n+1} - y_1^n)(z_2^{n+1} - z_1^n) - (z_1^{n+1} - z_1^n)(y_2^{n+1} - y_1^n)\} \\
& + \Delta t/6\{(y_2^{n+1} - y_1^n)(z_2^n - z_1^n) - (z_2^{n+1} - z_1^n)(y_2^n - y_1^n)\} \\
& + \Delta t/3\{(y_2^{n+1} - y_2^n)(z_3^{n+1} - z_2^n) - (z_2^{n+1} - z_2^n)(y_3^{n+1} - y_2^n)\} \\
& + \Delta t/6\{(y_3^{n+1} - y_2^n)(z_3^n - z_2^n) - (z_3^{n+1} - z_2^n)(y_3^n - y_2^n)\} \\
& + \Delta t/2\{(y_3^{n+1} - y_1^{n+1})(z_2^{n+1} - z_1^{n+1}) - (z_3^{n+1} - z_1^{n+1})(y_2^{n+1} - y_1^{n+1})\}
\end{aligned} \tag{8}$$

Here, at l = 4 and 5, the control volumes have only $\mathbf{n}_t$ component and correspond to the volumes in the (x, y, z)-space at time t^{n+1} and t^n themselves respectively. Thus, Eq.(7) can be expressed as,

$$q^{n+1}(n_t)_6 + q^n(n_t)_5 + \sum_{l=1}^{4}\left\{\left(E^{n+1/2}, F^{n+1/2}, G^{n+1/2}, q^{n+1/2}\right)\cdot \mathbf{n}\right\}_l = 0 . \tag{9}$$

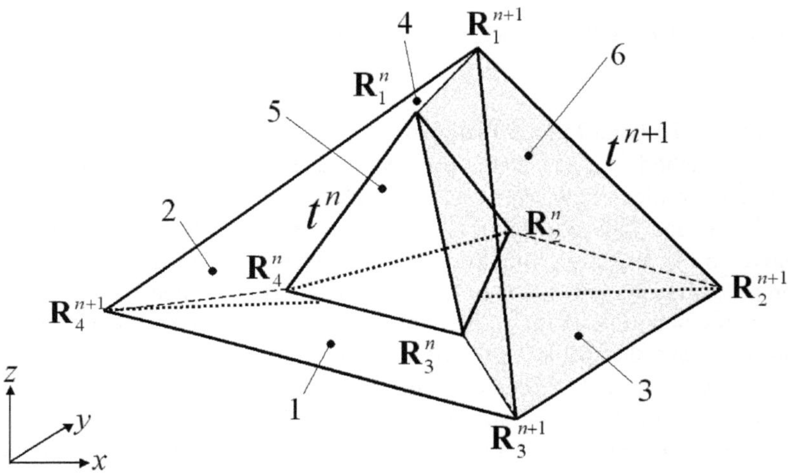

Figure 2. Normal vector ($l = 1$)

Here, the conservative variable vector and flux vector at (n+1/2)-time step are estimated by the average between n-time and (n+1)-time steps. Thus, for example, $E^{n+1/2}$ can be expressed as,

$$E^{n+1/2} = \left(E^n + E^{n+1}\right)/2 . \tag{10}$$

The flux vectors are evaluated using the Roe flux difference splitting scheme [2] with MUSCL approach, as well as the Venkatakrishnan limiter [3].
The method uses the variable at (n+1)-time step, and thus the method is completely implicit. We introduce sub-iteration strategy with a pseudo-time approach [4] to solve the implicit algorithm. Now by defining that the operating function L(q^{n+1}) as Eq.(11), the pseudo-time sub-iteration is represented as Eq.(12).

$$L\left(q^{n+1}\right) = \frac{1}{\Delta t(n_t)_6}[q^{n+1}(n_t)_6 + q^n(n_t)_5 + \sum_{l=1}^{4}\left\{\left(E^{n+1/2}, F^{n+1/2}, G^{n+1/2}, q^{n+1/2}\right)\cdot n\right\}_l], \tag{11}$$

$$\frac{dq^{n+1\langle\nu\rangle}}{d\tau} = -L\left(q^{n+1\langle\nu\rangle}\right), \tag{12}$$

where ν is index for iteration, and τ is pseudo-time step. To solve Eq.(12), we adopted the Runge-Kutta scheme to advance $\Delta\tau$ pseudo-time step. When ν inner iteration is converged, we can get (n+1)-time step solution, q^{n+1}.

3. A FLUID-STRUCTURE INTERACTION PROBLEM

3.1. Application To Gun-Tunnel Problem

We apply the method to a gun-tunnel problem as a fluid-structure interaction problem. The gun-tunnel problem means that a piston, high-pressure air and low-pressure are in a closed cylinder, and high and low pressure air is separated by the piston. Then, the piston moves by high-pressure air toward low-pressure air side, and the movement of the piston press the air again. So, the piston and the air affect mutually.

Figure 3. shows an outline of the gun-tunnel problem. In this paper, the piston is defined as a rigid body, and the piston is moved with only the difference pressure of air. The piston can move x, y and z direction, and the position of the piston is decided the balance with mass of the piston and the pressure of the wall of the piston. There are clearances from the piston and the wall of the cylinder. Then, the clearances are different x and y direction respectively as shown Fig.3. The initial condition of density, pressure, velocity components in the x, y and z directions are given by: ρ_L= 0.1, p_L = 1.0/γ(γ=1.4), u = 0.0, v = 0.0, w = 0.0 at a left side of the piston, and ρ_R= 0.1, p_R = 0.1/γ at a right side of the piston. And we calculate until t = 25.0 (50,000 time step with Δt = 0.0005). The initial grid is generated by dividing structure which is seen in Fig.4. The number of elements is 30,800. Due to the movement of the piston, the grid is deformed under keeping a topology. Thus, the total number of the element is constant at every time step.

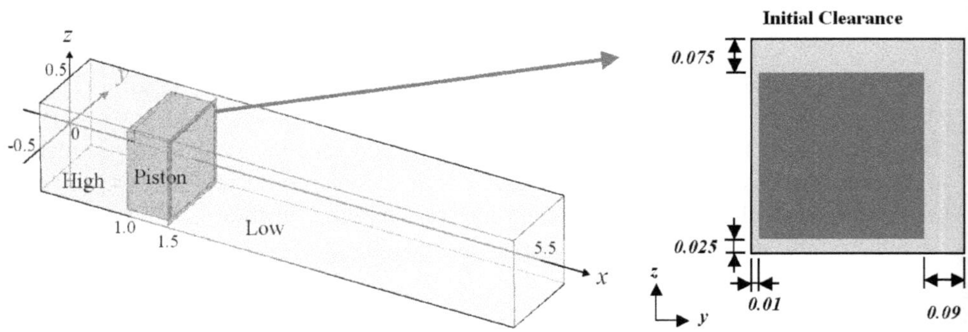

Figure 3. Outline of the gun-tunnel problem and the initial clearance

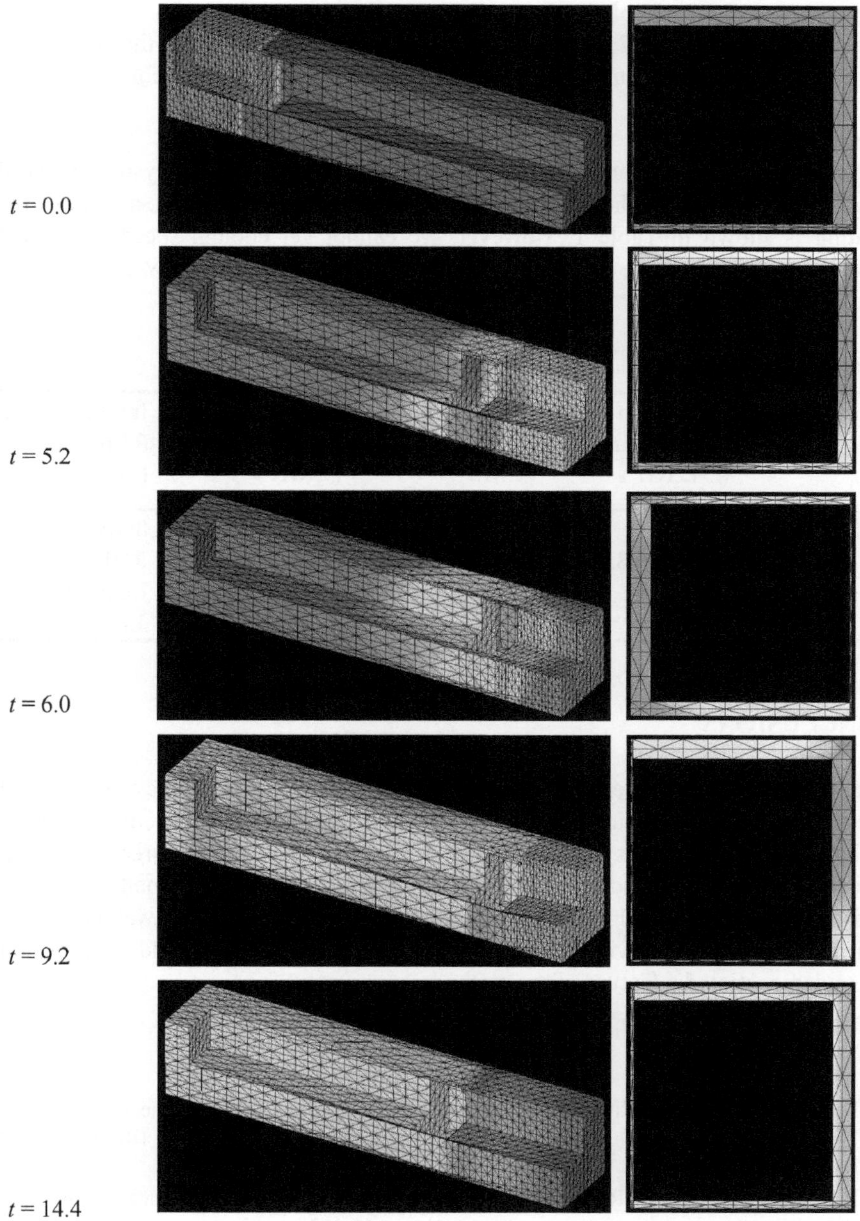

Figure 4. Results of the gun-tunnel problem
(Left: Pressure contours, Right: Velocity contours in *y-z* plane on the piston)

Figure 4 shows the result of the flow variable (pressure and velocity contours) and related grids at t = 0.0, 5.2, 6.0, 9.2 and 14.4.It can confirmed that the method can calculate a fluid-structure interaction problem in three-dimensional system.

3.2. Parallel Efficiency

To estimate a parallel computing on this method in three-dimensional system, OpenMP library was carried out for the gun-tunnel problem on several PC. Then, a results of the estimating is shown in Table 1. However, we cannot get enough speed up in these environment, as one of the most important reason, there are too few grid points to estimate in this OpenMP parallel environment.

Table 1. Environment and results (Parallel computing efficiency)

Processor (Clock speed) (Cashe)	Intel Xeon (3.06 GHz) (512KB)	Intel Xeon (3.6 GHz) (2MB)	Intel PentiumD (3.2 GHz) (1MB)	Intel Itanium2 (1.6 GHz) (3MB)
Main Memory	2GB	3GB	2GB	8GB
OS	RHLinux8.0	WinXP	WinXP	RHEnt.AS2.1
Compiler	Intel F 7.1	Intel F 8.1	Intel F 8.1	Intel F 9.1
Speed Up	0.98	1.00	1.01	1.00

4. CONCLUSIONS

To calculate a fluid-structure interaction problem accuracy and efficiency, we adopt the unstructured moving-grid finite-volume method in OpenMP parallel environment. In this paper, the method was applied to the gun-tunnel problem in three-dimensional coordinate system as a fluid-structure interaction problem, then the method was also extend to three-dimensional system. The efficiency of parallelization was not enough obtained in the parallel environment, however, the result of flow field in gun-tunnel problem demonstrated a feature of the method.

REFERENCES

1. M. Yamakawa and K. Matsuno, Unstructured Moving-Grid Finite-Volume Method for Unsteady Shocked Flows, *Journal of Computational Fluids Engineering*, 10-1 (2005) pp.24-30.
2. P. L. Roe, Approximate Riemann Solvers Parameter Vectors and Difference Schemes, *Journal of Computational Physics*, 43 (1981) pp.357-372.
3. V. Venkatakrishnan, On the Accuracy of Limiters and Convergence to Steady State Solutions, *AIAA Paper*, 93-0880, (1993).
4. C. L. Rumsey, M. D. Sanetrik, R. T. Biedron, N. D. Melson, and E. B. Parlette, Efficiency and Accuracy of Time-Accurate Turbulent Navier-Stokes Computations, *Computers and Fluids*, 25 (1996) pp. 217-236.

Parallel Computational Fluid Dynamics – Parallel Computing and Its Applications
J.H. Kwon, A. Ecer, J. Periaux, N. Satofuka and P. Fox (Editors)

Investigation of Turbulence Models for Multi-stage Launch Vehicle Analysis Including Base Flow

Soon-Heum Ko[a], Chongam Kim[a], Kyu Hong Kim[a] and Kum Won Cho[b]

[a] *Department of Mechanical and Aerospace Engineering,*

Seoul National University,

San 56-1 Shillim-Dong, Kwanak-Gu, Seoul, 151-742 Republic of Korea

E-mail: {floydfan,chongam,aerocfd1}@snu.ac.kr

[b] *Department of Supercomputing Applications,*

Korea Institute of Science and Technology Information(KISTI),

52 Eoeun-Dong, Yuseong-Gu, Daejeon, 305-333 Republic of Korea

E-mail: ckw@kisti.re.kr

Key Words: 2-Equation Turbulence Models; Base Flow; Chimera Overset Mesh; KSR-Ⅲ

1. INTRODUCTION

Accurate flight analysis of launch vehicle is one of the major topics on aeronautic and astronautic engineering. And the simulation of separated bodies in the atmosphere is an important part of launch vehicle researches for safe flight. For years, many researchers have simulated the detachment motion of strap-on boosters from the parent to predict the accurate trajectory of boosters numerically or experimentally[1~8]. However, many numerical researches failed to give the reliable results because of much simplification of present problem. At first, most analyses simply adopted the prescribed trajectory of strap-on boosters for dynamic analysis. Thus, they were unable to investigate the variation of detachment motions at different separation conditions. Secondly, researchers conducted fore-body simulation on aerodynamic-dynamic coupled anaysis, assuming that after-body flow gives minor effect on the motion of bodies. However, base flow effect cannot be neglected as expansion waves at the baseline of core rocket and the jet from core rocket change the trajectory of detached body a lot[9]. Finally, lots of works carried out inviscid analyses to reduce computing time and resources. But it cannot describe the detailed flow features around the bodies including accurate aerodynamic forces and moments of strap-ons.

The present paper focuses on applying the most appropriate turbulence model to launch vehicle simulations. Some recent works[10,11] accomplished flow analyses around the missiles and rockets using various turbulence models, but they just conducted fore-body flow analyses. Thus, present work will implement various turbulence models and conduct whole flow analysis including base region. The results from various turbulence models of k-ω SST[12], Spalart-Allmaras[13] and Craft-Launder-Suga[14] are to be compared and the best model for launch vehicle simulation will be utilized to steady-stae analysis of KSR-Ⅲ launch vehicle, which is researched in Korea.

2. GOVERNING EQUATIONS AND NUMERICAL TECHNIQUES

Since the present geometry contains massive flow separation regions in the base region, the three-dimensional compressible full Navier-Stokes Equations are adopted for an accurate flow analysis. Navier-Stokes equations can be written in general curvilinear coordinates ξ, η, ζ as follows:

$$\frac{\partial \hat{Q}}{\partial t}+\frac{\partial \hat{E}}{\partial}+\frac{\partial \hat{F}}{\partial}+\frac{\partial \hat{G}}{\partial}=\frac{1}{\mathrm{Re}}(\frac{\partial \hat{E}_v}{\partial}+\frac{\partial \hat{F}_v}{\partial}+\frac{\partial \hat{G}_v}{\partial}) \quad (1)$$

where $\hat{Q}$ is the conservative variable vector, $\hat{E}$, $\hat{F}$, $\hat{G}$ are the inviscid flux vectors and $\hat{E}_v$, $\hat{F}_v$, $\hat{G}_v$ are the viscous flux vectors.

For the accurate representation of turbulent flowfield, k-ω SST[12], Spalart-Allmaras[13] and Craft-Launder-Suga[14] turbulence models are implemented on the flow solver. Of those models, Craft-Launder-Suga is an algebraic stress model with a cubic stress-strain relationship. However, current research adopted their definition of turbulent viscosity only, leaving stress-strain relationship linear. That's because the cubic model requires unreasonably excessive computation time than linear model.

Dual time stepping is employed to obtain a second order accuracy for the temporal discretization of unsteady flowfield. Dual time stepping has the following form.

$$\frac{\partial \hat{Q}}{\partial} = -\hat{R}^{n+1,s+1} - \frac{3\hat{Q}^{n+1,s+1} - 4\hat{Q}^{n} + \hat{Q}^{n-1}}{2\Delta t} \tag{2}$$

where

$$\hat{R} = \frac{\partial (\hat{E} - \hat{E}_v)}{\partial} + \frac{\partial (\hat{F} - \hat{F}_v)}{\partial} + \frac{\partial (\hat{G} - \hat{G}_v)}{\partial} \tag{3}$$

Here, τ represents a pseudo time, n is the physical time level, and s is the pseodo time level. Equation (2) is discretized in pseudo time by the Euler implicit method and is linearized using the flux Jacobian. This leads to a large system of linear equations in delta form at each pseudo time step as

$$\left(\frac{I}{J\Delta} + \left[\frac{\partial R}{\partial Q} \right] + \frac{1.5I}{J\Delta t} \right) \Delta Q = -R^{n+1,s} - \frac{3Q^{n+1,s} - 4Q^{n} + Q^{n-1}}{2J\Delta t} \tag{4}$$

LU-SGS (Lower-Upper Symmetric Gauss-Seidel) scheme[15] is used for implicit time integration of equation (4). The viscous flux Jacobian is neglected in the implicit part since it does not influence the accuracy of a solution, and local time stepping is used.

AUSMPW+ (modified Advection Upstream Splitting Method Press-based Weight function)[16] has been applied as a spatial discretization technique. It was designed to remove the non-monotone pressure solutions of the hybrid flux splitting schemes such as AUSM and AUSM+ by introducing pressure weighting functions as a limiter at a cell interface. For a higher-order spatial accuracy, MUSCL(Monotone Upstream-centered Schemes for Conservation Laws)[17] approach is used. Primitive variables are extrapolated at a cell interface and the differentiable limiter is employed to suppress unphysical oscillations near physical discontinuities.

As for dynamic analysis, six degree-of-freedom rigid body equations of motion are integrated into the three-dimensional unsteady Navier-Stokes solution procedure to predict detachment motion of strap-on booster. Aerodynamic forces and moments from flow analysis are integrated to give linear and angular displacement.

For the efficient representation of relative motion among bodies, current study adopts Chimera overset grid technique by Steger[18]. The tri-linear interpolation technique with Newton-Raphson method is utilized for donor cell searching. Both main grid around core rocket and subgrid around booster are generated as multiblock to describe the fins and base regions of launch vehicle configuration. And tandem-type launch vehicle configuration is described as two-block mesh system.

3. NUMERICAL RESULTS

3.1. Flow Analysis around Tandem-type Launch Vehicle

Various turbulence models are applied to a two-stage launch vehicle analysis at NASA[19]. In the analysis of a launch vehicle, an advanced turbulence model is greatly required for the description of highly separated flow in the base. Thus, the inviscid and turbulent flow analyses are conducted and computed results are compared with the experimental data. Figure 1 shows a two-block mesh system and the pressure contour around the body. Of the abundant experimental data presented at Ref. 19, various angles of attack with Mach number 1.6 are simulated as they are most similar to the separation condition of KSR-Ⅲ, Mach number of 17. A two-block mesh has 121×31×101 and 121×31×201 mesh points, respectively. The baseline of the rocket is assumed to be wall and figure 1 shows the cross-sectional pressure contour of a launch vehicle at 6°, Reynolds number of 6.56×10^6.

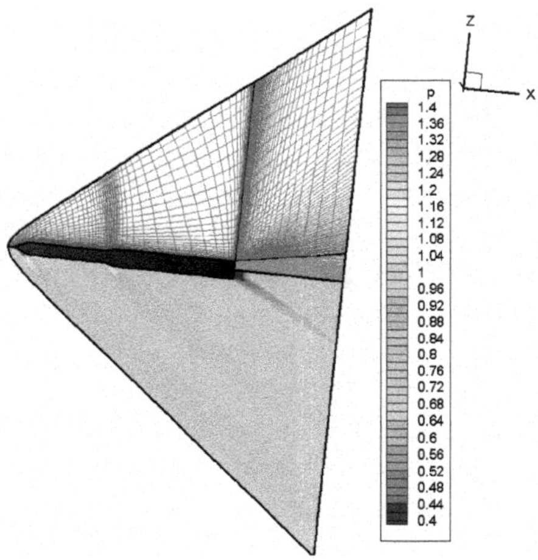

Figure 1. A two-block mesh and pressure contour around a tandem-typed launch vehicle

The results of aerodynamic coefficients at given conditions and computing time with total iterations until convergence are presented in table 1 and 2, respectively. From the results at two degree, both inviscid and turbulent results show good agreements with experimental data. But in high angle of attack, inviscid result becomes inaccurate especially in normal force. From all of the results, k-ω SST gives best solution in axial and normal forces, though it computes relatively larger pitching moment. In Craft-Launder-Suga model, lift coefficient was lower at all cases. It is presumed that the authors' simplification to Craft-Launder-Suga model caused this inaccuracy of the solution. But, original Craft-Launder-Suga model requires much more computation time if cubic stress-strain relation is implemented. So, current study maintains the linear relationship between stress and strain. As for Spalart-Allmaras, the result by Spalart-Allmaras was good for pitching moment computation, but even that result was insufficient to be called exact. The second merit of Spalart-Allmaras is that this one-equation model requires less computing time for one iteration than other two-equation turbulence models. However, Spalart-Allmaras showed poor convergency in all cases, including unconverged solution after 50,000 iterations at ten degree simulation, while other turbulence models are converged after 12,000 iterations. Thus, the authors adopted k-ω SST as main turbulence model for launch vehicle simulation and applied it to multi-stage launch vehicle simulations.

Table 1. Aerodynamic Coefficients of a Two-stage Launch Vehicle

AOA \ Turbulence Model		Inviscid	S-A	C-L-S	k-ω SST	Exp.
2.0	C_A	0.502551	0.509936	0.474216	0.496815	0.500
	C_N	0.103081	0.107124	0.106985	0.106282	0.102
	C_M	0.624332	0.632185	0.635362	0.631679	0.60
6.0	C_A	0.478736	0.524111	0.479773	0.506255	0.502
	C_N	0.331190	0.338175	0.345841	0.342224	0.335
	C_M	1.926793	1.939962	1.986300	1.955527	1.87
10.0	C_A	0.508151	DNC*	0.489721	0.515619	(0.515)
	C_N	0.633455	DNC*	0.647305	0.648954	0.682
	C_M	3.508933	DNC*	3.549657	3.541465	3.39

- DNC : Do Not Converge
- At ten degree simulation, Spalart-Allmaras model showed the residual of $1.5*10^{-3}$ in 2nd norm of density after 50,000 iterations
- At ten degree, experimental data on axial force is not presented clearly

Table 2. Convergence Criteria and Time for One Iteration at AOA 10 degrees

	No. of Iterations until Convergence	Time for One Iteration (sec.)	Total Computing Time (sec.)
k-ω SST	11,505	6.2309	71,686
C-L-S	10,733	6.5218	69,998
S-A	DNC	5.6521	-

3.2. Steady-state Analyses of Multi-stage Launch Vehicles

Validated flow solver is applied to flow analysis around multi-stage launch vehicle configuration. The aerodynamic characteristics of KSR-Ⅲ at initial stage are examined and the prediction is conducted with regard to the detachment motion of strap-ons during separation process. The free stream Mach number is 1.7 and the Reynolds number is 1.431×10^7. In the present research, jet will eject at the exit of the nozzle. The exit condition can be obtained by the chamber condition presented in the annual report of KSR-Ⅲ development. From the chamber condition and cold gas assumption, the properties at the exit are $p_{exit} = 1.084\times p_\infty$, $\rho_{exit} = 1.084\times\rho_\infty$ and $Ma_{exit} = 2.86$. Other properties are calculated from these relations. Additionally, as the exit diameter of a nozzle is 746.4mm, center region of the base plane is assumed to be the exit, while tip is assumed to be wall. Finally, the base plane of strap-on is assumed to be a flat wall as booster already completed exerting the plume gas. The boundary condition at the base plane can be seen at Figure 3 of current mesh system.

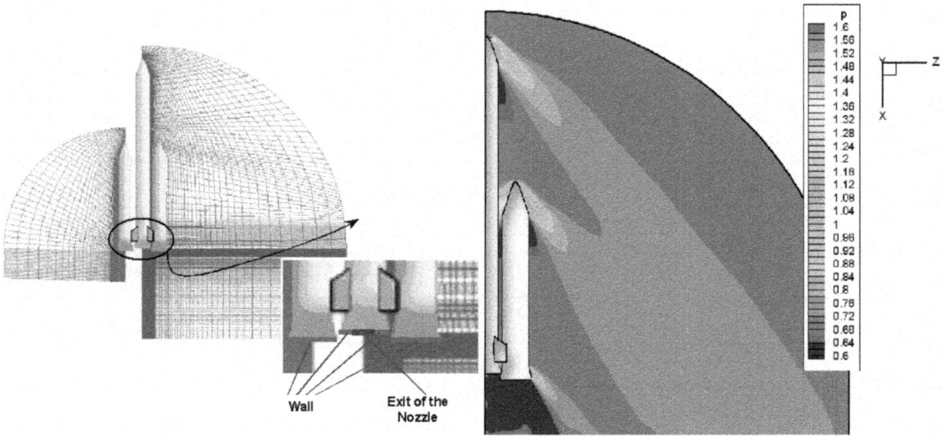

Figure 3. A Chimera Overlapping Mesh and Pressure Contour at Initial Stage

Pressure contours on and around the vehicle at the initial stage are presented in Figure 3, as well. From the pressure distributions in Figure 3, the aerodynamic characteristics of a launch vehicle can be investigated and a qualitative prediction of separation motion of booster can be made. The first factor which influences the detachment motion of strap-on is the bow shock from the nose of booster. Generated shock firstly propagates to the core rocket and the shock reflects to booster, consequently increasing pressure at near side of booster around the nose. This effect raises axial forces and it acts as a repulsive force between core rocket and booster. The next factors are the oblique shocks at fins and a skirt of core rocket. These shocks propagate to the bottom of booster and hit the near side, resulting in a negative pitching moment and the increase of the normal force. These two major mechanisms cause the strap-on to move away from core rocket, but it is not clear whether strap-on will not collide against core as rotating direction is still unclear.

For a detailed investigation of aerodynamic characteristics in real flowfield with a large flow separation at the base region, steady-state aerodynamic coefficients of a strap-on in turbulent and inviscid flowfield are computed and presented in table 3. From the result, normal force coefficient in the turbulent flow analysis is just a little bigger than inviscid analysis. However, the difference of pitching moment coefficient between these two cases is nearly 20%. It is mainly by the reason of the strength of the expansion wave from the base plane of core rocket. Most of the normal force of booster is induced by the oblique shock from the skirt and the expansion wave from the base plane of core rocket. In viscous analysis, the propagation range of the shock is less than the inviscid case. It means the increase of normal force by the shock will be less. Likewise, the area influenced by expansion wave is less in the viscous case. In the point of normal force, these two phenomena will cancel each other. However, in the point of moment coefficient, partial change of the force will be multiplied by the distance from the center of mass. Thus, the expansion wave will affect more to the pitching moment. Then it is natural that larger expansion region will bring about more positive pitching moment and it results in relatively more positive pitching moment in the inviscid analysis.

Table 3. Aerodynamic Coefficients of a Strap-on at Steady-state

	C_A	C_N	C_M
Inviscid	0.6730	0.1601	-0.1303
Viscous	0.6388	0.1626	-0.1544

4. CONCLUSION

Various turbulence models are implemented on a flow solver and the most suitable turbulence model for launch vehicle simulation is investigated. From the analysis of tandem-type launch vehicle, k-ω SST turbulence model was proven to be the appropriate turbulence model for launch vehicle simulation and applied to steady–state flow analysis of multi-stage launch vehicle. From flow analysis around KSR-Ⅲ rocket, turbulent analysis shows more normal force and less pitching moment. A relatively lower pitching moment in turbulent analysis implies that booster will show unsafer movement during separation than inviscid simulation. A dynamic analysis is highly demanded to investigate the effect of turbulence on motion of separated body.

ACKNOWLEDGEMENT

The authors would like to acknowledge the support from KISTI (Korea Institute of Science and Technology Information) under `The Sixth Strategic Supercomputing Support Program' with Dr. Kum Won Cho as the technical supporter. The use of the computing system of the Supercomputing Center is also greatly appreciated. And the authors also would like to appreciate the financial support from BK(Brain Korea)-21 research program.

References

1. G. Palmer and P. Buning, "Three-Dimensional Computational Analysis of Complex launch Vehicle Configurations," *AIAA J. of Spacecraft and Rockets*, Vol. 33, No. 1, 1996
2. S. Taylor and J. C. T. Wang, "Launch-Vehicle Simulations Using a Concurrent, Implicit Navier-Stokes Solver," *AIAA J. of Spacecraft and Rockets*, Vol. 33, No. 5, 1996
3. J. L. Azevedo and P. Moraes Jr., "Code Validation for High-Speed Flow Simulation Over Satellite Launch Vehicle," *AIAA J. of Spacecraft and Rockets*, Vol. 33, No. 1, 1996
4. R. L. Meakin and N. E. Suhs, "Unsteady Aerodynamic Simulation of Multiple Bodies in Relative Motion," *AIAA 89-1996-CP*, 1989
5. R. Lochan and V. Adimurthy, "Separation Dynamics of Strap-On Boosters in the Atmosphere," *AIAA J. of Guidance, Control and Dynamics*, Vol. 20, No. 4, 1997
6. L. M. Gea and D. Vicker, "CFD Simulations of the Space Shuttle Launch Vehicle with Booster Separation Motor and Reaction Control System Plumes," *Third International Conference on Computational Fluid Dynamics (ICCFD3)*, Jul. 2004
7. L. E. Lijewski and N. E. Suhs, "Time-Accurate Computational Fluid Dynamics Approach to Transonic Store Separation Trajectory Prediction," *AIAA J. of Aircraft*, Vol. 31, No. 4, 1994
8. S. J. Choi, C. Kim, O. H. Rho, and J. J. Park, "Numerical Analysis on Separation Dynamics of Strap-On Boosters in the Atmosphere," *AIAA J. of Spacecraft and Rockets*, Vol.39, No. 3, pp. 439-446, 2002
9. S. H. Ko, C. Kim, K. H. Kim and K. W. Cho, "Separation Analysis of Strap-ons in the Multi-stage Launch Vehicle Using the Grid Computing Technique," *Paper AIAA 2006-277*, 44th AIAA Aerospace Sciences Meeting and Exhibit, 2006
10. E. D. Bigarella and J. L. Azevedo, "Numerical Study of Turbulent Flows over Launch Vehicle Configurations," *AIAA J. of Spacecraft and Rockets*, Vol. 42, No. 2, pp. 266-276, 2005
11. X. Liu and S. Fu, "Numerical Simulation of Compressible Separated Turbulent Flows over Inclined Slender Body," *AIAA J. of Spacecraft and Rockets*, Vol. 42, No. 3, pp. 572-575, 2005
12. F. R. Menter, "Two-Equation Eddy-Viscosity Turbulence Models for Engineering Applications," *AIAA Journal*, Vol. 32, No. 8, pp. 1598-1605, 1994
13. P. R. Spalart and S. R. Allmaras, "A one-equation turbulence model for aerodynamic flows," *AIAA Paper 92-0439*, Jan. 1992
14. T. J. Craft, B. E. Launder and K. Suga, "Development and Application of a Cubic Eddy-Viscosity Model of Turbulence," *Int'l J. of Heat and Fluid Flow*, Vol. 17, pp. 108-115, 1996
15. S. Yoon and A. Jameson, "Lower-Upper Symmetric–Gauss-Seidel Method for the Euler and Navier-Stokes Equations," *AIAA Journal*, Vol. 26, No. 9, pp. 1025-1026, 1988
16. K. H. Kim, C. Kim and O. H. Rho, "Accurate Computations of Hypersonic Flows Using AUSMPW+ Scheme and Shock-aligned-grid Technique," *AIAA Paper 98-2442*, 1998
17. Van Leer, B., "Towards the Ultimate Conservative Difference Scheme. V. A Second Order Sequel to Godunov's Methods," *J. of Computational Physics*, Vol. 32, pp.101-136, 1979
18. J. L. Steger, F. C. Dougherty and J. A. Benek, "A Chimera Grid Scheme," *Advances in Grid Generation*, FED Vol.5, ASME, edited by Ghia. K. N., New York, pp. 59-69, 1983
19. R. D. Samuels and J. A. Blackwell, "Effects of Configuration Geometry on the Supersonic Aerodynamic Characteristics of a Simulated Launch Vehicle," *NASA TN D-3755*, 1966

Parallel Computational Fluid Dynamics – Parallel Computing and Its Applications
J.H. Kwon, A. Ecer, J. Periaux, N. Satofuka and P. Fox (Editors)

Path Optimization of Flapping Airfoils Based on NURBS

Mustafa KAYA[a] * and Ismail H.TUNCER[a] †

[a] Middle East Technical University
Department of Aerospace Engineering
06531 Ankara, Turkey

The path of a flapping airfoil undergoing a combined, non-sinusoidal pitching and plunging motion is optimized for maximum thrust and/or propulsive efficiency. The periodic flapping motion is described using Non-Uniform B-Splines (NURBS). A gradient based algorithm is then employed for the optimization of the NURBS parameters. Unsteady, low speed laminar flows are computed using a Navier-Stokes solver. The numerical evaluation of the gradient vector components, which requires unsteady flow solutions, and the flow solutions are all performed in parallel. It is shown that the thrust generation may significantly be increased in comparison to the sinusoidal flapping motion. For a high thrust generation, the airfoil stays at a high effective angle of attack during the upstroke and the downstroke.

1. INTRODUCTION

Based on observations of flying birds and insects, and swimming fish, flapping wings have been recognized to be more efficient than conventional propellers for flights of very small scale vehicles, so-called micro-air vehicles (MAVs) with wing spans of 15 *cm* or less. The current interest in the research and development community is to find the most energy efficient airfoil adaptation and flapping wing motion technologies capable of providing the required aerodynamic performance for a MAV flight.

Recent experimental and computational studies investigated the kinematics, dynamics and flow characteristics of flapping wings, and shed some light on the lift, drag and propulsive power considerations[1–5]. Anderson et al.[5] observed that the phase angle between pitch and plunge oscillations plays a significant role in maximizing the propulsive efficiency. A recent work by Lewin and Haj-Hariri[6] indicates that the aerodynamic forces generated by flapping insects are very sensitive to the wing kinematics. NavierStokes computations performed by Tuncer et al.[7–9] and by Isogai et al.[10,11] explore the effect of flow separation on the thrust generation and the propulsive efficiency of a single flapping airfoil in combined pitch and plunge oscillations.

Jones and Platzer[12] recently demonstrated a radiocontrolled micro air vehicle propelled by flapping wings in a biplane configuration (Figure 1). The experimental and numerical studies by Jones et al.[12–14] and Platzer and Jones[15] on flapping-wing propellers points at the gap between numerical flow solutions and the actual flight conditions

*Graduate Research Assistant, mkaya@ae.metu.edu.tr
†Prof., tuncer@ae.metu.edu.tr

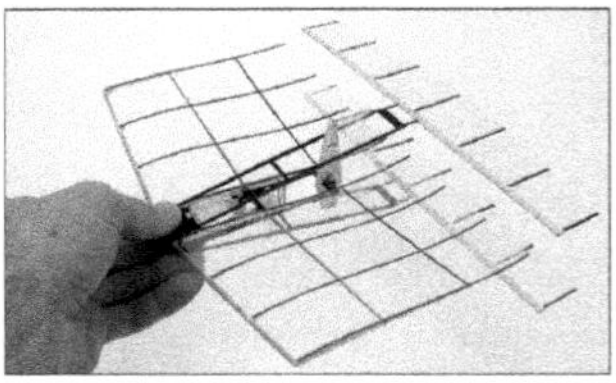

Figure 1. Flapping-wing MAV model (Jones and Platzer)

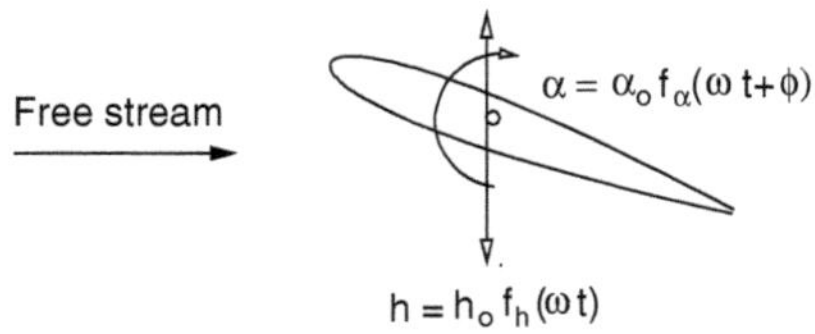

Figure 2. Flapping motion of an airfoil

over flapping wings.

The computational and experimental findings show that thrust generation and propulsive efficiency of flapping airfoils are closely related to the flow kinematics. In an earlier study[16], we employed a gradient based optimization of sinusoidal flapping motion parameters; flapping frequency, the amplitude of the pitch and plunge motions, and the phase shift between them to maximize the thrust and/or the propulsive efficiency of flapping airfoils. It should be noted that in the sinusoidal motion, the pitch and plunge positions are based on the projection of a vector rotating on a unit circle, and the maximum plunge and pitch velocities occur at the mean plunge and pitch positions. In a later study[17], the sinusoidal periodic motion was relaxed by replacing the unit circle with an ellipse, and introducing the flatness coefficient as the ratio of the axes of the ellipse. Optimization of flatness coefficient showed that the thrust generation of flapping airfoils may further be increased on an elliptic flapping path.

In the present study, the periodic flapping motion (Figure 2) is further relaxed using a closed curve based on a 3rd order Non-Uniform Rational B-Splines (NURBS). The parameters defining the NURBS is then optimized for maximum thrust and propulsive efficiency.

2. PERIODIC PATH BASED ON NURBS

The new closed, smooth curve describing the flapping path is produced by a 3rd order NURBS for an half stroke. It is defined by 3 parameters. The first parameter P_0 (y coordinate of the point) defines the center of the closed curve while the remaining two P_1 and P_2 (x coordinates of the points) define the periodic, smooth curve(Figure 3). The x and y coordinates on the curve may be obtained as a function of a nondimensional parameter u:

$$x = \frac{2x_{p1}u(1-u)^2 + 2x_{p2}u^2(1-u)}{(1-u)^3 + u(1-u)^2 + u^2(1-u) + u^3} \quad y = \frac{-(1-u)^3 - u(1-u)^2 + u^2(1-u) + u^3}{(1-u)^3 + u(1-u)^2 + u^2(1-u) + u^3}$$

3. NUMERICAL METHOD

2-D unsteady viscous flows around a flapping airfoil are computed by solving the Navier-Stokes equations on moving C-grids. Grids are partitioned into overlapping subgrids, and computations on subgrids are performed in parallel. A gradient based optimization is

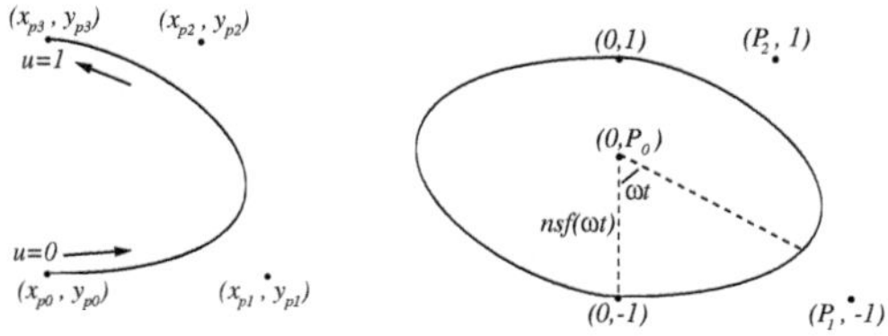

Figure 3. Flapping path defined by a 3rd order NURBS

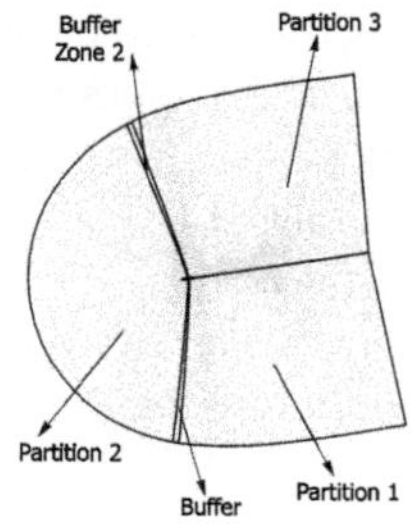

Figure 4. Domain decomposition with 3 partitions

employed for the optimization of flapping motion parameters to maximize the average thrust coefficient and/or the propulsive efficiency.

3.1. Navier-Stokes Solver

The strong conservation-law form of the 2-D, thin-layer, Reynolds averaged Navier-Stokes equations is solved on each subgrid. The convective fluxes are evaluated using the third order accurate Osher's upwind biased flux difference splitting scheme. The discretized equations are solved by an approximately factored, implicit algorithm[7,9].

The computational domain is discretized with a C-type grid around the airfoil. The flapping motion of the airfoil is imposed by moving the airfoil and the grid around it. Computations on subgrids are done in parallel.

3.2. Boundary Conditions

The flapping motion of the airfoil in combined plunge, h, and pitch, α, is specified by

$$h = -h_o\, f_h(\omega t), \qquad \alpha = -\alpha_o\, f_\alpha(\omega t + \phi) \tag{1}$$

where h_o and α_o are the plunge and pitch amplitudes, f is a periodic function based on NURBS, ω is the angular frequency which is given in terms of the reduced frequency, $k = \frac{\omega c}{U_\infty}$. ϕ is the phase shift between plunge and pitching motions. The pitch axis is located at the mid-chord.

At the intergrid boundaries of the overlapping subgrids, the conservative flow variables are exchanged among the subgrids.

3.3. Optimization

The objective function is taken as a linear combination of the average thrust coefficient, C_T, and the propulsive efficiency, η, over a flapping period:

$$O(C_T, \eta) = (1-\beta)\frac{C_T}{C_T + \varepsilon|\vec{\nabla} C_T \cdot \vec{D}|} + \beta\frac{\eta}{\eta + \varepsilon|\vec{\nabla}\eta \cdot \vec{D}|} \tag{2}$$

where

$$C_T = -\frac{1}{T}\int_t^{t+T} C_d\, dt, \qquad \eta = \frac{C_T\, U_\infty}{\dfrac{1}{T}\displaystyle\int_t^{t+T}\int_S p(\vec{V}\cdot d\vec{A})\, dt} \tag{3}$$

where T is the period of the flapping motion. The denominator in the efficiency expression accounts for the average work required to maintain the flapping motion. ε denotes the optimization step size. Note that $\beta = 0$ sets the objective function to the normalized thrust coefficient.

Optimization process is based on following the direction of the steepest ascent of the objective function, O. The direction of the steepest ascent is given by the gradient vector of the objective function:

$$\vec{\nabla} O(\vec{v}) = \frac{\partial O}{\partial V_1}\vec{v_1} + \frac{\partial O}{\partial V_2}\vec{v_2} + \cdots$$

where V_i's are the optimization variables, and the $\vec{v_i}$'s are the corresponding unit vectors in the variable space.

The components of the gradient vector is then evaluated numerically by computing the objective function for a perturbation of all the optimization variables one at a time. It should be noted that the evaluation of these vector components requires an unsteady flow solution over a few periods of the flapping motion until a periodic flow behavior is reached. Once the unit gradient vector is evaluated, a small step, $\Delta\vec{S} = \varepsilon\frac{\vec{\nabla} O}{|\vec{\nabla} O|}$, is taken along the vector. This process continues until a local maximum is reached. The stepsize ε is evaluated by a line search along the gradient vector.

3.4. Parallel Computation

A coarse parallel algorithm based on domain decomposition is implemented in a master-worker paradigm. The C-grid is partitioned into overlapping subgrids first, and the solution on each subgrid is assigned to a separate processor in the computer cluster. The flow variables at the overlapping subgrid boundaries are exchanged among the subgrid processes at each time step of the unsteady solution. PVM (version 3.4.5) library routines are used for inter-process communication. In the optimization process, unsteady flow solutions with perturbed optimization variables, which are required to determine the gradient vector of the objective function, are all computed in parallel. Computations are performed in a cluster of Linux based computers with dual Xeon and Pentium-D processors.

4. RESULTS AND DISCUSSION

In this optimization study, the optimization variables are chosen as the NURBS parameters defining the plunging and pitching paths, P_{0h}, P_{1h}, P_{2h}, $P_{0\alpha}$, $P_{1\alpha}$ and $P_{2\alpha}$ (Figure 3), the phase shift between the pitch and plunge motions, ϕ, and the pitch amplitude, α_o. As the airfoil is set to a flapping motion in plunge and pitch on the NURBS paths, the NURBS parameters are optimized for maximizing the thrust and/or propulsive efficiency. In an optimization process, parallel computations take about $30 - 100$ hours of wall clock time using $10 - 16$ processors.

In this study, the location of the points P_1 and P_2 are constrained within the range $0.2 — 5.0$, P_0 in the range $-0.9 — 0.9$ in order to define a proper periodic motion. The optimization variables are optimized for fixed values of the reduced flapping frequency, $k \equiv \frac{wc}{U_\infty} = 1.0$, the plunge amplitude, $h_0 = 0.5$. The unsteady laminar flowfields are

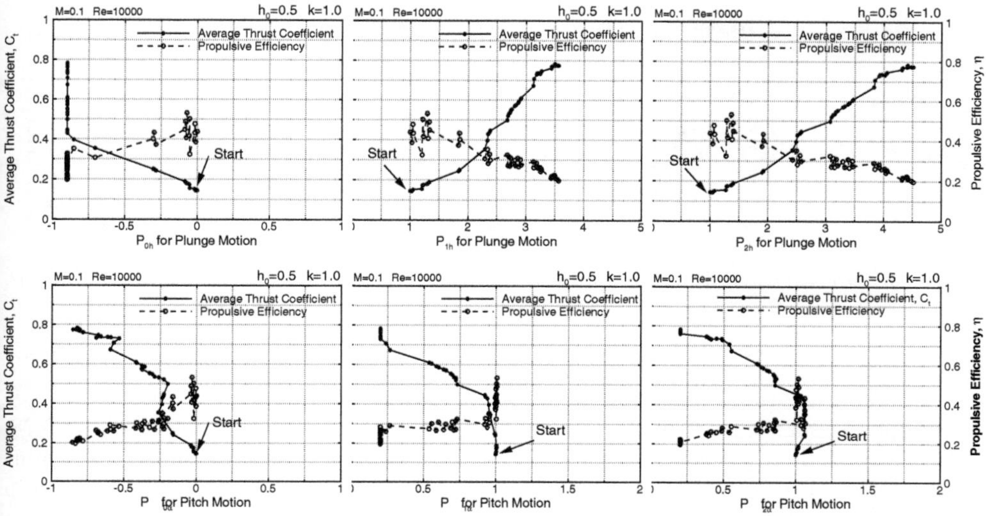

Figure 5. Variation of optimization parameters along the optimization steps

computed at a low Mach number of 0.1 and a Reynolds number of 10000. The optimization cases studied and the initial values for the optimization variables are given in Table 1. The initial values of the optimization variables correspond to the optimum sinusoidal flapping motion for the given β value. It should be noted that these initial values of the NURBS parameters define a sinusoidal flapping motion. The optimization results obtained are given in Table 2.

Table 1
Starting conditions for optimization case studies

Case	β	α_0	ϕ	P_{oh}	P_{1h}	P_{2h}	$P_{o\alpha}$	$P_{1\alpha}$	$P_{2\alpha}$	η	C_t
1	0.0	8.5°	90.3°	0.	1.	1.	0.	1.	1.	0.43	0.14
2	0.5	14.6°	89.7°	0.	1.	1.	0.	1.	1.	0.58	0.12
3	1.0	18.3°	81.5°	0.	1.	1.	0.	1.	1.	0.61	0.10

Table 2
Optimization results

Case	β	α_0	ϕ	P_{oh}	P_{1h}	P_{2h}	$P_{o\alpha}$	$P_{1\alpha}$	$P_{2\alpha}$	η	C_t
1	0.0	21.2°	41.4°	−0.9	3.5	4.4	−0.8	0.2	0.2	0.21	0.78
2	0.5	17.9°	80.8°	−0.4	1.3	1.6	−0.1	0.99	1.	0.57	0.18
3	1.0	21.5°	79.9°	−0.1	1.01	1.01	−0.01	1.	1.	0.64	0.09

Figure 5 shows the variation of NURBS parameters along the optimization process for Case 1. As all the optimization variables are incremented along the gradient vector, the average thrust coefficient increases gradually from $C_t = 0.14$, which is the maximum

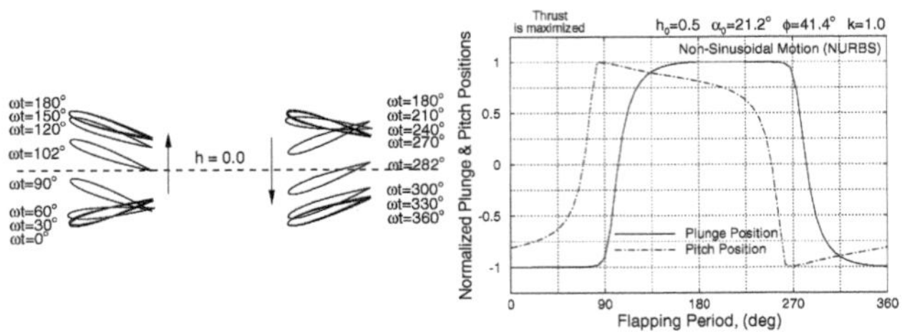

Figure 6. Non-sinusoidal flapping motion for Case 1

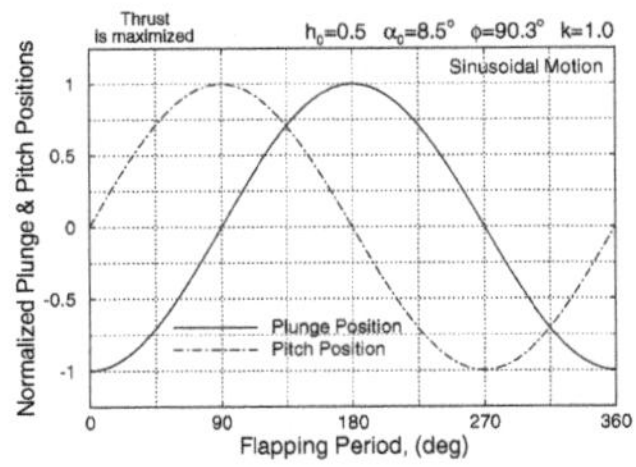

Figure 7. Sinusoidal flapping motion

value for the sinusoidal flapping motion, and reaches a maximum value of $C_T = 0.78$. The corresponding propulsive efficiency is about 21% in contrast to the starting value of $\eta = 43\%$. The corresponding NURBS based flapping motion is given in Figure 6 in comparison to the sinusoidal flapping motion (Figure 7).

It is noted that the non-sinusoidal, NURBS based optimum flapping motion differs significantly from the sinusoidal motion. The pitching motion now mostly occurs at the minimum and maximum plunge positions, and the airfoil stays at about a constant angle of attach during the up and down strokes. Such a flapping motion produces a much higher thrust, which is more than 5 times that of the sinusoidal motion.

The unsteady flow for Case 1 is depicted in Figure 8 in terms of particle traces shed along a vertical line at the leading edge of the airfoil. As seen the flowfield is highly vortical with strong leading edge vortices formed during the upstroke and the downstroke and shed into the wake.

For Cases 2 and 3, the optimized flapping motions are given in Figures 9 and 10. In Case 2, the thrust and propulsive efficiency is optimized simultaneously. The resulting motion still deviates from the sinusoidal motion. However, in Case 3, where the efficiency is optimized, the optimization process does not deviate from the sinusoidal motion. It appears that the sinusoidal motion may provide the best efficiency without producing much thrust.

5. CONCLUSIONS

In this study, the flapping motion of an airfoil is defined by a constrained 3rd order NURBS. The NURBS parameters are successfully optimized for maximum thrust and/or efficiency using a gradient based optimization with an unsteady Navier-Stokes solver.

The study shows that the thrust generation of a flapping airfoil may be increased significantly on a non-sinusoidal flapping path at the expense of propulsive efficiency. Thrust producing flows are observed to be highly vortical with the formation of strong leading edge vortices during the upstroke and the downstroke. The propulsive efficiency is improved as the flapping path converges to a sinusoidal one.

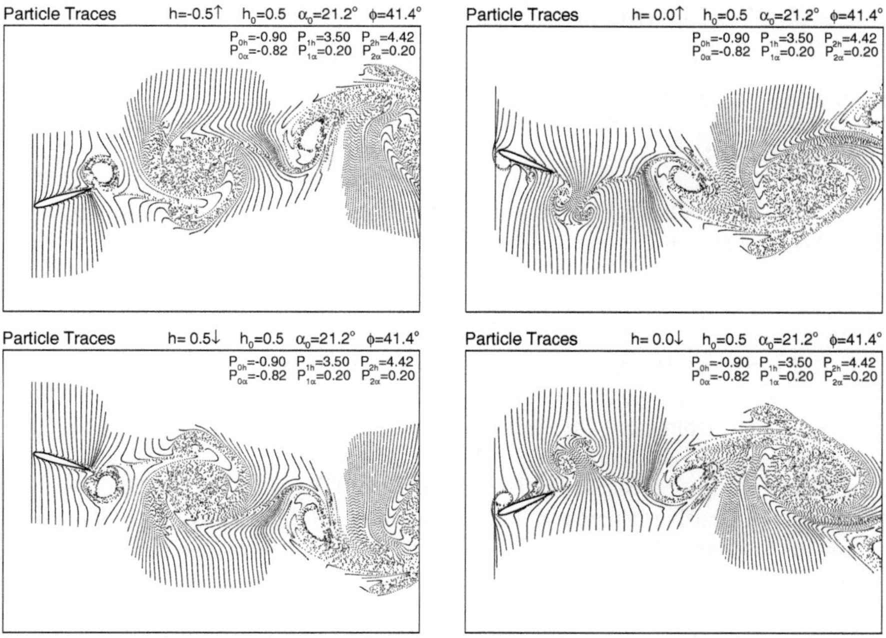

Figure 8. The unsteady flowfield for Case 1

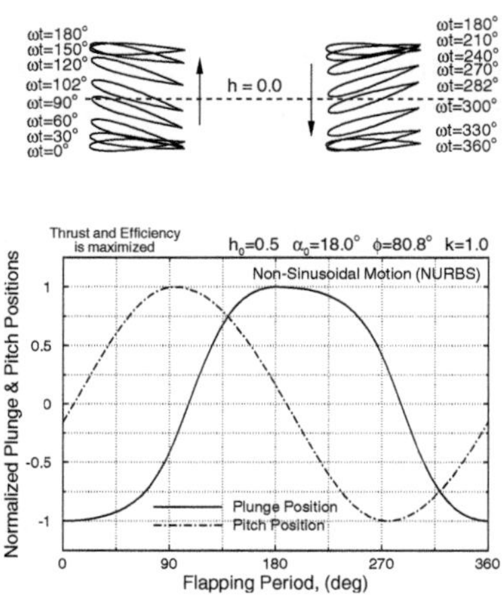

Figure 9. Non-sinusoidal flapping motion for Case 2

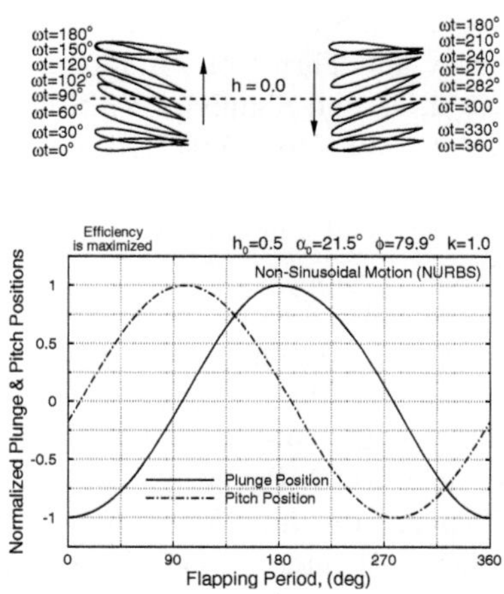

Figure 10. Non-sinusoidal flapping motion for Case 3

6. ACKNOWLEDGMENT

The present work was partially supported by NATO Research and Technology Organization under the project TX01-02.

REFERENCES

1. Mueller, T.J. (editor), Fixed and Flapping Wing Aerodynamics for Micro Air Vehicles, AIAA Progress in Aeronautics and Astronautics, Vol 195, Reston, VA, 2001.
2. Shyy, W., Berg, M. and Lyungvist, D., Flapping and Flexible Wings for Biological and Micro Air Vehocles, Pergamon Progress in Aerospace Sciences, Vol 35, p: 455-505, 1999.
3. Lai, J.C.S. and Platzer, M.F., The Jet Characteristics of a Plunging Airfoil, 36th AIAA Aerospace Sciences Meeting and Exhibit, Reno, NV, Jan 1998.
4. Jones, K.D., Dohring, C.M. and Platzer, M.F., An Experimental and Computational Investigation of the Knoller-Beltz Effect, AIAA Journal, Vol 36, p: 1240-1246, 1998.
5. Anderson, J.M., Streitlen, K., Barrett, D.S. and Triantafyllou, M.S., Oscillating Foils of High Propulsive Efficiency, Journal of Fluid Mechanics, Vol 360, p: 41-72, 1998.
6. Lewin, G.C. and Haj-Hariri, H., Reduced-Order Modeling of a Heaving Airfoil, AIAA Journal, Vol 43, p: 270-277, 2005.
7. Tuncer, I.H. and Platzer, M.F., Thrust Generation due to Airfoil Flapping, AIAA Journal, Vol 34, p: 324-331, 1996.
8. Tuncer, I.H., Lai, J., Ortiz, M.A, and Platzer, M.F., Unsteady Aerodynamics of Stationary/Flapping Airfoil Combination in Tandem, AIAA Paper, No 97-0659, 1997.
9. Tuncer, I.H. and Platzer, M.F., Computational Study of Flapping Airfoil Aerodynamics, AIAA Journal of Aircraft, Vol 35, p: 554-560, 2000.
10. Isogai, K., Shinmoto Y., Watanabe, Y., Effects of Dynamic Stall on Propulsive Efficiency and Thrust of a Flapping Airfoil, AIAA Journal, Vol 37, p: 1145-1151, 2000.
11. Isogai, K. and Shinmoto Y., Study on Aerodynamic Mechanism of Hovering Insects, AIAA Paper, No 2001-2470, 2001.
12. Jones, K.D. and Platzer, M.F., Experimental Investigation of the Aerodynamic Characteristics of Flapping-Wing Micro Air Vehicles, AIAA Paper, No 2003-0418, Jan 2003.
13. Jones, K.D., Castro, B.M., Mahmoud, O., Pollard, S.J., Platzer, M.F., Neef, M.F., Gonet, K., Hummel, D. A Collaborative Numerical and Experimental Investigation of Flapping-Wing Propulsion, AIAA Paper, No 2002-0706, Jan 2002.
14. Jones, K.D., Duggan, S.J., Platzer, M.F., Flapping-Wing Propulsion for a Micro Air Vehicle, AIAA Paper, No 2001-0126, Jan 2001.
15. Platzer, M.F. and Jones, K.D., The Unsteady Aerodynamics of Flapping-Foil Propellers, 9th International Symposium on Unsteady Aerodynamics, Aeroacoustics and Aeroelasticity of Turbomachines, Ecole Centrale de Lyon, Lyon, France, Sept 2000.
16. Tuncer, I.H. and Kaya, M., Optimization of Flapping Airfoils For Maximum Thrust and Propulsive Efficiency, AIAA Journal, Vol 43, p: 2329-2341, Nov 2005.
17. Kaya, M. and Tuncer, I.H., Path Optimization of Flapping Airfoils for Maximum Thrust Based on Unsteady Viscous Flow Solutions, 3rd Ankara International Aerospace Conference, Ankara, Aug 2005.

Parallel Computational Fluid Dynamics – Parallel Computing and Its Applications
J.H. Kwon, A. Ecer, J. Periaux, N. Satofuka and P. Fox (Editors)

Aerodynamic Optimization Design System for Turbomachinery Based on Parallelized 3D Viscous Numerical Analysis

Zhirong Lin[a], Bo Chen[a]and Xin Yuan[a]

[a]Key laboratory for Thermal Science and Power Engineering of Ministry of Education, Tsinghua University, Beijing 100084, China, E-mail: yuanxin@tsinghua.edu.cn

This paper has developed an aerodynamic optimization design system for blade design optimization based on a parallel computation platform. The optimization system includes three sub-systems: parametric modeling, evaluation system and optimization strategies. In parametric modeling sub-system, Non-Uniform Rational B-Splines (NURBS) technique is successfully applied. In evaluation system, an in-house parallelized CFD code and an commercial code NUMECA fine/turbo are both adopted. The optimization strategies sub-system is provided by a commercial software named iSIGHT. The iSIGHT software is also the integration platform of the three sub-systems mentioned above. To reduce the time cost by optimization process, a parallel computation platform including hardware and software was used in this paper. Two kinds of parallel mode can be applied during the optimizing process, one is using parallelized CFD code, another is using parallelized optimization strategy. Both of them can obtain good parallel speed-up coefficient. Finally, three test cases were optimized to improve the aerodynamic efficiency by using the present system. High efficiency and performance were confirmed by comparing the analytic results with those of the previous designs.

1. INTRODUCTION

In aerospace and power generation industries, engineers are challenged to design the highest quality systems while reducing the cost and the duration of the design cycle in order to achieve reactivity to the market demands and business changes.

Turbomachinery blade has complex design problems[1]. A considerable amount of research has been conducted on blade design for turbomachinery. 2D/3D blade-to-blade inverse methods employing specified flow quantities have been commonly used in the past few decades[2–4]. But inverse design is lack of direct control on parameters that indicate the improvement, and the specification of the aerodynamic duty in terms of local variables to achieve global aerodynamic optimum is not clear and needs more deliberation. In the meantime, Owing to the sophistication of internal flows in turbomachinery, many flow features are fully 3D and can not be predicted by quasi-3D approach[5]. With the development of the in-house 3D Navier-Stokes solver, full 3D blade design using optimization techniques is becoming desirable and practical.

In optimization course, one problem is how to choose the design variables and reduce their number while maintaining the freedom and quality of the blade representation. The design variables can be derived from the geometric parameterization of blade curves and surfaces. Bezier and B-Spline curves of third- or fourth-order have been chosen for 2D blade profile parameterization which ensured good continuity of the blade surface, and the parameters defining the geometry have intuitive meanings that facilitate imposing constraints on their variance and allow restricting the design space[6,7].

Another problem is optimization strategy. Because many turbomachinery design optimization are multimodal and discontinuous problems, the use of gradient based numerical optimization algorithms perform inefficiently and drop into local minimum prematurely. Hence, the exploratory algorithms such as Genetic Algorithms (GA), Adaptive Simulated Annealing (ASA) are required for global exploration[8].

2. AERODYNAMIC OPTIMIZATION PLATFORM

The optimization platform constructed in this paper can be divided into three sub-systems: Parametric Modeling, Evaluation System and Optimization Strategy. The detailed information about these sub-systems will be described in the following subsections.

2.1. Parametric Modeling Based on NURBS Technique

The geometric representation of curves in 2D space, or surfaces in 3D space, is an important part of any shape optimization procedure. Researchers always want to represent desired shape with the least possible number of parameters for a given accuracy. Turbine blades are usually highly cambered and have round leading and trailing edges, whereas compressor blades are usually thin, low cambered and have a round leading edge and a sharp trailing edge, sometimes also have a round trailing edge. Therefore, the geometric representation of such blades with only one type of functions is quite a challenge[9].

In past decades, Bezier curves were widely used in representing airfoil, probably because of their ease of implementation. However, they have two limitations, first they are global in nature, i.e. when a control point is moved the entire blade shape is modified, which results in less control over the local blade profile; second they cannot represent conics (e.g. leading and trailing edge of circular arc profile) exactly. To overcome the global nature of the Bezier polynomials, B-splines use the concept of control points introduced by Bezier, but with more complex interpolation functions that can capture local characteristics such that the displacement of a control point introduces a local modification of the curve near that point. However, B-splines still cannot represent conics exactly so that e.g. leading and trailing edge cannot be described exactly. To alleviate the shortcomings of Bezier curves and B-splines, i.e. to allow for local control of the curve and represent conics exactly, the NURBS technique can be introduced to represent the blade curves.

NURBS is becoming an industry standard tool for representation and design of geometry. The reasons are[10]: represent exactly conics, e.g. circles, and provide flexibility to design a large variety of shapes; be evaluated reasonably fast by numerically stable and accurate algorithms; be invariant under affine as well as perspective; be generalizations of B-splines and Bezier curves and surfaces. For turbine blade parametric modeling, there are three key points for parametric modeling process[11]: Bigger design space for all possibilities of surface transformation; Less design variables for accelerating optimization; More accurate and smooth for evaluation. Figures 1, 2 and 3 shows NURBS curve to control 2D blade profiles for turbine, compressor and fan respectively. Considering these pictures, we

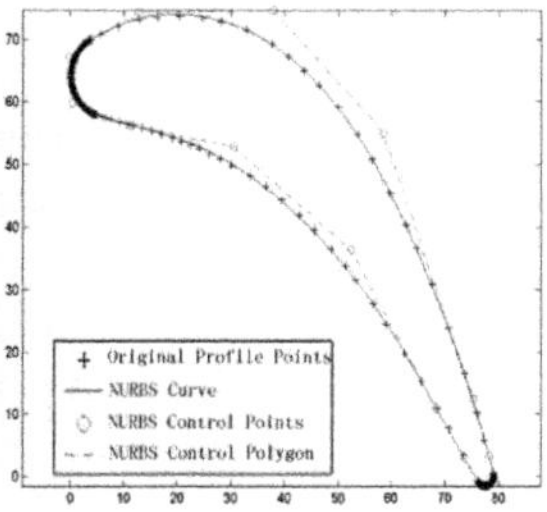

Figure 1. Turbine blade profile based on NURBS curve

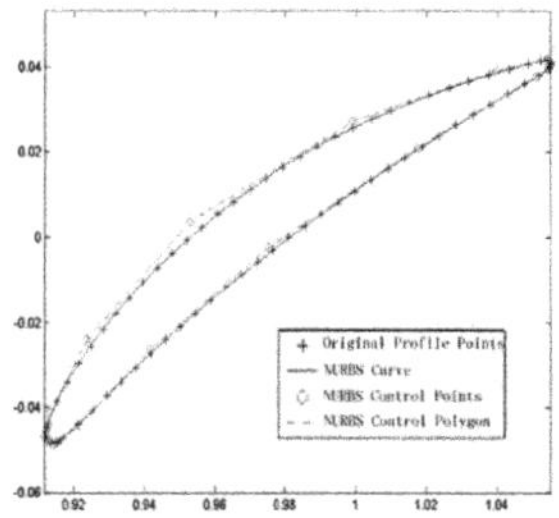

Figure 2. Compressor blade profile based on NURBS curve

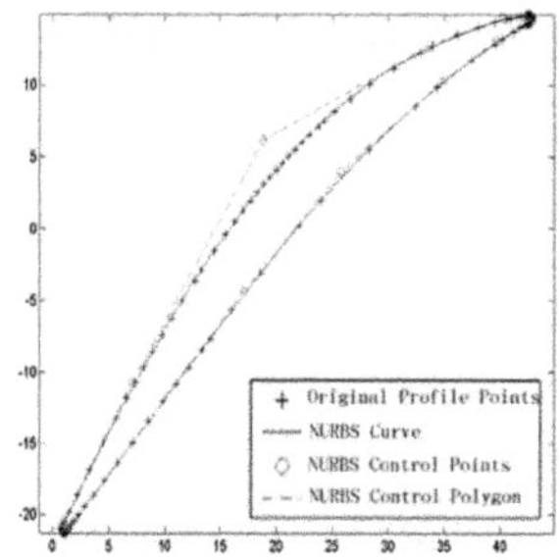

Figure 3. Fan blade profile based on NURBS curve

can say that NURBS curves are able to represent various blade profiles of turbomachinery. To form a complete 3D blade shape, 2D blade profiles must be stacked by specific stacking line. Where Figure 4 shows the stacking line represented by NURBS curve. By using this kind of stacking line, the characteristic of blade bend/lean/sweeping/torsion can be achieved. For example, supposing the meaning of x-coordinate in Figure 4 is torsion angle of blade profile, then the curve in Figure 4 can control the torsion characteristic of blade shape. By integrating 2D blade profiles and stacking lines based on NURBS technique, the complete blade can be shaped as Figure 5 shows.

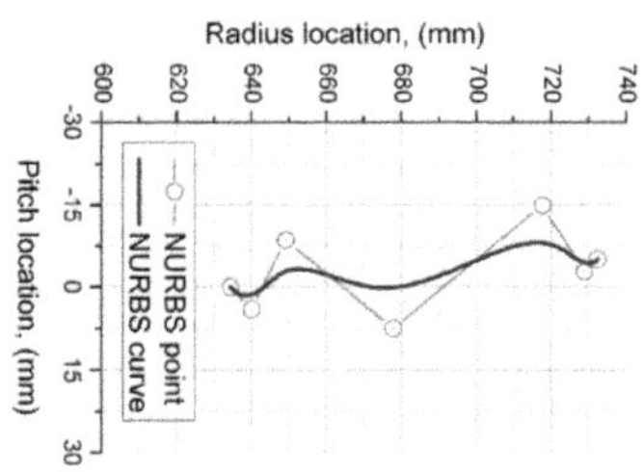

Figure 4. Turbine blade profile based on NURBS curve

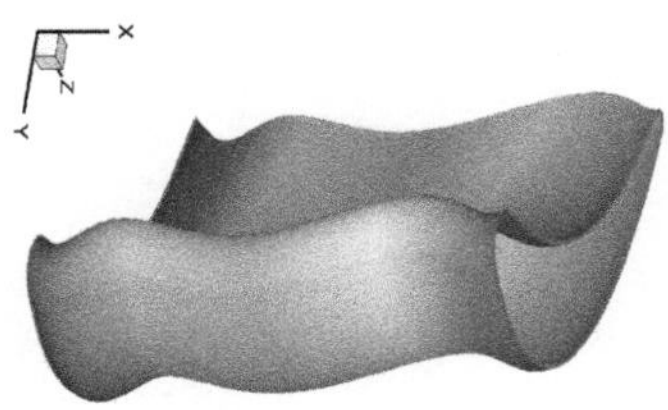

Figure 5. Complete 3D blade shape based on NURBS technique

2.2. Evaluation System

A 3D viscous flow analysis code proposed by one of authors, which solves the Reynolds-Averaged Navier-Stokes equations and the low-Reynolds-number q-ω two-equation turbulence model, is used to predict the effect of full 3D viscous flow through turbomachinery. The code combines LU-SGS-GE implicit scheme and modified fourth-order MUSCL TVD scheme[12]. On the other hand, the commercial CFD software, for instance NUMECA FINE/Turbo, can also be used to evaluating the viscous flow fields.

3. PARALLEL COMPUTATION PLATFORM

The parallel computation platform consists of both software and hardware. In this paper, two kinds of network data transformation software platform are used, MPI and Windows Socket. A High Performance Computer Cluster (HPCC) were used in this paper to accomplish the heavy computation task during optimization process. The HPCC consists of 32 nodes, and each node was made up of two AMD Opteron 246 CPUs and 2 gigabytes memory. These nodes were connected through 1 gigabit/s fast ethernet.

3.1. Parallelization based on domain decomposition

The most natural way to achieve parallelization for numerical simulation is by domain decomposition. This is relatively easy for explicit schemes because the explicit schemes are defined pointwisely and are inherently parallel. In order to achieve parallelization using implicit scheme, a special interface conditions called time-lagging interface conditions with overlapping sub-grid was introduced. In previous study, the numerical results show that sometimes the grid overlapping technique will influence the numerical convergence speed and accuracy[13]. But during doing the numerical simulation of steady and unsteady cases for some turbine blade flow using overlapping multi-block grids and time-lagging interface conditions, we have found it hardly influence the convergence speed and the accuracy of the numerical results with the present LU-SGS-GE implicit scheme and higher-order, high resolution MUSCL TVD scheme[12]. A suitable overlapping width of the sub-grid for domain decomposition is a key point for the present method of parallelization. Here overlapping width is defined in terms of the number of grid points in the overlap. Figure 6 shows the decomposition scheme of blade computation domain. Figure 7 compares linear speed-up, real speed-up and idea speed-up coefficient.

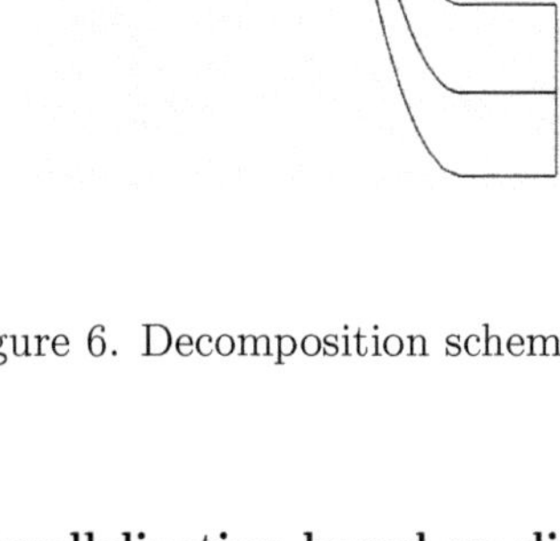

Figure 6. Decomposition scheme

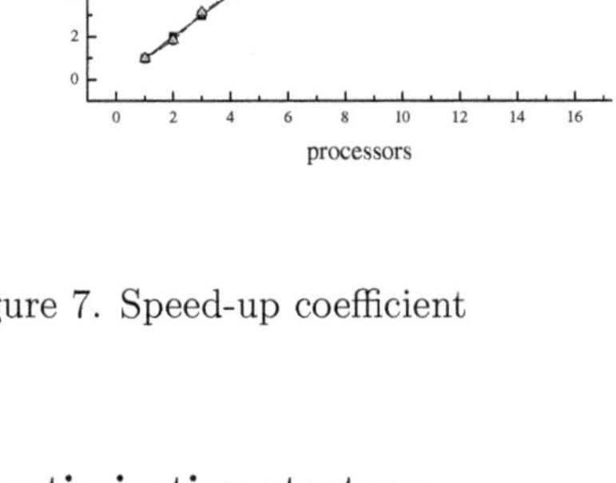

Figure 7. Speed-up coefficient

3.2. Parallelization based on distributed optimization strategy

When using some of the optimization strategy, perform the different design attempts simultaneously can really accelerate the optimization course. Figure 8 shows the diagram of parallelized optimization strategy.

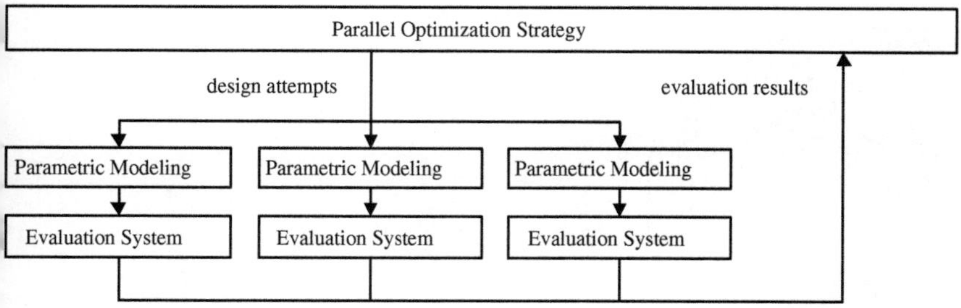

Figure 8. Parallelized optimization strategy

4. OPTIMIZATION CASES AND DISCUSSION

In this section, three test cases of optimization design system are performed. Case I and II are about a single row of turbine blade, case III is about a turbine stage. In these cases, the objective function is the total sum of the circumferential mass flow averaged loss at different radial heights at the exit plane of the cascade. The main information of them is listed in Table 1.

Table 1
Brief summary of optimization cases

	Mesh points	CPUs	Improvement	Elapsed time
case I	158,025	8	0.85%	250 Hours
case II	134,113	12	0.56%	200 Hours
case III	423,108	2	0.53%	128 Hours

4.1. Optimization of turbine blade based on parallelized CFD code

In this case, a single row of turbine blade was considered to be optimized. The geometry modeling techniques used in this case included NURBS blade profile and bend/lean stacking line. In more detail, the number of design parameters for 2D blade profile is 16, and the number of design parameters for bend/lean stacking line is 11. So, totally there are 27 design parameters in this case. For evaluating system, the in-house code presented by one of the authors was applied. The code was already parallelized by domain decomposition technique and running on 8 CPUs for parallel computation. ASA method was adopted as the only optimization strategy in this case. And the optimization target is maximum aerodynamic efficiency. Figure 9 shows the evolution history of this case. Figure 10 shows the original and optimized blade shape respectively.

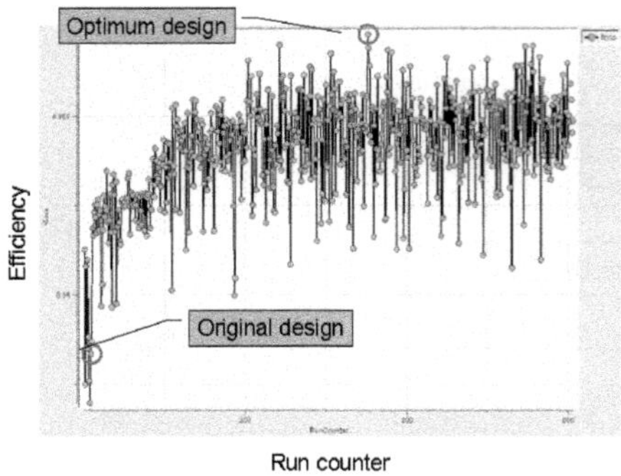

Figure 9. Optimization evolution history

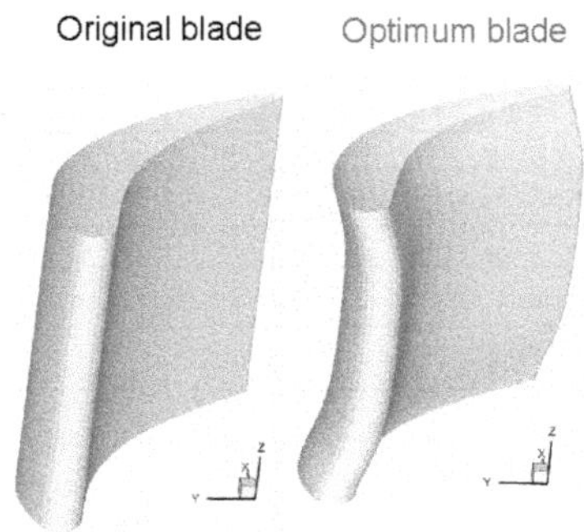

Figure 10. Blade shapes

4.2. Optimization of turbine blade using parallelized optimization strategy

This case is almost the same as the last one, a single row of turbine was optimized. The geometry modeling in this case included NURBS blade profile and bend/lean stacking line. In more detail, the number of design parameters for 2D blade profile is 16, and three sections at different position was treated as key section to control the 3D blade shape more subtly. So the total number of design parameters of profile is 16×3=48. The number of design parameters for bend/lean stacking line is 11 just as last case. So, totally there are 59 design parameters in this case. For evaluating system, the commercial software NUMECA Fine/Turbo was applied and running under single-thread mode. Multi-Island GA and Sequence Quadratic Programming method was combined as the optimization strategy, and both of them were running under parallel mode. The optimization target is also maximum aerodynamic efficiency. Figure 11 shows the evolution history of this case. Figure 12 shows the original and optimized blade shape respectively.

4.3. Optimization of a turbine stage

This case is about a stage of turbine blade. The geometry modeling techniques used in this case included NURBS blade profile and bend/lean stacking line. In more detail, the number of design parameters for 2D blade profile is 16, and 5 sections at different position was treated as key section to control the 3D blade shape for stator. So the total number of design parameters of profile is 16×5=80. The number of design parameters for bend/lean stacking line is 11 for stator and 8 for rotor. So, totally there are 99 design parameters in this case. For evaluating system, the commercial software NUMECA Fine/Turbo was applied and running under parallel mode. ASA method was adopted as the only optimization strategy in this case. The optimization target is also maximum aerodynamic efficiency. Figure 13 shows the evolution history of this case. Figure 14 shows the original and optimized blade shape respectively.

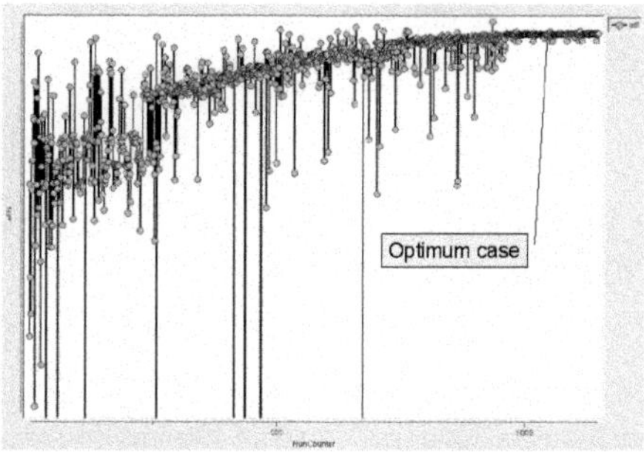

Figure 11. Optimization evolution history

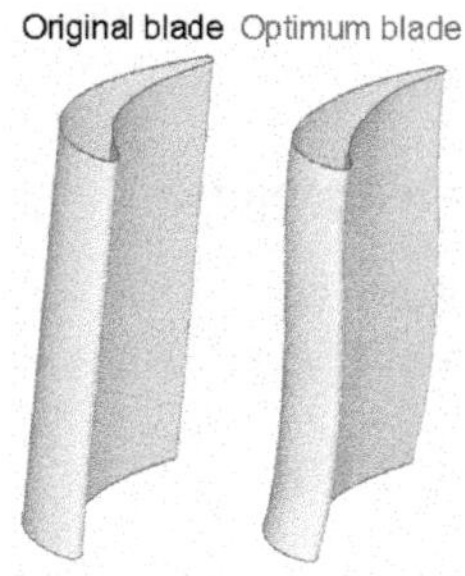

Figure 12. Blade shapes

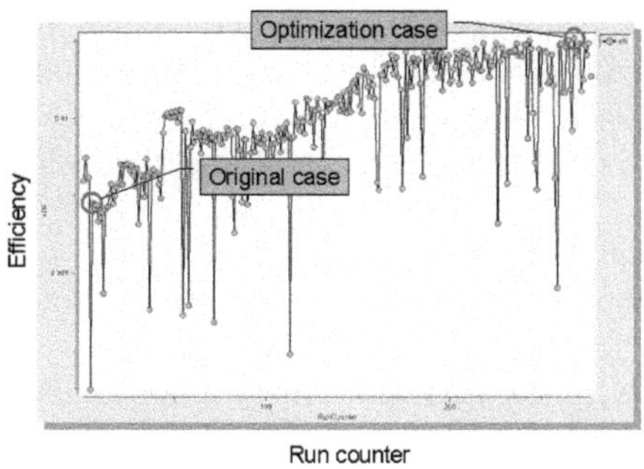

Figure 13. Optimization evolution history

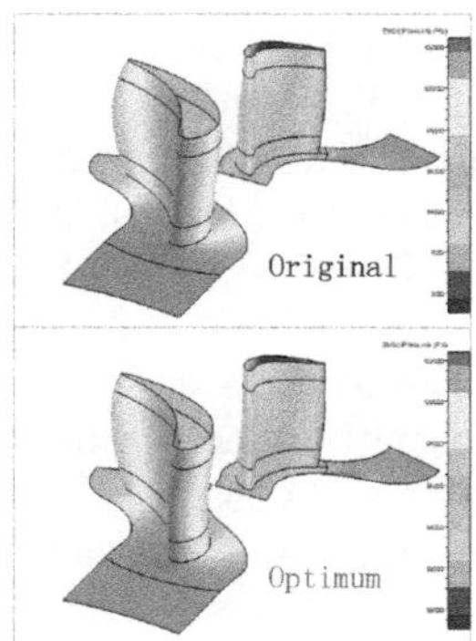

Figure 14. Blade shapes

5. CONCLUSION AND PROSPECT

An optimization design system has been used for the blade design optimization, which include three modules: parametric modeling, evaluation system and optimization strategy.

The advanced NURBS technique has been successfully applied in turbomachinery blade parametric modeling system. A higher-order high resolution parallel CFD code was introduced as 3D viscous numerical evaluation system for turbomachinery. An automatic aerodynamic optimization system based on commercial software iSIGHT was also proposed in the paper.

Three cases of turbine blade were optimized to improve the efficiency by using the present system. Reasonably high efficiency and high performance were confirmed by comparing the CFD results with those of the previous designs.

The present aerodynamic optimization design system can be an efficient and robust design tool to achieve good aerodynamic blade design optimization in a reasonable time.

We expect this optimization system can become an efficient and robust design tool to achieve better blade performance in the near future for manufactory.

Acknowledgements

This work was partly supported by the Specialized Research Fund for the Doctoral Program of Higher Education (Grant No. 20050003063).

REFERENCES

1. S. Havakechian and R. Greim. Recent advances in aerodynamic design of steam turbine components, VGB conference, 1999.
2. Schwering, W., Design of Cascades for Incompressible Plane Potential Flows with Prescribed Velocity Distribution, J Eng Power Trans ASME, Vol. 93 Ser A, No. 3(1971) 321-329.
3. Meauze, G., An Inverse Time Marching Method for the Definition of Cascade Geometry, J ENG Power Trans ASME, Vol. 104, No. 3(1982) 650-656.
4. Dang, T., Damle, S., Qiu, X., Euler-Based Inverse Method for Turbomachine Blades, Part 2: Three-Dimensional Flows, AIAA Journal, Vol. 38, No. 11(2000) 2007-2013.
5. Denton, J.D., Dawes, W.N., Computational Fluid Dynamics for Turbomachinery Design, Proc Instn Mech Engrs, Part C: Journal of Mechanical Engineering Science, Vol. 213, Iss. 2(1999) 107-124.
6. T Korakianitis and G I Pantazopoulos, Improved turbine-blade design techniques using 4th-order parametric-spline segments, Computer-Aided Design, Vol. 25, No. 5(1993) 289-299.
7. Pierret.S., and Van den Braembussche, R.A., Turbomachinery Blade Design Using a Navier-Stokes Solver and Artificial Neural Network, Journal of Turbomachinery, Vol.121(1999) 326-332.
8. Shahpar, S., A Comparative Study of Optimisation Methods for aerodynamic Design of Turbomachinery Blades, ASME Paper 2000-GT-523(2000).
9. Wahid S. Ghaly and Temesgen T. Mengistu. Optimal Geometric Representation of Turbomachinery Cascades Using NURBS. Inverse Problems in Engng, Vol. 11, No. 5 (2003) 359-373.
10. Temesgen T. Mengistu and Wahid S. Ghaly. A Geometric Representation of Turbomachinery Cascades Using NURBS. 40th Aerospace Sciences Meeting & Exhibit. 14-17 January 2002.
11. Yuyang Lai, Xin Yuan. Blade Design With Three-demensional Viscous Analysis And Hybrid Optimization Approach. 9th AIAA/ISSMO Symposium on Multidisciplinary Analysis and Optimization, Atlanta, USA, Sept 4-6, 2002.
12. Xin Yuan, H. Daiguji, A Specially Combined Lower-Upper Factored Implicit Scheme For Three-Dimensional Compressible Navier-Stokes Equations, Computers and Fluids, Vol.30, No.3(2001) 339-363.
13. Wu Z. N., Zou H., Grid overlapping for implicit parallel computation of compressible flows, Journal of Computational Physics, Vol. 157, No. 1(2000) 2-43.

Parallel Computational Fluid Dynamics – Parallel Computing and Its Applications
J.H. Kwon, A. Ecer, J. Periaux, N. Satofuka and P. Fox (Editors)

Genetic Algorithm Optimisation of Fish Shape and Swim Mode in Fully-Resolved Flow Field

Sho KUSUDA[a], Shintaro TAKEUCHI[a] and Takeo KAJISHIMA[a]

[a]Dept. of Mechanical Engineering, Osaka University,
2-1 Yamada-oka, Suita-city, Osaka 565-0871, Japan

Flow field around a single 2-D fish is analysed by solving the full Navier-Stokes equations, and optimisation of fish shape and swim mode with genetic algorithm (GA) is attempted. To solve the flow field around the moving fish meshed with time-dependent body-fitted coordinate system, arbitrary Lagrangian Eulerian (ALE) method is employed. A study with specific parameters shows that swim mode with/without phase-lag causes distinct differences in fluid's velocity and pressure fields past a fish, suggesting that the phase-lag parameter plays an important role in the interaction between the fluid and fish's global movement. For pursuing good performances of fish, five characteristic parameters are optimised in conjunction with the surrounding fully resolved flow field. The GA results show that fish exhibits different ideal shapes and manoeuvres for different desired fitnesses. Fish of efficient energy consumption qualitatively agrees with the prediction by Lighthill[1]. On the other hand, the best-performing fish attaining the furthest distance in a given time exhibits totally opposite trends for shape and swim mode with little consideration on efficiency. The study shows that difference in energy consumption for those two extreme cases is about 15 times.

1. INTRODUCTION

Finned aquatic animals, like fish and dolphin, often exhibit extraordinary physical capabilities for swimming speed, efficiency or steep turn, and the performances have encouraged a number of scientists to study the potential capacities of them from many aspects. Meanwhile, ship and undersea vessel have been using rotary propeller to get thrust. If the above fish-like propulsion system is applied to man-made vessels, it may drastically improve efficiency or noise level. In reality, several prototypes of robot fish have been developed already, and in the future the competition in development and improvement of robot fish will be further accelerated. However, it is impossible to manufacture and test all possible types of robot to search good performance, because one robot fish has many adjustable parameters for identifying shape and swim mode.

Therefore, a sophisticated optimisation method is needed to seek superior samples efficiently. There is one particular technique for optimisation, genetic algorithm (GA), suitable for problems of multi-dimensional multi-peaked solutions. GA has learnt its basic idea, i.e. *reproduction*, *crossover* and *mutation*, from the natural evolution system of individuals in a group of families, and it can be easily modified to search optimal solution through a various situations. Due to its controlled non-linear stochastic process, GA solves the optimisation problem efficiently, and it has gained popularity for predicting optimal solution and efficient machine learning process. For robot fish, Barrett[2] conducted some experiments to optimise fish locomotion by directly incorporating the experimental results into GA, and reported that GA works effectively to achieve desired robot-fish functions. Kuo and Grierson[3] employed GA to search efficient prospective gait based on computational results. Five parameters characterising the carangiform locomotion were optimised in their work. However, their model for fluid-object interaction was rather simplified and restricted by employing an empirical fluid drag model. As object's global motion is affected by fluid motion passing by it and vortical structures in the wake[4], fully-resolved flow field is of importance in predicting good performance of fish by numerical optimisation.

Due to the recent increase in computational power, large scale numerical simulations of fluid dynamics are carried out. Generally, numerical simulation of fluid-solid interaction needs treatments of both fluid and solid phases in a coupled fashion. Arbitrary Lagrangian-Eulerian (ALE) method is one of the approaches to solve the interaction problems of fluid and deformable object [5]. The method resolves a flow field at a desired scale along object surface, and has found applications in a number of flow fields accommodating moving object. For example, Liu et al[6] applied their finite volume ALE approach to a problem of tadpole swimming, and studied the effect of morphological development of the tadpole on the hydrodynamic forces and efficiency.

In the present work, interaction between a simplified model fish and fluid flow is studied. The fish locomotion in a fully-resolved flow field is facilitated by our ALE method efficiently combined with a mesh generation procedure. GA optimisation is applied for five parameters characterising the fish's shape and swim mode, and the effect of the parameters on the fish movement and the flow field are discussed. Also, parallel processing is attempted to the GA optimisation, and our approach is shown to exhibit a remarkable applicability for the large scale parallel computation.

2. GOVERNING EQUATIONS AND NUMERICAL METHODS

2.1. Fluid phase

Body-fitted curvilinear coordinate system is employed for fluid phase. In the present study, the coordinates alter to fit the object surface which changes the shape with time in a prescribed manner.

Governing equation of fluid motion is Navier-Stokes (N-S) equation of ALE form[5] and the equation of continuity for incompressible fluid, which are described as follows:

$$\frac{\partial u_i}{\partial t} + \frac{\partial u_i}{\partial \xi^j}(U^j - V^j) = -\frac{\partial \xi^j}{\partial x_i}\frac{\partial P}{\partial \xi^j} + \frac{1}{Re}\frac{\partial \xi^j}{\partial x_k}\frac{\partial}{\partial \xi^j}(\frac{\partial \xi^l}{\partial x_k}\frac{\partial u_i}{\partial \xi^l}) \tag{1}$$

$$\frac{1}{J}\frac{\partial JU^j}{\partial \xi^j} = 0 \tag{2}$$

where u_i is velocity, p pressure, Re Reynolds number, x_j Cartesian coordinates, ξ^j curvilinear coordinates, U^j and V^j contravariant velocities of fluid and grid, respectively, and J is Jacobian of coordinate transformation between x_j and ξ^j.

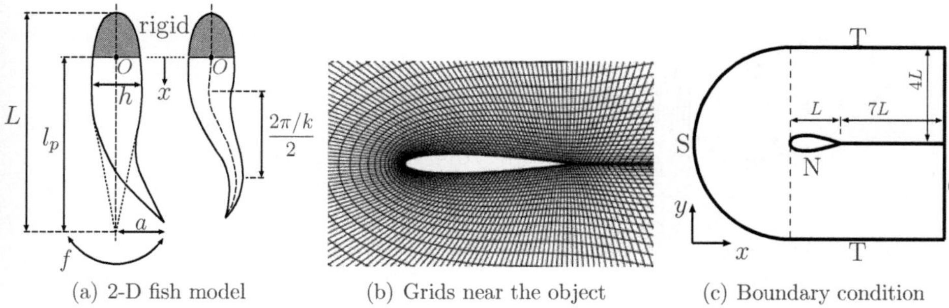

Figure 1. Schematics of (a)neutral position (dotted line) and two typical deformed positions (solid line), (b)grid system and (c)boundary condition.

2.2. Solid phase

In the present study, the authors employ a simplified 2-D fish model with a limited number of parameters to easily isolate the effect of each parameter (from the others) on fish performance and flow structure. The fish is allowed three fundamental degrees of freedom (two for translating motion and one for rotating motion) and some additional degrees for the joints on the vertebra. The three fundamental degrees of freedom represents the global behaviour of the fish movement, and they are governed by the following equations:

$$\frac{d}{dt}\begin{pmatrix} m\dot{G}_i \\ L_i \end{pmatrix} = \int_{\Omega}\begin{pmatrix} F_i \\ M_i \end{pmatrix} d\Omega \qquad (3)$$

where m is mass, $\dot{G}_i$ velocity of gravity centre, F_i fluid force, $d\Omega$ surface element of the object, L_i angular momentum, M_i first order moment of the fluid force.

A NACA0012 hydrofoil is used here as a fish model at the neutral position. Five characteristic parameters are considered to identify individuals. Figure 1(a) shows a schematic of the neutral and deformed positions of the fish and the symbols for the parameters. Within the entire body length L, the fish is assumed to have a deformable rear part (length l_p) and a rigid front part ($L-l_p$). The thickness h at the neutral position is $0.12L$. In the present study, amplitude a is selected in a range including $0.2L$, which is reported as a typical amplitude of a real fish's movement by Bainbridge[7]. Wavenumber k of undulatory motion (Fig. 1(a), right) is tested in the range between 0 and 0.4. Here, zero wavenumber means that the every point of the backbone is set in phase (in-phase motion), and non-zero wavenumber causes some phase-lag between arbitrary two points on the deformable part (phase-lag motion). For both extreme vales of k, volume deviations during the swim are found to be less than 10^{-5} with respect to the initial value. Frequency f determines the speed of the undulatory motion propagating down towards the tail tip as $2\pi f/k$. The all five parameters are kept constant for each fish, except that amplitude a is gradually increased in the first one unit time to avoid unnecessary numerical instability. Density ratio of fish to fluid is set to 1.0.

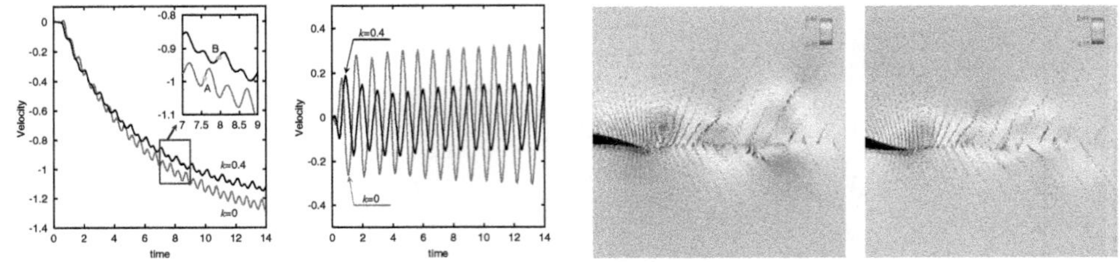

(a) x velocity (b) y velocity (c) k=0 (In-phase; Point B in Fig. 2(a)) (d) k=0.4 (Phase-lag; Point A in Fig. 2(a))

Figure 2. Fish velocities and flow fields. (a)(b) time evolution of fish velocities for k=0.4 and 0; (c)(d) instantaneous pressure and velocity fields in slip streams of k=0 and 0.4.

2.3. Numerical and boundary conditions

In the present study, fish is set in an open space. C-type mesh system is used for the simulation of the flow field. Figure 1(b) shows a mesh near the object, and Fig. 1(c) outlines the computational domain and boundary conditions. The computational domain is a semi-circle of diameter $8L$ in the front of the hydrofoil and $7L$ in length in the behind. Boundary condition of fluid velocity is the non-slip condition at the surface of the fish (N). At the inlet (S) and out-flowing boundaries (C), stationary fluid and convective boundary condition are assumed, respectively. Neumann's condition is applied for pressure at the inlet boundary. A traction-free boundary condition is employed at the upper and lower straight sections of the domain (T).

The number of grid points is 240×40. There are 120 grid points on the fish surface. A computational mesh fitted to a deformed fish is given every time step by the elliptic mesh generation algorithm.

3. SPECIFIC STUDY OF PROPULSIVE STAGE

Some specific cases of typical fish characteristics are studied to investigate the fundamental dynamics of fluid and the fish-like deformable hydrofoil. In this section, a, l_p and f are set to $0.2L$, $0.8L$ and 1, respectively. Two wavenumbers k=0 and 0.4 are tested to compare the fluid behaviours past the object.

Figures 2(a) and 2(b) compare diagrams of fish velocities and instantaneous flow fields in the initial propulsive stage for the different k values. From the velocity diagrams, the fishes experience large accelerations in both x and y directions after t=0.5. Then, for both k values, x components of velocity develop exponentially with small fluctuations, which eventually attain stable oscillations around the constant cruising velocities at about t=20. On the other hand, rather stable oscillations are observed for the y components around small non-zero constant y velocities, suggesting that the fishes travel in specific directions.

The above fish motions may be related to the fluid motions. Figures 2(d) and 2(c) compare instantaneous pressure and velocity fields in the slip streams for k=0 and 0.4, respectively. The corresponding moments are depicted in Fig. 2(a) as points A and B.

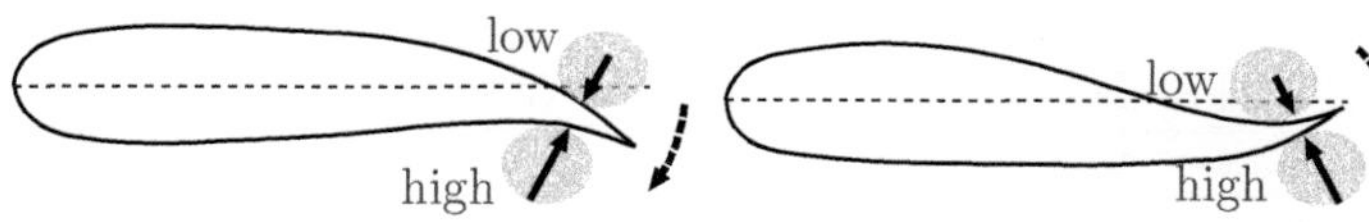

(a) k=0 (In-phase motion) (b) k=0.4 (Phase-lag motion)

Figure 3. Schematic of pressure distributions near the tailing edges in Figs.2(c) and 2(d)

In both snapshots, fish tail is in the movement towards the largest displacement in $-y$ direction, generating a vortex of positive circulation in the upper side and a high pressure region in the lower side of the tail.

The difference in wavenumber has caused a formation of stronger vortex in the slip stream of k=0 case (Fig. 2(c)) than k=0.4 (Fig. 2(d)). Generally, fish's total gain in momentum is dependent on the momentums held by the discharged vortices. Therefore, terminal travelling velocity of fish is dominated by the strength of the series of vortices. This may explain the difference in terminal velocities observed in Fig. 2(a).

However, the fish of phase-lag motion overcomes this weakness of generated vortices by its efficient swim mode. The phase-lag fish develops a positive pressure gradient across the convex and concave sides of the trailing edge, as compared in Fig. 3(a)(b) for the respective instances of Fig. 2(c)(d). This adverse pressure gradient helps the phase-lag fish to be less decelerated in each fluctuating cycle and over the initial propulsive stage (Fig. 2(a)), and consequently, allows the fish to swim efficiently at large. The vortex dynamics induced by the efficient swim mode is the subject of ongoing research by the present authors.

The above observation suggests that the fish with non-zero wavenumber is more efficient in converting the hydrodynamic force into propulsion, while the fish with little phase-lag effectively transfer kick power to fluid to get strong thrust.

Also the results demonstrates that only fully-resolved flow field enables analysis and prediction of fish-fluid dynamics at this detailed level.

4. OPTIMISATION WITH GENETIC ALGORITHM

It would be interesting to ask what fish shape and swim mode give optimal solution for a given objective. For this question, a genetic algorithm (GA) is employed in the present study, and fish's characteristics and performances optimised for travelling distance and efficiency are compared.

4.1. Fitness functions

The following five parameters of fish is varied in search for optimised performance of fish: a, l_p, h, f and k. Each parameter has 16 equally-divided quanta in the respective range, shown in Table 1, including the values studied in the previous section.

To qualify the adaptivity of the individual fish to the environment, the present study employs the following two fitness functions; distance travelled in the first 10 unit time

Table 1
Ranges of the GA parameters

Parameter	h	a	f	l_p	k
Minimum	$0.08L$	$0.08L$	0.60	$0.5L$	0
Maximum	$0.23L$	$0.23L$	1.35	$1.0L$	0.4

Table 2
Numerical condition of GA process

The number of individuals(N)	6
The number of generations	30
Probability of crossover(p_c)	0.3
Probability of mutation(p_m)	0.01

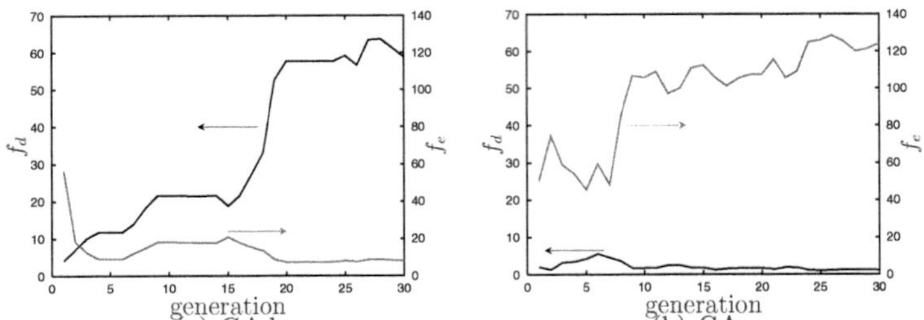

Figure 4. Comparison of f_d and f_e histories for GAd and GAe

and efficiency (travel distance per unit energy input) in the same duration of time. The fitnesses functions, f_d and f_e, are defined as follows:

$$\text{GAd} \ : \ f_d = x\text{-displacement at } t = 10 \tag{4}$$

$$\text{GAe} \ : \ f_e = \frac{f_d}{\text{cumulative workload}} \ , \tag{5}$$

where GAd and GAe are GA session names targeting optimisation of f_d and f_e, respectively. For both fitnesses, higher value means better performance. The conditions used for both sessions are summarised in Table 2.

4.2. Results and discussion

Figure 4 compares the progresses in distance and efficiency of GAd and GAe. Figure 4(a) shows that GAd acquires a remarkable capability of long range travel as generation goes, paying little attention to the optimisation of efficiency. Similarly, GAe improves the efficiency progressively despite the poor travel distances. The ratio of energy consumptions of GAd and GAe at 30th generation is about 15 times. This result suggests that an individual optimised in one environment is not always suited for another environment.

Trends of development of each parameter are shown in Figure 5. Experimental results of a real mackerel by Bainbridge[7] and predictions by Lighthill[1] are also included for some parameters. Significant differences are observed between the trends of the solutions of GAd and GAe.

The overall trend for GAd session is summarised as follows. Amplitude, deformable body length and frequency are optimised to near-maximum values of the respective parameter ranges as shown in Figs. 5(a), 5(b) and 5(d). Also, from Fig. 5(e), smaller

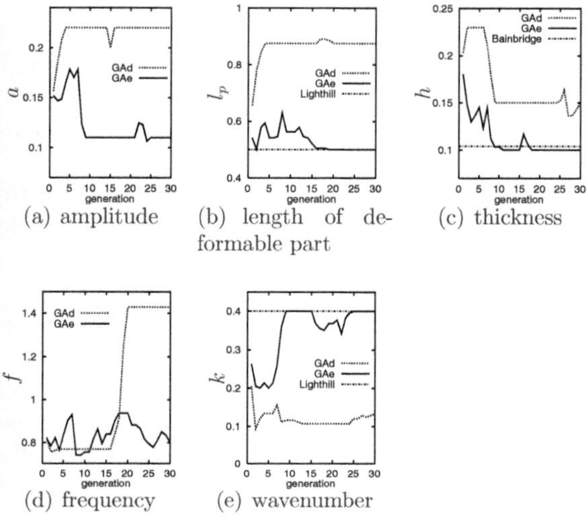

(a) amplitude (b) length of deformable part (c) thickness

(d) frequency (e) wavenumber

Figure 5. Evolution of fish parameters.

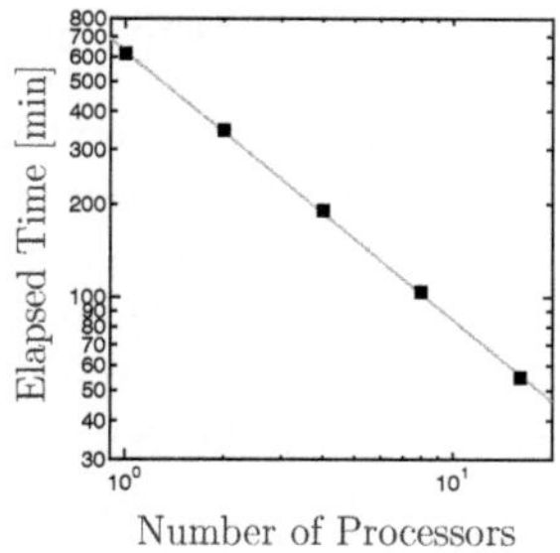

Figure 6. Trend of computational time for one generation (containing 16 individuals) against number of processors. A maximum number of 16 processors are used.

wavenumber is suitable for the purpose, which is in agreement with the result of the in-phase motion presented in the previous section. The result suggests that, to attain a longer distance in a certain duration of time, the fish twists its entire body as quickly as possible with a large amplitude and little phase-lag within the body.

On the other hand, the optimised fish in GAe has totally different traits; a small amplitude, a shortest possible deformable section, a very thin body, and a largest possible phase-lag but a slow wave propagation. The fish characteristics predicted in GAe agree with the prediction by Lighthill[1] for real fish, and this result may point to the fact that real fish give first priority to efficiency, not to travel distance or speed, in a usual situation.

4.3. Optimisation with multiple processors

Genetic algorithm requires a long run over generations. Particularly, the present case involves fully-resolved numerical simulation of fluid phase, and computational time for the whole process becomes a serious issue. For this problem, we attempted a parallelised processing within each generation by the message passing interface (MPI) library employing 16 processors at the maximum. The computer used is Dell Precision 350 with processor of 2.66 GHz Pentium4 inter-connected by 5224 Gigabit switches. The trend of computational time for one generation is plotted in Fig. 6. The elapsed time is inversely proportional to the number of processors, suggesting that the MPI parallel computation is suited for this type of problem.

5. CONCLUDING REMARKS

Two dimensional simulation of flow field past a fish-like deformable hydrofoil was carried out with arbitrary Lagrangian-Eulerian method. The flow field was fully resolved from the boundary layer to the far-field of the object by time-dependent body-fitted coordinate system. Also fish shape and swim mode are optimised by genetic algorithm (GA).

Our preliminary study on fluid-fish interaction showed that the flow structures in the fish's near and far fields are dependent on wavenumber of the undulatory motion of the fish and that the swim modes affect the hydrodynamic forces acting on the fish surface and terminal velocity of the fish.

It was found that GA solution for one objective is not always suited for another objective. The GA result showed that the fish attaining the furthest range in a given time exhibits totally opposite trends for shape and swim mode to the least energy-consuming fish in unit travel distance. The shape and swim mode optimised with energy show qualitative agreement with the theoretical and experimental predictions reported in the literatures.

And in the present study, our method demonstrated the capability of solving this type of multi-objective-multi-solution problem by including the fully-resolved flow field surrounding the object. And the computation time for the large scale GA was found to be effectively reduced by parallel computation.

The authors are currently testing more realistic fish models with an increased number of characteristic parameters (e.g. adaptive undulatory motion) and aiming for further improvement of fish performance with an original acceleration technique of GA. A series of new attempts and results will be presented as a separate paper in a relevant journal.

REFERENCES

1. M.J. Lighthill, Note on the swimming of slender fish, *J. Fluid Mech.* **9** (1960) 305
2. D. Barrett, Optimization of swimming locomotion by genetic algorithm, *NeuroTechnology for Biomimetic Robots* MIT Press (2002) 207
3. P.D. Kuo and D. Grierson, Genetic algorithm optimization of escape and normal swimming gaits for a hydrodynamical model of carangiform locomotion, *Genetic and Evolutionary Comp. Conf. Late Breaking Papers*, Chicago, USA (2003) 170
4. S. Takeuchi, T. Yamazaki and T. Kajishima, Study of solid-fluid interaction in body-fixed non-inertial frame of reference, *J. Fluid Science and Technology* **1**-1 (2006) 1
5. C.W. Hirt, A.A. Amsden and J.I. Cook, An arbitrary Lagrangian-Eulerian computing method for all flow speeds, *J. Comp. Phys.* **14** (1974) 227
6. H. Liu, R.J. Wassersug and K. Kawachi, A computational fluid dynamics stdy of tadpole swimming, *J. Exp. Biol.* **199** (1996) 1245
7. R. Bainbridge, The speed of swimming of fish as related to the size and to the frequency and amplitude of the tail beat, *J. Exp. Biol.* **35** (1958) 109

ACKNOWLEDGEMENT

The authors gratefully acknowledge the Merit Allocation Scheme Grant of the Australian Partnership for Advanced Computing (APAC).

www.ingramcontent.com/pod-product-compliance
Ingram Content Group UK Ltd.
Pitfield, Milton Keynes, MK11 3LW, UK
UKHW020219041026
15470UKWH00012B/664

9 780444 53035